Silt Sand Slurry

Dredging, Sediment, and the Worlds We Are Making

Silt Sand Slurry

Dredging, Sediment, and the Worlds We Are Making

Rob Holmes, Brett Milligan, Gena Wirth

With contributions by Sean Burkholder, Brian Davis, Justine Holzman

Dredge Research Collaborative

APPLIED RESEARCH +DESIGN

Published by Applied Research and Design Publishing, an imprint of ORO Editions.
Gordon Goff: Publisher

www.appliedresearchanddesign.com
info@appliedresearchanddesign.com

Authors: Rob Holmes, Brett Milligan, Gena Wirth.
Contributors: Sean Burkholder, Brian Davis, Justine Holzman.
Book Design: Pablo Mandel, Silvina Synaj / circularstudio.com
Project Manager: Jake Anderson

Cover image: Restored Napa River marshes in California's San Francisco Bay.
Image by Brett Milligan

10 9 8 7 6 5 4 3 2 1 FIRST EDITION

ISBN: 978-1-954081-84-0

Color Separations and Printing: ORO Group Inc.
Printed in China.

AR+D Publishing makes a continuous effort to minimize the overall carbon footprint of its publications. As part of this goal, AR+D, in association with Global ReLeaf, arranges to plant trees to replace those used in the manufacturing of the paper produced for its books. Global ReLeaf is an international campaign run by American Forests, one of the world's oldest nonprofit conservation organizations. Global ReLeaf is American Forests' education and action program that helps individuals, organizations, agencies, and corporations improve the local and global environment by planting and caring for trees.

Table of Contents

*Humans move tremendous amounts of soil and rock.
By some measures, the amount of earth that we move exceeds
that of any other geomorphic agent.*

Roger LeBaron Hooke, "On the History of Humans as Geomorphic Agents"

*We will either have justice, sustainability, and peace together or we will
descend into ecological catastrophe, social chaos, and conflict.
Soil, not oil, offers a framework for converting the ecological catastrophe
and human brutalization we face into an opportunity to reclaim our humanity
and our future.*

Vandana Shiva, *Soil, Not Oil: Environmental Justice in an Age of Climate Crisis*

*The mud should be dredged out. This maintenance operation could
be treated in terms of art, as a "mud extraction sculpture"...
The mud could be deposited on a site in the city that needs "fill."
The transportation of mud would be followed from point of extraction
to point of deposition. A consciousness of mud and the realms
of sedimentation is necessary in order to understand the landscape
as it exists.*

Robert Smithson, "Frederick Law Olmsted and the Dialectical Landscape"

*To imagine other forms of human existence is exactly the challenge that is
posed by the climate crisis: for if there is one thing that global warming has
made perfectly clear it is that to think about the world only as it is amounts
to a formula for collective suicide. We need, rather, to envision what it might be.*

Amitav Ghosh, *The Great Derangement*

Credits and Acknowledgments

Gena Wirth, Brett Milligan, and Rob Holmes authored *Silt Sand Slurry* together. Gena led development of the visual essays, while Rob and Brett led development of the text chapters. We are deeply grateful for the contributions of every colleague and collaborator who has made this book possible: Justine Holzman, Brian Davis, and Sean Burkholder each contributed essays, in addition to much dialogue and comradery across seven years of book development. Tim Maly helped us conceptualize the project, structure the chapters, and contributed early editing; both the shape of this book and the way in which it is written owe a great deal to Tim's work and sensibilities. Josh Wallaert did both developmental and line editing at crucial points in the development of the manuscript. Jake Anderson, who managed the project for AR+D, provided editorial review throughout. Justine Holzman provided editorial review for "Disappearing Delta," and Sean Burkholder provided editorial review for "Inland Chokepoint." Lisa Baron, Peter Weppler, Juanita Irizarry, Todd Bridges, Brenda Goeden, and Phoenix Armenta generously agreed to be interviewed, and Helena Starnes transcribed their interviews.

The drawings in the visual essays were largely produced by our intern-collaborators in the summers of 2015, 2016, and 2017: Alex Hill, Gena Morgis, and Inwa Park (2015); Michael Biros, Kate Lenahan, Nick Shannon, and Crystal Xing (2016); Michelle Benoit, Nathalie Mitchell, and Gavin Zeitz (2017). Yinjia Gong, Miriam Grunfeld, Yuzhou Jin, Ian Miller, Anna Mitchell, and Maria Elena Vanegas Perez also contributed to the drawings in the visual essays. The graphic design and layout of the visual essays was led by Gena Wirth with these intern-collaborators, and further refined by Pablo Mandel and Circular Studio. Graphic design and layout for the remainder of the book was done by Pablo Mandel and Circular Studio.

From the start of this project in summer 2015, we benefited from many conversations, discussions, and reviews with friends, colleagues, and collaborators. Review discussions with Kate Orff, Catherine Seavitt Nordenson, Stephen Becker, Ben Mendelsohn, Alex Chohlas-Woods, Chris Dols, Anne Weber, Glen Cummings, Pippa Brashear, Mary Kimball, Neil Freeman, Patrick Thrasher, and many others came at particularly key points in the project. We are deeply appreciative of each and every person who spoke with us and helped shape the development of *Silt Sand Slurry*.

Specific credits for photographs and other images follow the chapters in which they appear. A portion of "Choreographing Sediment" was adapted from

Rob and Brett's 2013 essay "Feedback: Designing the Dredge Cycle," published online in *Scenario 03: Rethinking Infrastructure* (eds. Stephanie Carlisle and Nicholas Pevzner). "Why Design?" would not exist without the efforts of every member of the Public Sediment team, including but not limited to Kate Orff, Pippa Brashear, Gena Morgis, Nans Voron, Nick Shannon, Sophie Riedel, Wright Yang, Lee Stickles, Brett Snyder, Claire Napawan, Beth Ferguson, Cy Keener, Adam Marcus, Evan Jones, Margaret Ikeda, Chris Devick, and Jingting Li. We would like to extend special thanks to our colleagues at SCAPE Landscape Architecture for supporting the DRC over many years as project collaborators, reviewers, and intellectual advisors.

We would also like to thank everyone who contributed to, participated in, and attended each of the four DredgeFests, as the DredgeFests were crucial in forming our understanding of the worlds of sediment.

For DredgeFest NYC, we would particularly like to thank: Geoff Manaugh and Nicola Twilley, for hosting us at Studio X-NYC (twice), for helping to conjure the DRC, and for naming DredgeFest (Geoff) and the DRC (Nicky); panelists and speakers Lisa Baron, Andrew Genn, Roger Hooke, Michael Ezban, Bill Murphy, Douglas Pabst, Edgar Westerhof, Vicki Ginter, Catherine Seavitt Nordenson, Kate Orff, Phillip Orton, Dave Arvin, Hans Hesselein, and Debbie Mans; exhibition contributor Seth Denizen; Alex Chohlas-Woods and Ben Mendelsohn for producing the event trailer; sponsors Arcadis, TenCate, and TWFM Ferry; and Tom Palladino and the crew of the *American Princess*.

For DredgeFest Louisiana, we would particularly like to thank: Brad Cantrell, Jeff Carney, and the Coastal Sustainability Studio, for hosting us at Louisiana State University; event partners the Robert Reich School of Landscape Architecture at LSU, the Center for Land Use Interpretation, Gulf Coast Public Lab, and *Scenario Journal*; panelists and speakers Eugene Turner, Edward Creef, Paul Aucoin, Travis Bost, Kees Lokman, Richard Hindle, Clint Willson, Ehab Meselhe, Robert Twilley, Jeff Carney, Kyle Graham, Derek Hoeferlin, Matthew Coolidge, Susan Testroet-Bergeron, Charlie Hailey, Windell Curole, Becki Chall, Scott Eustis, Karen Westphal, Sarah Cowles, Stephen Hall, and Brad Cantrell; workshop leaders Brad Cantrell, Alex Robinson, Rich Hindle, Casey Lance Brown, Jeff Carney; and film-makers Ben Mendelsohn and Alex Cholas-Wood.

For DredgeFest Great Lakes, we would particularly like to thank: Ozayr Saloojee and Vincent deBritto, for hosting us at the University of Minnesota, as well as the School of Architecture and Department of Landscape Architecture there; speakers and panelists Peter Annin, Neeraj Bhatia, Ted Smith, Eli Sands, Jen Holmstadt, Chris Bennett, Patrick Phenow, James White, William Hanson, Jim Sharrow, David Knight, María Arquero de Alarcón, Jen Maigret, Matthew Tucker, Ozayr Saloojee, Vincent deBritto, Michael Ezban, Margaux Valenti, Karen Lutsky, and Catherine de Almeida; the St. Anthony Falls Laboratory and Ellicott Dredge Technologies, for the tours; and sponsors University of Minnesota Imagine Chair in the Arts, Design, and Humanities, University of Minnesota Imagine Special Events Fund, University of Minnesota College of Design, University of Minnesota School of Architecture, *Landscape Architecture Magazine*, Great Lakes Dredge and Dock, Bay West, Duluth Seaway Port Authority, and the Minnesota chapter of the American Society of Landscape Architects.

For DredgeFest California, we would particularly like to thank: the Department of Landscape Architecture & Environmental Planning at the University of California, Berkeley, for hosting the event and workshops; workshop leaders Neeraj Bhatia, Richard Hindle, Halina Steiner, Forbes Lipschitz, Nicholas Pevzner, Stephanie Carlisle, Meg Studer, Brennan Cox, Kees Lokman, Matthew Seibert, Benjamin Wellington, and Derek Hoeferlin; contributing experts Lester McKee, Jeremy Lowe, Warner Chabot, Brenda Goeden, John Durand, Bill Fleenor, Richard Norgaard, Elizabeth Murray, Joe Gailiani, and Diana Sokolove; and sponsors Great Lakes Dredge and Dock, *Landscape Architecture Magazine*, UC Davis Hellman Fellows Program, the UC Davis Department of Human Ecology, Groundworks Office, the Delta Protection Commission, and the Dutra Museum Foundation.

We would also like to thank the many friends, colleagues, and collaborators who have supported the DRC and its work during the writing of this book, particularly Jennifer Reut and Brad McKee at *Landscape Architecture Magazine*; Jeff King and Todd Bridges at Engineering With Nature, as well as many other EWN advocates; the Council of Educators in Landscape Architecture, Ball State University, the Exploratorium, Rhode Island School of Design, and Washington University's Sam Fox School of Design for hosting exhibitions of DRC material; Mason White, Lola Sheppard, Maya Przybylski, and Neeraj Bhatia for organizing *Bracket [Goes Soft]*; Tess Ruswick, project designer for Healthy Port Futures and key member of the EWN-LA Design Research Initiative; and Stephen Becker, founding member of the Dredge Research Collaborative, current board member, co-organizer of DredgeFest NYC, and the only member to ever be organized enough to order business cards for the DRC.

Finally, thank you to Gordon Goff, Jake Anderson, Pablo Mandel, and the team at AR+D Publishing for supporting and helping us to realize this project.

Support

Silt Sand Slurry is published with financial support from the Seed Fund grants program, the New York State Council on the Arts grants program, and Auburn University's College of Architecture, Design, and Construction's Seed Grants program.

The Dredge Research Collaborative

The Dredge Research Collaborative is an independent 501c3 nonprofit organization that investigates human sediment handling practices, through publications, events, and other projects. Our mission is to improve sediment management through design research, building public knowledge, and facilitating transdisciplinary conversation.

The DRC was established in 2012, by Stephen Becker, Rob Holmes, Tim Maly, and Brett Milligan. Gena Wirth joined the DRC after DredgeFest NYC, Sean Burkholder and Brian Davis joined for DredgeFest Great Lakes, and Justine Holzman joined for DredgeFest California. For DredgeFest Great Lakes, DredgeFest California, and the production of this book, DRC members were supported by three teams of intern-collaborators, Alex Hill, Gena Morgis, and Inwa Park (2015); Michael Biros, Kate Lenahan, Nick Shannon, and Crystal Xing (2016); Michelle Benoit, Nathalie Mitchell, and Gavin Zeitz (2017). Stephen Becker and Tim Maly are currently inactive members of the DRC.

Our first major initiative was the DredgeFest event series, which began with DredgeFest NYC in fall 2012, and continued through DredgeFest California in summer 2016. The aims of the DredgeFests included building awareness of the practices and significance of sediment management through symposia, public landscape tours, and exhibitions; fostering conversations between sediment management professionals, designers, and members of the public; and exploring the potential of sediment design through workshops.

As the DredgeFests drew to their conclusion, DRC members' efforts shifted toward more direct design engagement with sediment. To name a few of these: Gena Wirth, Brett Milligan, Justine Holzman, and Rob Holmes led the DRC's participation in the 2017–18 Resilient by Design Bay Area Challenge with the Public Sediment team; that proposal is documented in detail in the final chapter of this book. Sean Burkholder and Brian Davis, together with their collaborator Tess Ruswick, launched the Healthy Port Futures initiative in the Great Lakes, supported by the Great Lakes Protection Fund. Two of their built projects are documented in this book's Great Lakes visual essay, "Inland Chokepoint." Brett Milligan took on a leadership role in Franks Tract Futures, a co-design effort in the California Delta, which is discussed in both "Aesthetics and Publics" and the visual essay "Sediment Shortfall." In summer 2017, the DRC helped the US Army Corps of Engineers' Engineering With Nature program convene a workshop at the Engineer Research and Development Center in Vicksburg, Mississippi, to explore ways in which landscape architects might engage with EWN's work. That workshop led to the on-going EWN+LA Design Research Initiative, which all of the active DRC members have contributed to.

Sean Burkholder is the Andrew Gordon Assistant Professor of Landscape Architecture at the University of Pennsylvania's Weitzman School of Design. As a member of the DRC, Sean's work has focused particularly on the Great Lakes region, and with Brian Davis, he co-coordinated Dredgefest Great Lakes in 2015. Since then, he has participated continually in the ongoing collaborative work as part of the USACE Engineering with Nature program. In addition to his work with the DRC, Sean is also the co-founder of the Environmental Modeling Lab at the Weitzman School of Design, which works with local state and federal agencies to model, monitor, and design dynamic coastal landscapes. With fellow DRC member Brian Davis, Sean also co-directs both the Great Lakes Protection Fund-supported Healthy Port Futures Project and the office Proof Projects, both established to investigate novel methods of coastal research and design that place sediment and sediment-based processes as central agents in the design process.

Brian Davis is an associate professor of landscape architecture at the University of Virginia School of Architecture. For the DRC, he co-coordinated Dredgefest Great Lakes in 2015 and has helped lead collaborations with the Engineering with Nature program of the US Army Corps. With Sean Burkholder, Brian directs both the Great Lakes Protection Fund-supported Healthy Port Futures Project and the office Proof Projects. At UVA, he co-directs the Natural Infrastructure Lab, where his research focuses on ports, rivers, and cities, and their formal interrelations through time. Prior to joining UVA, he was an assistant professor at Cornell University and has practiced landscape architecture in Buenos Aires and New York City; he is also a registered landscape architect. In 2019 he was the American Academy Rome Prize recipient in Landscape Architecture.

Brett Milligan is an associate professor of landscape architecture at the University of California, Davis. With the DRC, he developed and co-organized the *DredgeFest* event series, was a core member of the Public Sediment team for the Bay Area Resilient by Design Competition, and consults on projects with the USACE *Engineering with Nature* program. At UC Davis, he is the director of the Metamorphic Landscapes Lab, dedicated to advancing multi-benefit adaptations to accelerated landscape change through design research. Recent projects include *Franks Tract Futures*, a transdisciplinary effort funded by California's Department of Fish and Wildlife to restore tidal marshes, attenuate salinity intrusion, and create new recreational opportunities in California's Delta; and *Delta Island Adaptations*, which is piloting landscape strategies to capture and store carbon, reverse land subsidence, restore terrestrial and aquatic habitats, and create new opportunities for recreation, eco-cultural restoration, and co-stewardship with indigenous tribes on reclaimed lands in the central delta.

Rob Holmes is an associate professor and chair of the undergraduate landscape architecture program at Auburn University. With the DRC, he co-organized the DredgeFest event series, participated in the Resilient By Design Bay Area Challenge as part of the Public Sediment team, and, since 2017, has been engaged in a range of collaborations with the US Army Corps of Engineers' Engineering With Nature program. Other current projects include study of landscape infrastructure designs for coastal communities in Mobile Bay, funded by NOAA's Effects of Sea Level Rise program, and "Feral," an investigation of unruly designed landscapes with David Hill. He is also organizing a Landscape Infrastructure Pilot Studio on the Gulf Coast with the Auburn landscape architecture program's Alabama Lab, funded by the National Academies' Gulf Research Program.

Justine Holzman is a landscape researcher, designer, and educator with a background in landscape architecture. Holzman is currently training as a historian of science at Princeton University where her doctoral work focuses on how knowledge is produced about environments and how landscapes are designed and transformed for scientific research. With the DRC, Holzman co-organized DredgeFest California, was a member of the Public Sediment Team, and has led and progressed funded research and collaboration with the US Army Corps of Engineers. Outside the DRC, Holzman's interdisciplinary scholarship and creative works have contributed to conversations surrounding climate adaptation, green infrastructure, and environmental monitoring. Holzman is co-author of *Responsive Landscapes: Strategies for Responsive Technologies in Landscape Architecture* (2016), and has previously taught in departments of Landscape Architecture, Architecture, and Urban Design at Columbia University, the University of Toronto, the University of Tennessee, and Louisiana State University.

Gena Wirth is a partner and design principal at SCAPE Landscape Architecture. With the DRC, she co-organized the Dredge Fest event series, led the DRC intern-collaborator program, participated in the Resilient by Design Bay Area Challenge as part of the Public Sediment team, and is an advisor to the US Army Corps of Engineers' Engineering with Nature program. At SCAPE, Gena translates research into practice, leading the design and implementation of complex, multi-stakeholder landscapes—including public and private waterfronts, regional trail systems, parks, plazas, and climate adaptation plans. Her portfolio includes Living Breakwaters, a 2,400-linear-foot necklace of nearshore breakwaters designed to reduce risk to coastal communities and regenerate aquatic ecosystems in Staten Island, and the Hudson Highlands Fjord Trail, a 7.5-mile walkable, bikeable linear park connecting people with the sublime landscapes of the Hudson River. Gena has taught at Harvard University, Columbia University, Syracuse University, and Rutgers University.

Introduction: Dredge Matters

1. Where the Dredging Never Stops

No place in North America is dredged more than the deep-draft navigation channel of the Lower Mississippi River, which connects the ports of Baton Rouge, South Louisiana, and New Orleans to the Gulf of Mexico. Along two hundred miles of twisting river, crews of cutterhead and hopper dredges—specialized ships with attached tools for removing sediment from river bottoms—fight a Sisyphean battle to keep the ports open for cargo ships.[1]

Consider that the Mississippi's upland watershed drains about 40 percent of the land area of the lower 48 states, and you can imagine how much sediment ends up here. Some 150 million tons of silt, sand, and clay flow through the channel every year. Much of it is pushed furiously into the Gulf of Mexico, shooting out through the Birdsfoot Delta at river's end, toward the underwater cliff at the edge of the continental shelf.[2] But some of the sediment, particularly the heavier sands, slows down enough to deposit on the river bottom. As spring rains and snowmelt in the American Midwest swell the river with sediment-laden water, the channel gets clogged up.

The US Army Corps of Engineers, which manages most of the nation's shipping channels, is authorized by Congress to maintain the Lower Mississippi at a depth of 50 feet. Shippers plan their cargos accordingly. But when deposition outpaces dredging, shoals form in the river and the channel grows shallower.[3] Sometimes this happens because the river carries more sediment than expected. Sometimes it's because dredging budgets are tight. Either way, shipping is disrupted, companies lose money, ports lose money, upriver sellers of grain and petroleum products lose money, and, because maintaining this navigation channel is a federal project, ultimately dependent on federal funding, congressional representatives start getting phone calls.[4] Sooner or later, more dredges are assigned to the channel. Their crews work twenty-four hours a day, seven days a week, until the standard depth is re-established.[5]

The crews on these dredge ships lead unusual lives. From home ports in places like Boston, Newark, and New York, they are summoned to Gulf channels like the Lower Mississippi, and to farther destinations such as the Panama Canal. Like oil rig crews, they work in rhythms of months on-ship and weeks at home, pulling long shifts while aboard. Their workdays (and the workdays of their compatriot shore crews) are segmented into specialized tasks: extracting chewed-up cable and other debris from the enormous drill bit that is a dredge

1 Dredging is not spread equally along the length of this channel, as the river shoals primarily in the entrance channel, between Venice and the Gulf, and upriver at locations between Baton Rouge and Convent known as "the crossings." The entrance channel has one dredging season, usually from February to July, and the crossings have another, usually from May to September.

2 The dynamics of sediment movement at the advancing limb of the delta are quite complex and disputed. Some sediment settles in the delta itself, while other sediment is transported west along the Louisiana coast. Scientists also believe that some sediment slumps toward the abyssal plain beyond the continental shelf, where it is lost to Louisiana completely.
Muth, "The Once and Future Delta," 14–18.

3 In recent history, this usually happens two or three times per decade.

4 Schleifstein, "Shippers Say."

5 This standardization of a riverbed in service of the efficient flow of commerce is typical of the dominant paradigm for landscape infrastructure design throughout the twentieth century. This paradigm was shaped by many intellectual and practical forces, notably including Taylorism's scientific management, which emphasizes planning, predictability, and control.
Bélanger, "Is Landscape Infrastructure?," 200–1.
O'Neill, *Rivers by Design*.
Shallat, *Structures in the Stream*, 2–4.

cutterhead; monitoring the slow progress of the cutterhead as it cuts a predetermined grade along the river bottom, while the ship creeps forward; riding ATVs beside the long thick black pipes that carry sediment-laden slurry to spotlight-illuminated deposition sites on freshly-built land emerging from the marshes of the Birdsfoot; standing watch on those deposition sites, where slurried sand and silt gushes continuously out of the end of a pipe.

That last moment, where sediment spills out of a pipe to form new land, connects the small drama of dredging to the larger drama of a continent eroding, a river disconnected from its floodplain, and a delta imperiled. Until the twentieth century, the Mississippi River was, in the words of geographer Richard Campanella, a land-making machine. The historic delta plain was formed over thousands of years by sediments that washed down the river, as much as 400 million tons annually.[6] Erosion across a wide swath of the continent provided the sand, silt, and clay to make the salt marsh and muddy channels, the cypress swamps and tidal flats of the Mississippi River Delta. But over the past century, south Louisiana's wetlands have been disappearing. Nearly 2,000 square miles of land have been lost, an area about the size of Delaware.[7] The causes are multiple, complex, interdependent, and disputed. They include sea-level rise, subsidence, severe storms, channels cut into the marshland for oil and gas extraction, and the isolation of the delta from the Mississippi by levees built to control flooding and protect communities.

Today it is the dredge crews that make land. The sediment they remove from navigation channels has to be put somewhere, and using it to shore up disappearing wetlands makes more sense than dumping it in open water.[8] Sometimes dredged material is carried by hopper dredges to a temporary aquatic dump site, where the dredges' split hulls swing open on giant hinges, and the sediment spills out to settle on the river bottom. Later, other dredges, working on their own contracts, arrive at the dump site, and re-dredge the material. At other times, cutterhead dredges can be positioned close enough to marshes that are selected for placement. In either case, slurried sediment is pumped into a long pipeline that floats on the river's surface and then snakes across soft land. Eventually, the slurry reaches a new edge of land and water, where it sprays, settles, and builds. Plants seed themselves on this young soil, and over time spongy marsh emerges from what was once open water.[9]

This book is about stories like that. It starts with a simple premise. We—people within the political territory of the United States—are manipulating sediments at a tectonic scale. That manipulation matters because it shapes the current and future conditions of life. And so it is crucial that we make our manipulations—that we *design with sediment*—intelligently, democratically, and equitably. To do that, we need to look closely at the many terrestrial and aqueous landscapes where sediment is manipulated, examining multiscalar landscape processes and associated systems of management. We need to study the shifting entanglements of economic and ecological systems. We need to consider social innovation and political aesthetics, that is, what is accessible to various publics and what is concealed within institutions and expert knowledge. And through this work, we need to sharpen a critical assessment of technologies, policies, and machines, reckoning with current and past applications that can inform adaptation in the face of multiple crises—not least, climate change.

6 Meade and Moody, "Causes for the Decline."

7 CPRA, *Louisiana's Comprehensive Master Plan*, 2017.

8 This calculation is not just about benefit, though the Corps has used dredged sediment to build up banks and marshes along the delta passes since it assumed responsibility for navigation there in the early twentieth century. It is also a matter of cost and logistics. Dredged material is typically placed as close to the location of its removal as is feasible. Thus, when material dredged from the ten miles upriver and eleven miles downriver of the aquatic dump site (the Head of Passes Hopper Dredge Disposal Area) cannot be directly reused, it is placed at the dump site, while when material dredged further down the Southwest Pass cannot be directly reused, it is placed just beyond the delta in the Ocean Dredged Material Disposal Site.
USACE, New Orleans District, "Current and Future Dredging."

9 Portions of the description in this and preceding paragraphs are based on a tour of dredging operations at Head of Passes facilitated by USACE, New Orleans District in February 2013 and direct communication with Jeff Corbino, Environmental Specialist, USACE, New Orleans District in March 2022.

These concerns are intense within the domain of sediment, but they are not uniquely found there. *Silt Sand Slurry* also speaks generally to the contemporary landscapes of the United States—to their late-capitalist peculiarities, their brokenness, their mix of ugliness and exquisite beauty. It asks, how do we design within the ruins of contemporary landscape infrastructure, while seeking justice, equality, and dignity? How can we shape better futures together? Lessons learned from designing with sediment are often lessons about design and world-making broadly.

2. Moving Earth

Over the last 6,000 years, the quantity of earth moved by human actions has grown continuously. For most of that time, earthmoving was a side effect of activities like plowing fields and clearing forests, which accelerated erosion. Technologies changed slowly, and human influence over sediment expanded in a more or less linear fashion, driven primarily by population growth, which was itself relatively slow. In the past two hundred years, though, the world population grew exponentially, technologies advanced with unprecedented speed, and the rate of earthmoving correspondingly exploded. Picture a "hockey stick" graph—horizontal through most of human history, now rising vertically toward some unknown and novel future.[10]

People today intervene in hydrogeological processes at every scale, accelerating and decelerating flows of sediment, both above and below the waterline, while altering their physical and chemical composition.[11] As sediment moves around on the inclined surfaces of the planet, above and below the waterline, humans play roles both grand (networks of river dams trap enough sediment to produce seismic events) and humble (ubiquitous orange silt fences prevent loose dirt from escaping construction sites).[12] In the United States, an estimated 30 tons of earth is moved annually for every person. This is the highest per capita volume in the world, although only a fraction of the 120 billion tons moved globally.

Roughly one-third of this earthmoving is intentional. Mining companies dig holes, level mountains, and scrape plains. Urban developers level, pile, and sculpt land to accommodate buildings, roads, and other infrastructure. Coasts from the Netherlands to Singapore are shaped by extensive land reclamation, in which sediments are piled up to make dry terrain out of wet.[13] Entire cities are built on dredge. That leaves more than 80 billion tons of earth moved annually through unintentional, unplanned actions, seemingly beyond the grasp of design.[14] Agriculture and deforestation, often working in concert, strip away vegetative cover and loosen soils, accelerating erosion. (Indeed, the earth disturbed and transported by agriculture may be roughly ten times that carried by the world's rivers.[15]) Meanwhile, water control infrastructures like dams and levees decelerate and trap sediments, restricting their flow. Sediment fails to accumulate where it is needed, in places like marshes cut off from rivers, and deposits in places where it is not wanted, like in underwater "deltas" behind dams. All these earthmoving roles, intentional and not, have placed humans in the ethically awkward and highly unprepared position of being the earth's *preeminent geologic agents*.[16]

10 Hooke, "On the History," 843–46.

11 It has frequently been rightly noted that people, meaning global humanity, are not equally implicated in these processes of transformation. Some people—generally wealthy, disproportionately Europeans and their descendants—have benefited greatly from these transformations. Other people are often harmed by them. With these critiques in mind, we use the second-person plural in *Silt Sand Slurry* as an invitation to solidarity at many scales, including that of a planetary humanity, recognizing that every scale of analysis is,an abstraction and requires interrogation from other scales and perspectives.
Chakrabarty, "The Climate of History," 197–222.
McAfee, "The Politics of Nature," 65–72.
Chakrabarty, "Whose Anthropocene?," 103–13.

12 McNeill, *Something New Under the Sun*.
Syvitski and Kettner, "Sediment Flux," 957–75.
Wilkinson, "Humans as Geologic Agents," 161–64.
Wolman, "The Human Impact," 81–98.

13 Masoud, *Terra-Sorta-Firma*.

14 Hooke, "On the History of Humans as Geomorphic Agents."

15 Wolman, "The Human Impact," 85.

16 Cooper et al., "Humans," 222–29.

Still, human agency should not be understood as separate from hydrological and geological cycles; it is coevolutionary. Our earthmoving is interwoven with dynamic landscape processes, feedback loops, and emergent novelty.[17] Landscapes are not passive canvases. They actively and elastically respond to what humans do to them. This in turn affects our subsequent action, which provokes a new response, and so on through time.[18] Agency is unevenly shared between actors as disparate as bulldozers and dredgers, concrete storm channels, corporations, political lobbyists, citizens who live beside hulking containment dikes, rain, streams, tectonic plates, and the molten core of the earth itself, exerting force via gravity.[19]

Dredging is a decisive and forceful moment within this cloud of activity. It is a rapid displacement of sediment that marks a switching point between processes driven primarily by gravity, like erosion, water transport, and siltation, and the co-option of sediment into mechanized systems of uplift, placement, and transformation. Sediments which had been migrating down river beds and along bay bottoms, which might eventually have been compressed into sedimentary rock, are forcefully removed by drill, suction, blast, and claw, to be transported into a novel realm of pipes, barges, dewatering sites, diked storage facilities hundreds of acres large, and wetlands augmented with emplaced sediments.

Within the United States alone, roughly 300 million cubic yards of sediment is annually dredged and redistributed along the nation's coastlines, lakes, rivers, harbors, constructed channels, bays, deltas, and estuaries.[20] This is equivalent to digging a new Panama Canal every year,[21] and the direct bidded cost of these infrastructural projects averages over one billion US dollars.[22] Around the world, dredging shapes waterways from Rotterdam to Singapore, from Lagos to Dubai. A handful of multinational corporations, including the Dutch Boskalis, Belgian Jan de Nul, American Great Lakes Dredge and Dock, and China Harbour Engineering Company, operate fleets of enormous dredgers that rove every ocean, competing for the biggest contracts. Countless smaller operators deal with smaller but more numerous projects.

The impact of dredging can be seen not only in the sheer volume of material removed from navigation channels, but also in the land masses that are produced. In the United States, most navigable channels are accompanied by diked islands and leveed peninsulas that store and dewater dredged material. These "confined disposal facilities" are often quite large. Virginia's Craney Island is composed of three cells each as large as New York City's Central Park. And port cities have relied for centuries (indeed, millennia) on artificial fill—dredged to match excavated materials dumped inside purpose-built dikes—to add valuable new land at their edges. This process has entangled dredging with the political economy of real estate markets,[23] reaching a ludicrous extreme in the construction of Dubai's Palm and World islands, where vast quantities of sand and rock were moved to build landforms in the shape of symbols (a palm tree and a world map) that are legible only from the sky.

Historically, dredging has been perceived in isolation from other landscape processes. Dredged material has been seen as a waste product, as "spoils." The people tasked by society to deal with such material have been asked to think mainly about the most economical way to get rid of it. But a new paradigm is emerging that connects dredging to a broader realm of sediment design. Sediment is increasingly valued as a material that makes landscapes and fosters

17 Holmes, "Design with Change."
Milligan, "Landscape Migration."

18 Time itself, as an abstract measure of duration, is made, fashioned, and experienced as different temporalities, which are a core element of sediment design.
Milligan, "Accelerated and Decelerated Landscapes."

19 The agency of such varied actors in the making of landscapes has been broadly discussed in both landscape architecture and related fields in recent years.
Barnett, *Emergence in Landscape Architecture*.
Bennett, *Vibrant Matter*.
Davis, "Landscapes and Instruments."
Seibert, *Atlas of Material Worlds*.
Tsing et al., *Feral Atlas*.

20 Childs, *Dredged Material Management Categories*.

21 Castro, "Isthmus in the World," 36.

22 USACE, *Navigation Data Center*.

23 Maly, "A City Built on Dredge."

24 These different kinds of agents—as well as bodies and diverse material assemblages—exist in co-evolutionary relationships. All have discernible effects on the others, but none are fully determinative. As a result, their relationships are not fixed; there are points of slippage and decouplings, which enables each of them to act semi-autonomously, at times.
DeLanda, *A New Philosophy of Society*.
DeLanda, *Assemblage Theory*.

25 Critiques of monofunctionality have been an important part of discourse on landscape infrastructure, both within landscape architecture and in related conversations like geographers' critiques of modern infrastructure or the engineering and scientific discussion of "nature-based infrastructure."
Amidon, "Big Nature."
Bélanger, *Landscape as Infrastructure*.
Bridges et al., *Use of Natural and Nature-based Features*.
Graham and Marvin, *Splintering Urbanism*.
Orff and SCAPE, *Toward an Urban Ecology*.

life. Meanwhile, the sheer volume of material being moved and managed by humans has grown exponentially. As fossil fuel emissions accelerate climate change to a dizzying, world-threatening pace, and as ecological communities like beaches, marshes, and floodplains are destroyed on a planetary scale, we should see dredging in this greater context.

Dredging is a clarifying act: it reveals how people intervene in the material flow and deposition of sediments, and it makes visible the technologies, cultural practices, and values that are active in those material effects and feedbacks.[24] Through dredging, we alter, destroy, and create landscapes at the interface between land and water—often dramatically. But this clarifying act should not be seen as wholly exceptional. Dredging is just one type of human activity affecting the flow of sediments. It is a good place to begin understanding sediment design, but we need to consider the much greater extent of human influence. We need to look not just at ports, navigation channels, and spoil islands, but also at urbanized estuaries, upland forests and "sedimentsheds," dune complexes, water control infrastructures like levees and dams, beaches, offshore sand reservoirs, braiding rivers, and more: the full geography of sediment, in its rich diversity and life-supporting dynamism.

3. The State of Sediment Design

Since 2011, we—as members of the Dredge Research Collaborative, with colleagues and partners from a broad range of organizations and institutions—have been working to understand the current state of sediment design in the United States. We've focused on four distinct regions: the Mississippi River Delta; the New York-New Jersey Harbor Estuary; the Great Lakes; and California's Bay-Delta. These regions sample geographic differences between the four coasts of the lower 48 states. Through a roving, transdisciplinary event series called "DredgeFest," we brought together researchers, public agencies, private industry, and members of the general public to investigate the human manipulation of sediments, using mapping and visualization, symposiums, design workshops, and landscape tours.

One thing we learned is that many sediment systems are formatted for conditions that no longer apply, values that have changed, and performance criteria that are outdated. Like most human activities, sediment design has significant institutional inertia. For example, the government policies and institutions that set the regulatory framework for sedimentary management in California's Bay Area evolved during a period known as "mudlock," when the bay received greater supplies of sediment from upstream sources than it could handle. In the mid-twentieth century, environmental activists mobilized against developers who hoped to build on the bay's muddy shallows. Their efforts protected the bay from the major threat of that era, that the bay would be filled in. But that activism combined with the sedimentary excess of mudlock to shape a Bay Area regulatory framework that is ill-equipped to cope with a reversal in conditions. Plans to rebuild lost wetlands and protect against sea-level rise are now stymied by a sediment shortage.

Like many infrastructures, sediment systems have been designed according to narrow criteria prioritizing an efficient mono-functionality.[25] For instance, the "confined disposal facilities" (CDFs) of the Great Lakes were precisely engineered to dewater sediment efficiently; any recreational or ecological value they achieved was (almost always) accidental, even though CDFs are such promising bird habitats that their operations are often disrupted by endangered birds.[26] Similarly, the monumental levees, dams, and concrete channels that encase rivers like the Sacramento and Mississippi were engineered to achieve strict floodwater conveyance criteria; the distribution of sediment by those floodwaters, which is crucial to the growth of estuaries and deltas, was not considered.

For the most part, technocratic design goals have been pursued through decision-making processes that only rarely emerge into public discourse or awareness. If infrastructure in general tends to be metaphorically "black-boxed"[27]—assumed rather than interrogated, hidden rather than explained—this is even more true of sediment. For one thing, it is often literally submerged. And even when it is above the waterline, piled on land or packed into a dump truck, sediment often disappears by virtue of its quotidian nature. It is merely "dirt." The invisibility of sediment makes it a difficult infrastructure to mobilize around or advocate for, and that lack of public interest has translated into disinvestment, like the underfunding of navigation channels in the Great Lakes. Even more worryingly, sediment systems see little to no pressure toward equitable design; this often means, for instance, that contaminated dredged material is placed in diked ponds near marginalized communities.

So, what would better sediment design look like? In Louisiana, nonprofits have banded together to empower citizen-scientists who monitor marshes with cameras hanging from DIY balloon rigs and citizen-dredgers who build marshes with personal mini-dredges. In California, an entrepreneur has formed partnerships with local ports, the US Army Corps of Engineers, and environmental regulators to build hundreds of acres of new wetlands on land drained for agriculture in the nineteenth century, using sediment dredged from San Francisco Bay.[28] In New York, after Hurricane Sandy, government agencies, communities, and designers have invented and are now implementing a new generation of coastal infrastructure that is multifunctional, adaptive, and biologically fecund.[29]

Containment facilities for dredged material can be sited where new land is most desirable, not merely where the facilities are cheapest to build. The land can be sculpted to perform in new ways: to create nesting habitat for endangered shorebirds, or host aquaculture,[30] or phytoremediate toxins from contaminated slurries. Other facilities can be used as temporary way-stations, landscape machines for dewatering and processing sediment before it is excavated, hauled off, and redeployed. And containment can be avoided entirely through new techniques for placing sediments to build wetlands, shallows,[31] offshore sand reservoirs,[32] and other valuable landforms, bolstering ecosystems and harnessing processes like wave action as co-shapers of new landscapes.

Similar opportunities exist within other domains of sedimentary infrastructure, management, and manipulation. New dams can be designed and existing dams retrofitted to better convey sediment, to ameliorate impacts on ecological systems, and to provide for the public, communal experience of sediment at grand scales. Levees can be made more adaptive to environmental conditions—more membrane than barrier—and can serve multiple ecological and recreational purposes. Breakwaters, groins, and other coastal structures can be

26 Burkholder, "Designing Dredge."

27 Hinchliffe, "Technology, Power, and Space," 665.
Star, "The Ethnography of Infrastructure."

28 Camhi, "How Dredging the Bay is Helping."

29 Ovink and Boeijenga, *Too Big*.

30 USACE Waterways Experiment Station, *Site Selection, Acquisition, and Planning*.

31 Thin-layer placement of dredged material has recently become an important technique for adding sediment to existing shallows, flats, and wetlands.
US Army ERDC, *Thin Layer Placement*.

32 There have been several decades of experimental landform design utilizing the unconfined "strategic placement" of dredged material. The Dutch Zandmotor is perhaps the best known of these; one early American antecedent was the National Berm Demonstration Program, initiated off the Alabama coast in 1987.
McFall et al., "Evaluation techniques."

33 Suedel et al., "Creating Horseshoe Bend Island."

34 Nassauer, Santelmann, and Scavia, *From the Corn Belt to the Gulf.*

35 Orff, "What is Design Now?"

designed to admit the inherently dynamic structure of coasts, accommodating change rather than seeking to fix the shoreline in place.

New sediment design should not be limited to making the conventional infrastructure of the twentieth century more ecologically workable. Systemic change is possible, even necessary. Radical experiments are underway in places like southern Louisiana, where thickly forested Horseshoe Bend Island is growing, fed by sediment dredged from an adjacent navigation channel and strategically placed where currents will carry it to the island.[33] Sites of sedimentary surplus can be linked to sites of deficit, through pipelines, new regulatory policies, and sediment trading exchanges. Intentionality can be introduced into matrices of unintentional effects, by altering practices such as farming techniques, or by developing new infrastructural networks, such as wetlands strategically distributed among farm fields in Iowa to reduce fertilizer nutrient loads downstream.[34]

Ideas like this have shaped Public Sediment, a proposal we helped develop for the Resilient by Design Bay Area Challenge. The plan aims to relink upland sediment sources with the muddy Baylands that ring San Francisco Bay. It would unmake obsolete infrastructure and create space for sediment to facilitate dynamic interactions between water, human communities, and non-human species.[35] This is the challenge and opportunity of sediment design today: to address the many issues that sediment systems are entangled with while cultivating the value latent in them.

4. Our Guide to Designing with Sediment

In this book, we explore how sediment systems came to be the way they are, how they relate to other concerns, and how those systems and relationships might be redesigned.

We take as our starting point the four coasts studied by the Dredge Research Collaborative, which are surveyed here in five visual essays. Immediately following this introduction, *Dredge 101* describes the why, where, and how of dredging, situating the four coasts in geographic and material relationship. If you are not familiar with dredging, you'll want to begin there. The other four visual essays each focus on one coast, showing the sedimentary challenges facing these regions, alongside hopeful examples that point toward the future of sediment design. *Dredge City* looks at the harbors of New York and New Jersey, where a shipping economy is intertwined with the Hudson River sedimentshed, shaping disposal landscapes and opening opportunities to deploy sediment to renew vanishing marsh islands. The Great Lakes are the focus of *Inland Chokepoint*. There, sediment managers have been grappling with how and where to place an expanding sediment surplus as budgets shrink, demanding innovation and producing novel landscape types. *Disappearing Delta* examines the Mississippi River and coastal Louisiana, where rapid land loss—tied in part to enormous water-control infrastructures—is prompting crucial experiments in moving sediment to build land. The final coast is documented in *Sediment Shortfall*. The California Bay-Delta has been dramatically transformed by colonial settlement, infrastructure-building, and urbanization. Climate change is now further accelerating that transformation, and sediment will be crucial to adaptation.

Text chapters alternate with these visual essays. Like the visual essays, each text chapter can be read on its own; collectively, they lay out a program for designing sedimentary landscapes.

Choreographing Sediment explores the human alteration of sediment at a grand scale. We discuss how flows of sediment have been refashioned through material, economic, social, and political networks. These networks alter distribution patterns across space and time, producing sediment surpluses and shortages across landscapes. Many environmental and social problems are generated through these networked displacements and emplacements—for example, rapid land loss in the Mississippi River Delta. We argue that these pressing concerns can be addressed through choreography: the intentional, holistic, and dynamic orchestration of surpluses and deficits across entire *sedimentsheds*. We use the example of New York City's Jamaica Bay to show why accounting for change is central to the task of choreography—and to show how current restoration practices fall short. Choreography involves working without perfect knowledge and in the context of disagreements over goals and values, as seen in Louisiana's Coastal Master Plan, America's largest ongoing sediment design project.

Mud is Vital looks closely at how and where sediment can be ecologically and economically valuable. We begin in the Great Lakes, where Duluth's Erie Pier has been transformed into a machine for sorting dredged sediments. This transformation is part of a broader shift toward sediment being valued as a resource and a means of ecological reconciliation. We delve into the history of American sediment management, revealing the consequences of the nation's historical failure to value sediment. We consider the example of the underwater Mud Dump Site, off the coasts of New York and New Jersey, which was a wasteful component of an economic system that treated sediment as a disposal problem, to reveal the ecological consequences of that approach: contamination, degradation, and habitat loss. We then trace the factors that are propelling a shift toward valuing sediment, including the recognition of the problems produced by wasteful disposal, the expanding volume of sediment being dredged, and the accelerating environmental crises that sediment is linked to—particularly climate change. We conclude by looking at a few more places where innovative sediment practices are engendering new ecological systems. We discuss how sediment is entangled with some of the biggest issues of political economy and ecological imagination facing humanity in the twenty-first century.

Aesthetics and Publics examines the many ways that sedimentary landscapes are experienced and proposes ideas for how a wider range of participants might engage with and value these landscapes. Sediment is stealthy and often invisible. It is often in motion, but its movements usually occur underwater and out of sight. Additionally, the institutional and bureaucratic processes through which sediment is managed are orchestrated behind closed doors by actors who rarely represent the diversity of impacted communities or the interests of society as a whole. Grappling with these invisibilities and associated environmental justice issues is crucial to improving the design of landscapes. Aesthetics—by which we mean not merely appearances, but an entire realm of sensorial and socio-political experiences—are central to the processes of design. We advocate for expanding and democratizing the actors engaged in the making of

36 The work of our colleagues in design firms like SCAPE Landscape Architecture and Mahan Rykiel Associates who have begun to design with sediment, is, unfortunately, an exception that proves the rule. Much broader and more expansive engagement by landscape architects is necessary.
O'Connell, "Dredging up the Future."

sedimentary landscapes. We detail efforts at fostering new aesthetic sensibilities for *sediment as infrastructure*, emphasizing process, performance, and political access. We conclude by describing our efforts to build *sediment publics* through the DredgeFest series.

Earthmoving techniques vary widely, from embodied physical action, to industrial machines like clamshell dredgers, to policies and protocols that structure behavior and acceptable practices. For nearly every landscape technology that has been designed and deployed, intentional effects are accompanied by unintended effects that place humans and other species at risk. The complexity and indeterminacy of landscapes nearly always exceeds human intention and technological foresight. The accelerated landscape change of the current global ecological crisis means that the need for socio-technical adaptation is great, yet the unknowns and risks of failure are great too. In *Technology, Pace, and Adaptation*, we call for a humble approach to technology, guided by public and transparent design processes. We probe how adaptation pathways might be safely tested and refined through small-scale prototyping of experimental landscapes. We focus on the pre-colonial and colonial landscape technologies that made and remade California's Sacramento-San Joaquin Delta and the downstream San Francisco Bay, and we discuss how policies and practices can adapt to meet new conditions and evolving values.

Interspersed with the text chapters and the visual essays are contributions from other voices. The visual essays include interviews with practitioners working with sediment, including both experts from within key institutions like the US Army Corps of Engineers and community advocates who are working to change design practices. Three essays by our Dredge Research Collaborative colleagues Sean Burkholder, Brian Davis, and Justine Holzman offer further perspective on sediment design. Sean considers how the dredged landscapes of the Great Lakes teach us to design with fluctuation and change, rather than against it. Brian argues for river sediment as a public good, a source of rich confluence between ecological need, urban life, and aesthetic purpose. And Justine looks closely at the history of the science of sediment, where techniques of measurement, modeling, and detection have shaped how sediment is used as a design material.

The challenges we have outlined—choreographing sediment flows amid landscape change, seeing sediment as an ecological resource, making sediment a public and accessible commons, and dealing with the transformations wrought by landscape technologies—indicate the need for a paradigm shift in how landscapes are made. The final chapter, *Why Design?*, synthesizes these lessons through a close examination of Public Sediment. That plan responds to the bay's sediment shortfall by reconfiguring channelized Alameda Creek to prioritize sediment, people, and fish. Public Sediment links sediment to social and ecological needs in a region marked by stark inequality and faced with rapid environmental change, and it demonstrates how design can unlock more inclusive decision-making.

Although earth and soil underlie all landscape architecture, designers have not yet assumed a significant and sustained role in working with sediment.[36] Designers must make a case that they belong at the table alongside the many scientists, policy makers, public advocates, industry practitioners, and other stakeholders who are already working to bring new sedimentary worlds into being. This book is our effort to consider what designers need to learn, and how they might best contribute.

Humans are already redesigning the planet, both deliberately and inadvertently.[37] In this sense, we have no choice but to design. Altered sediment systems are just one piece of a larger process, which Jedediah Purdy has called "a kind of collective landscape architecture" that enrolls all humanity as we "shape the world by living."[38] Our designs make many worlds on the surface of the Earth.[39] Agency in how these worlds are shaped is neither equally nor equitably distributed, but we all must live within them, and so we all have a stake in how they are being made. Thus, sediment design should not belong exclusively to professionals. The challenges we face in designing better sediment systems, from learning to design infrastructures that foster dynamic, healthy landscapes to democratizing this work, are challenges that reverberate through many dimensions and instances of our collective world-shaping. When we look closely at sediments—observing why they move, where they deposit, and how they relate to all that they touch—we see many worlds more clearly.

It's time to get muddy.

37 The breadth of this influence across topics beyond sediment has been well-documented in multiple recent sources, such as John Robert McNeill's environmental history of the twentieth century *Something New Under the Sun*, the work of Erle Ellis's Laboratory for Anthropogenic Landscape Ecology, Globaïa's Cartography of the Anthropocene, the International Geosphere-Biosphere Programme, and HKW's Anthropocene project.

38 Purdy, *After Nature*, 22.

39 These worlds are thick with material, ecological, social, and political dimensions. Escobar, *Designs for the Pluriverse*. Fry, *Design Futuring*.

Bibliography:

Jane Amidon, "Big Nature," in *Design Ecologies: Essays on the Nature of Design*, eds. Lisa Tilder and Beth Blostein (New York: Princeton Architectural Press, 2010): 165–81.

Rod Barnett, *Emergence in Landscape Architecture* (New York: Routledge, 2013).

Pierre Bélanger, "Is Landscape Infrastructure?," in *Is Landscape...?*, eds. Gareth Doherty and Charles Waldheim (New York: Routledge, 2016).

Pierre Bélanger, *Landscape as Infrastructure: A Base Primer* (New York: Routledge, 2017).

Jane Bennett, *Vibrant Matter: A Political Ecology of Things* (Duke University Press, 2010).

Todd S. Bridges, Kelly Burks-Copes, Matthew Bates, Zachary Collier, J. Craig Fischenich, Candice Piercy, Edmond Russo, et al., *Use of Natural and Nature-based Features (NNBF) for Coastal Resilience*, Vicksburg: US Army Engineer Research and Development Center, Coastal and Hydraulics Laboratory, 2015.

Sean Burkholder, "Designing Dredge: Engaging the Sediment Landscapes of the Great Lakes Basin," *Journal of Landscape Architecture* 11, no. 1 (2016): 6–17.

Tiffany Camhi, "How Dredging the Bay is Helping Restore Delta Wetlands," *KEQD*, August 15, 2018, https://www.kqed.org/news/11686939/how-dredging-the-bay-is-helping-restore-delta-wetlands.

Guillermo Castro, "Isthmus in the World: Elements for an Environmental History of Panama," *Global Environment* 1 (2008).

Dipesh Chakrabarty, "The Climate of History: Four Theses," *Critical Inquiry* 35, no. 2 (2009): 197–222.

Dipesh Chakrabarty, "Whose Anthropocene? A Response" in: "Whose Anthropocene? Revisiting Dipesh Chakrabarty's 'Four Theses,'" eds. Robert Emmett and Thomas Lekan, *RCC Perspectives: Transformations in Environment and Society* 2 (2016), 103–13.

John Childs, *Dredged Material Management Categories for Tracking Beneficial Use*, DOER Technical Notes Collection, ERDC TN-DOER-R22, Vicksburg, MS: US Army Engineer Research and Development Center, https://budm.el.erdc.dren.mil/guidance/ERDC-TN-DOER-R22.pdf.

Coastal Protection and Restoration Authority of Louisiana, *Louisiana's Comprehensive Master Plan for a Sustainable Coast*, 2017, ES-2.

Anthony H. Cooper, Teresa J. Brown, Simon J. Price, Jonathan R. Ford, and Colin N. Waters, "Humans are the Most Significant Global Geomorphological Driving Force of the 21st Century," *The Anthropocene Review* 5, no. 3 (2018): 222–29.

Brian Davis, "Landscapes and Instruments," *Landscape Journal* 32, no. 2 (2013): 293–308.

Manuel DeLanda, *A New Philosophy of Society: Assemblage Theory and Social Complexity* (New York: Continuum, 2006).

Manuel DeLanda, *Assemblage Theory* (Edinburgh University Press, 2016).

Arturo Escobar, *Designs for the Pluriverse: Radical Interdependence, Autonomy, and the Making of Worlds* (Durham: Duke University Press, 2018).

Tony Fry, *Design Futuring: Sustainability, Ethics, and New Practice* (New York: Bloomsbury Visual Arts, 2018).

Stephen Graham and Simon Marvin, *Splintering Urbanism: Networked Infrastructures, Technological Mobilities and the Urban Condition* (New York: Routledge, 2001).

Steve Hinchliffe, "Technology, Power, and Space—the Means and Ends of Geographies of Technology," *Environment and Planning D: Society and Space* 14 (1996): 665.

Rob Holmes, "Design with Change," in *Design with Nature Now*, eds. Frederick Steiner, Richard Weller, Karen M'Closkey, and Billy Fleming (Cambridge, MA: Lincoln Institute of Land Policy, 2019).

Roger Hooke, "On the History of Humans as Geomorphic Agents," *Geology* 28, no. 9 (2000): 843–46.

Tim Maly, "A City Built on Dredge," *Urban Omnibus*, December 12, 2012, https://urbanomnibus.net/2012/12/a-city-built-on-dredge/.

Fadi Masoud, *Terra-Sorta-Firma: Reclaiming the Littoral Gradient* (New York: Actar Publishers, 2021).

Kathleen McAfee, "The Politics of Nature in the Anthropocene" in: "Whose Anthropocene? Revisiting Dipesh Chakrabarty's 'Four Theses,'" eds. Robert Emmett and Thomas Lekan, *RCC Perspectives: Transformations in Environment and Society* 2 (2016), 65–72.

Brian McFall, Katherine Brutsché, Anthony Priestas, and Douglas Krafft, "Evaluation techniques for the beneficial use of dredged sediment placed in the nearshore," *Journal of Waterway, Port, Coastal, and Ocean Engineering* 147, no. 5 (2021).

J. R. McNeill, *Something New Under the Sun: An Environmental History of the Twentieth-century World* (New York: W. W. Norton & Company, 2001).

Robert H. Meade and John A. Moody, "Causes for the Decline of Suspended-Sediment Discharge in the Mississippi River System, 1940–2007," *Hydrological Processes: An International Journal* 24, no. 1 (2010): 35–49.

Brett Milligan, "Accelerated and Decelerated Landscapes," *Places Journal* (2022), https://placesjournal.org/article/accelerated-and-decelerated-landscapes/.

Brett Milligan, "Landscape Migration," *Places Journal* (2015), https://placesjournal.org/article/landscape-migration/.

David P. Muth, "The Once and Future Delta," in *Perspectives on the Restoration of the Mississippi Delta*, eds. John W. Day, G. Paul Kemp, Angelina M. Freeman, and David P. Muth (Springer: 2014), 14–18.

Joan Nassauer, Mary Santelmann, and Donald Scavia, eds., *From the Corn Belt to the Gulf: Societal and Environmental Implications of Alternative Agricultural Futures* (Washington, DC: Resources for the Future Press, 2007).

Kim O'Connell, "Dredging up the Future," *Landscape Architecture Magazine* 107, no. 12 (2017): 74–91.

Karen O'Neill, *Rivers by Design: State Power and the Origins of U.S. Flood Control* (Durham: Duke University Press, 2006).

Kate Orff, "What is Design Now? Unmaking the Landscape," *Architectural Design* 90, no. 1 (2020): 94–99.

Kate Orff and SCAPE, *Toward an Urban Ecology* (New York: The Monacelli Press, 2016).

Henk Ovink and Jelte Boeijenga, *Too Big: Rebuild by Design: A Transformative Approach to Climate Change* (Rotterdam: nai010 publishers, 2018).

Jedediah Purdy, *After Nature: A Politics for the Anthropocene* (Cambridge, MA: Harvard University Press, 2015).

Mark Schleifstein, "Shippers Say Mississippi River Sediment Again Costing Them Millions," *The Times-Picayune*, April 16, 2016, http://www.nola.com/environment/index.ssf/2016/04/mississippi_river_sediment_aga.html.

Matthew Seibert, ed., *Atlas of Material Worlds: Mapping the Agency of Matter for a New Landscape Practice* (New York: Routledge, 2021).

Todd Shallat, *Structures in the Stream: Water, Science, and the Rise of the U.S. Army Corps of Engineers* (Austin: University of Texas Press, 2010).

Susan Star, "The Ethnography of Infrastructure," *American Behavioral Scientist* 43, no. 3 (1999): 377–91.

Burton Suedel, Jacob Berkowitz, Sung-chan Kim, Nathan Beane, Elizabeth Summers, Darrell Evans, and Jeffrey Corbino, "Creating Horseshoe Bend Island, Atchafalaya River, Louisiana," *Terra Et Aqua* 140 (2015): 26–31.

James Syvitski and Albert Kettner, "Sediment Flux and the Anthropocene," *Philosophical Transactions of the Royal Society A: Mathematical, Physical and Engineering Sciences* 369, no. 1938 (2011): 957–75.

Anna Tsing, Jennifer Deger, Alder Keleman Saxena, and Feifei Zhou, *Feral Atlas: The More-than-human Anthropocene* (Stanford University Press, 2020).

US Army Corps of Engineers, *Navigation Data Center*, http://www.navigationdatacenter.us/dredge/drgcorps.htm.

US Army Corps of Engineers, New Orleans District, "Current and Future Dredging Along the Lower Mississippi River" by Mark Wingate, presentation to Association of Levee Boards of Louisiana 79th Annual Meeting (December 4, 2019), http://albl.org/wp-content/uploads/2019/12/Mark-Wingate.pdf.

US Army Corps of Engineers Waterways Experiment Station, Site Selection, *Acquisition, and Planning for Aquaculture in Dredged Material Containment Areas*, by Jonathan Wilson, Jurij Homziak, and Richard Coleman, Technical Report EL-93-13, Vicksburg, MS: US Army Corps of Engineers, 1993.

US Army Engineer Research and Development Center, *Thin Layer Placement: Technical Definition for U.S. Army Corps of Engineers Applications*, by Jacob Berkowitz, Candice Piercy, Tim Welp, and Christine VanZomeren, Technical Note ERDC TN-19-1, Vicksburg, MS: US Army Corps of Engineers, 2019.

Bruce Wilkinson, "Humans as Geologic Agents: A Deep-time Perspective," *Geology* 33, no. 3 (2005): 161–64.

Gordon Wolman, "The Human Impact: Some Observations," *Proceedings of the American Philosophical Society* 146, no. 1 (2002): 81–98.

Emergency dredging on the Mississippi River (US Army Corps of Engineers, George Stringham, 2014)

Dredge 101

Dredging is the active removal of sediment from the bottoms of lakes, rivers, harbors, and other water bodies. Dredging enables much of modern life, yet the practice is little known, except to technical experts and engineers. When you eat a banana shipped from South America or sunbathe on a beach replenished by offshore sand, you benefit from landscapes of dredge. Dredge 101 is an introduction to the mucky and murky practice of dredging.

Why dredge?

We dredge to control the movement of sediment—to take sediment from where it is not wanted and place it where it is desired. The practice of dredging reveals conflicts between dynamic landscape processes and fixed patterns of urbanization.

In the United States, most dredging occurs to enable navigation. Shipping channels are constructed and maintained to allow the safe passage of commercial and recreational vessels that account for billions of dollars in commerce and transportation. The US alone has more than 25,000 miles of dredged channels feeding more than 400 ports.[1] Many ports are concentrated at the mouths of the nation's muddiest rivers, ensuring a consistent demand for dredging that extends from the coasts to inland waterways.

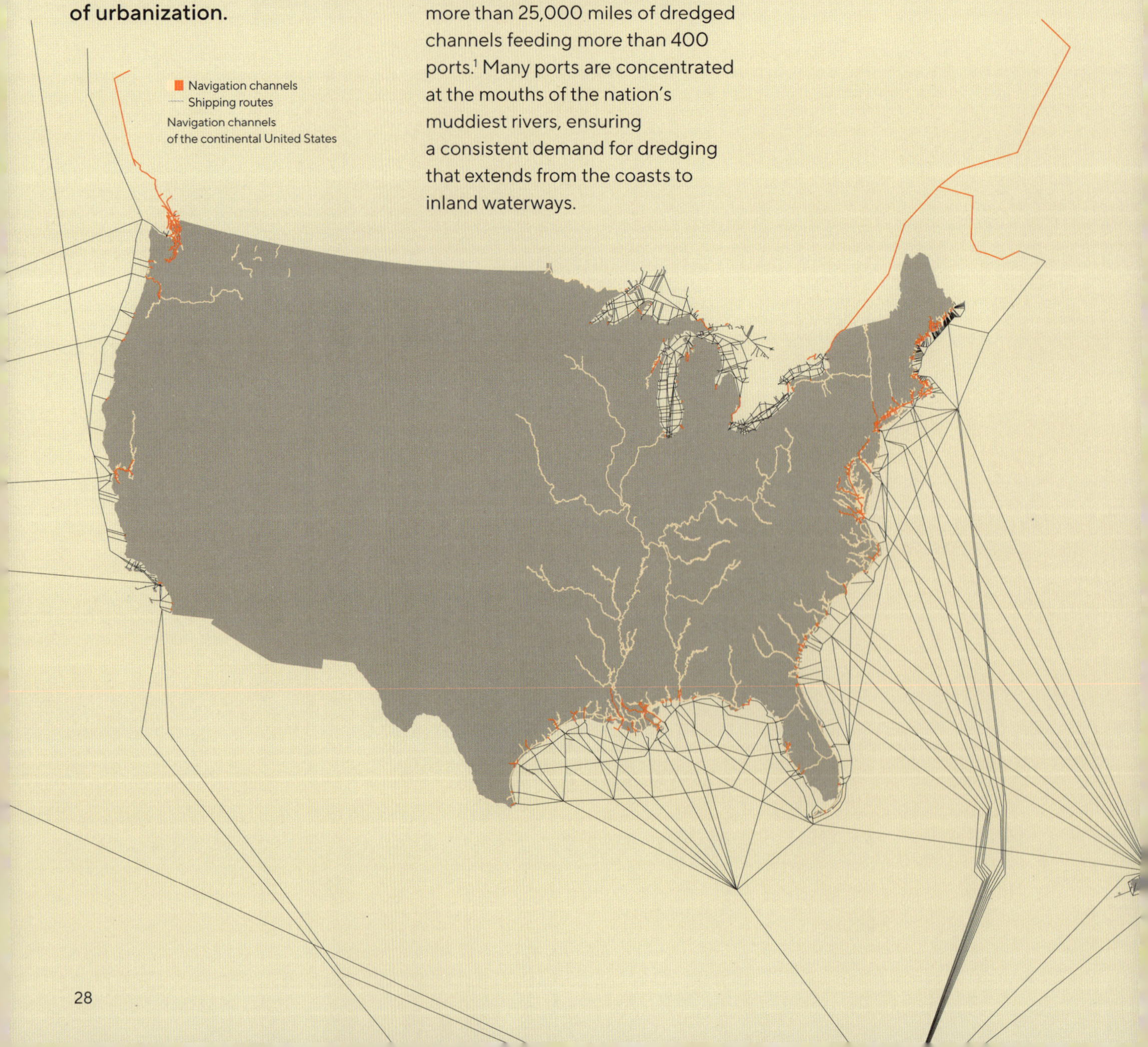

Navigation channels of the continental United States

Channel deepening in New Jersey (Vince Elias, US Army Corps of Engineers)

Why dredge, cont.

Shipping channels are periodically cleaned to enable continuous access, and often deepened to accommodate larger vessels. But there are other reasons to dredge besides navigation. Dredging is used for environmental projects, such as removing legacy contaminants from water bodies or placing sand and silt for marsh and mudflat restoration.[2] Dredging removes sediment from channels that manage floodwater, especially in areas where hydrologic flows and patterns of sediment transport are altered. Sometimes dredging occurs simply to excavate the raw material of sand, gravel, and silt for another use. Increasingly, sediment is a valuable commodity harvested to replenish recreational beaches; build coastal protection features like dikes, dunes, and levees; and protect vulnerable neighborhoods from sea level rise.

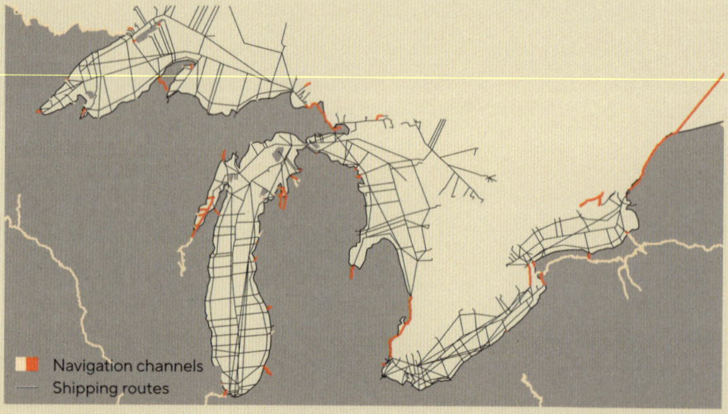

Dredging for navigation in the channels of the Great Lakes

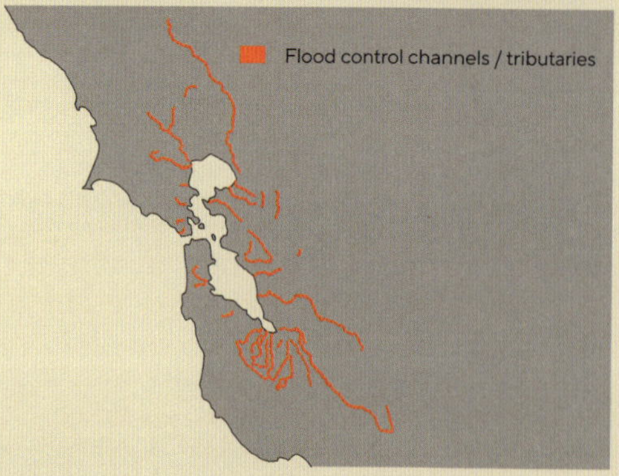

Dredging for flood control in the tributaries of the San Francisco Bay

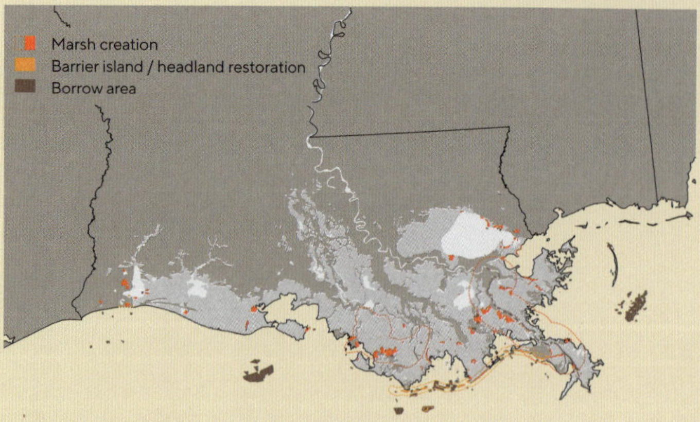

Dredging for marsh building in Southern Louisiana

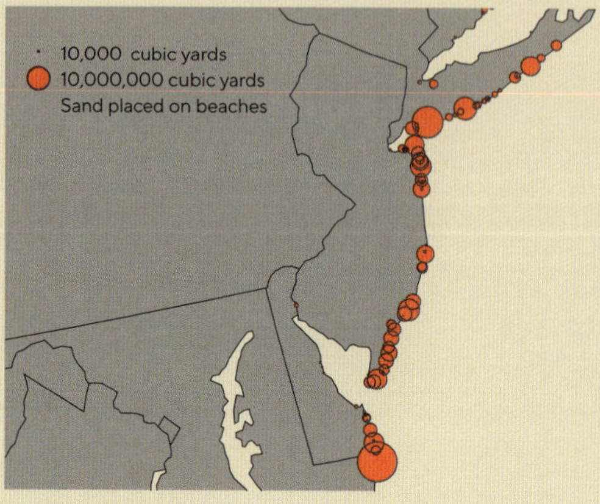

Dredging for beach nourishment in the New York / New Jersey region, 1923-2016

Dredging sediment from the Mississippi River for marsh creation in Louisiana

Dune construction and beach berm replenishment in New Jersey with dredged sand (US Army Corps of Engineers, 2015)

What is dredged material?

Dredged material is excavated sediment removed from the bottoms of rivers, lakes, and oceans. Dredging pulls up material deposited by short- and long-term processes of sediment transport.

Fine-textured silts and clays become suspended in water and can move along with the slightest current or slowest tide. Coarser, heavier sands and cobbles move downstream in higher energy storm events, like flash floods or hurricanes, pushed around by fast water and crashing waves. Sediment settles to the bottom as the water slows, often in navigation channels and harbor entrances, triggering a desire for dredging.

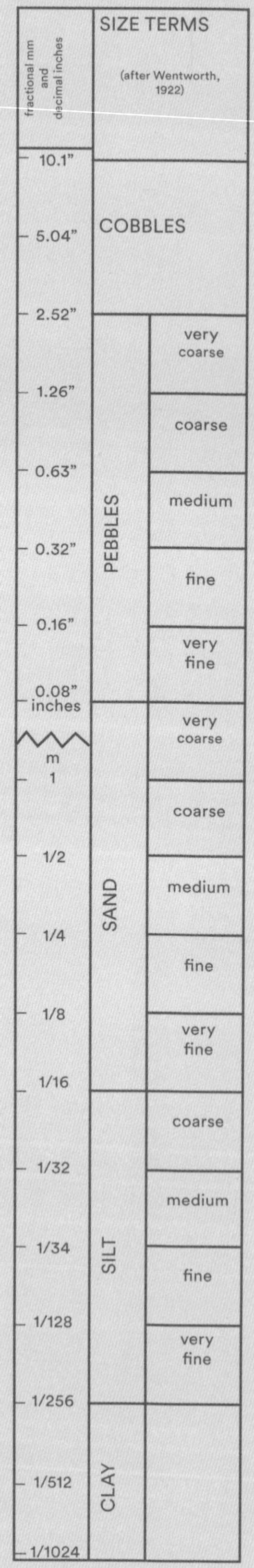

Wentworth grain size chart

Beach replenishment after Hurricane Sandy with dredged sand at Rockaway Beach, NY
(Chris Gardner, US Army Corps of Engineers, 2013)

Where does dredged material come from?

Every water body has a sedimentshed, a geographic area where sediment erodes off mountains and hillsides and moves into the water system.

Dredging frequently occurs at the downstream end of the system. When water bodies meet the ocean, the water slows, depositing sediments which form deltas and other shallow wetland environments that have historically been ideal sites for agriculture, trade, and settlement. As these environments urbanize over time, edges harden, boats and channels grow larger, and more material is removed through dredging.

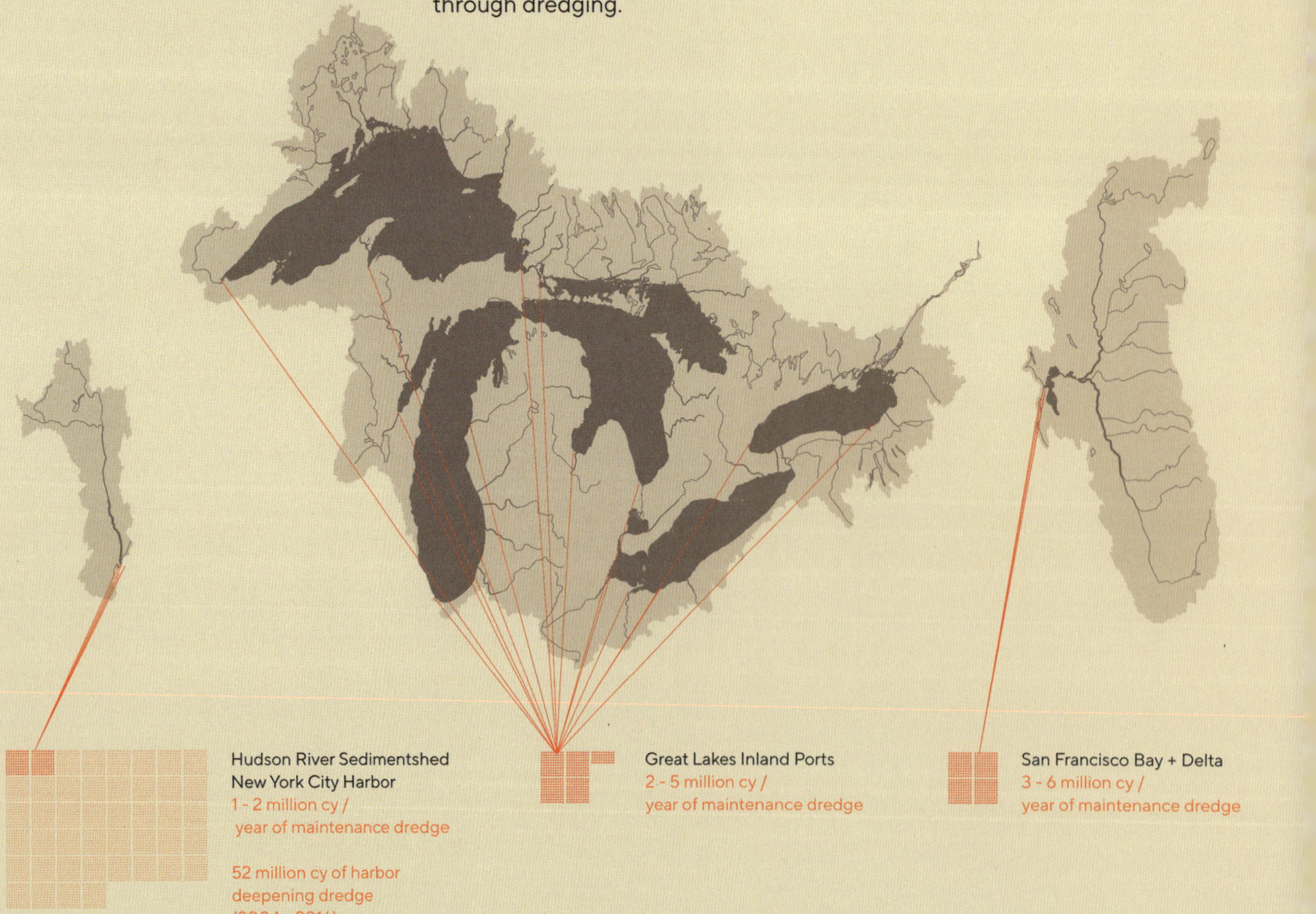

Sedimentsheds and average annual dredging volumes across the four coasts

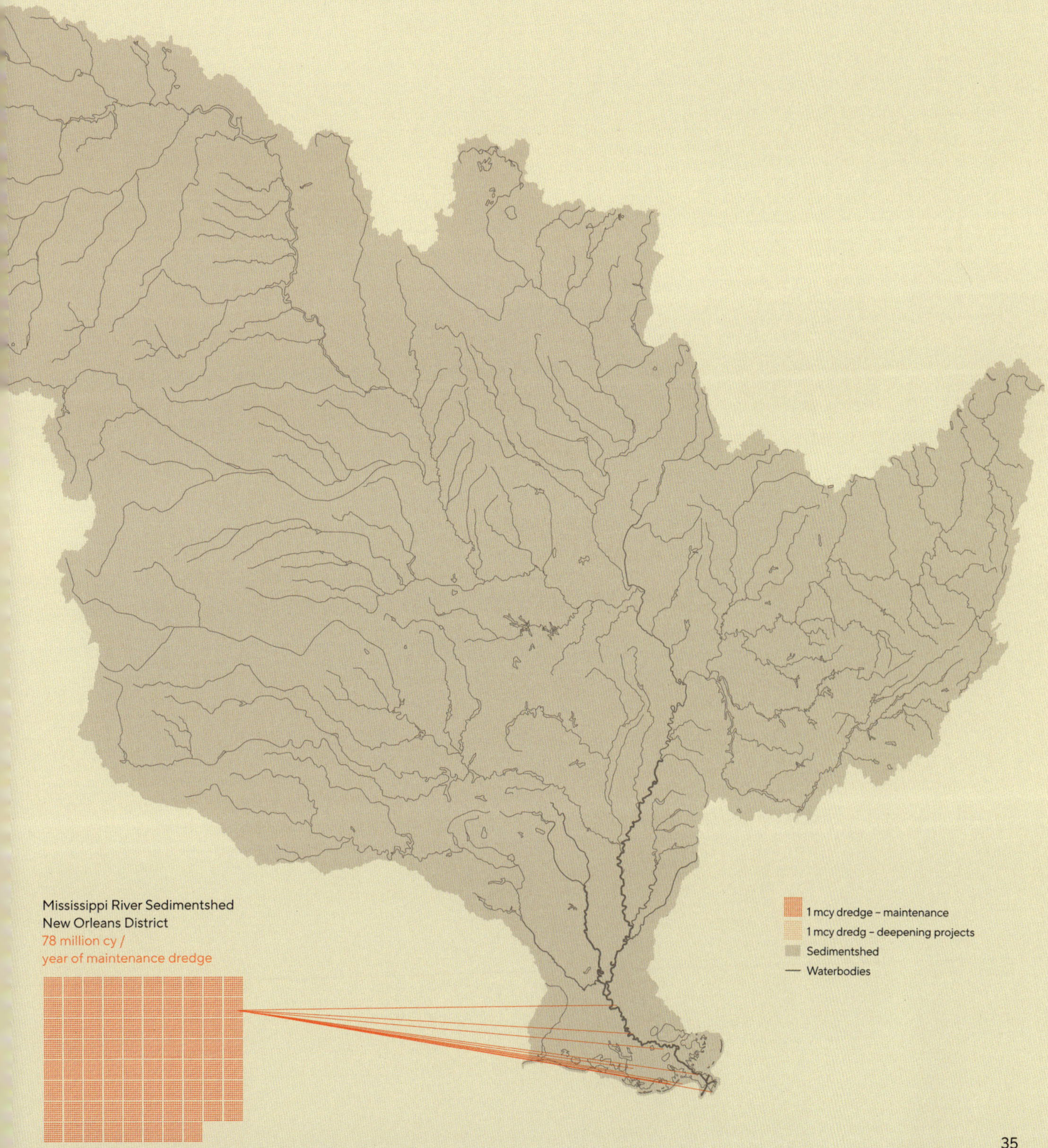
Mississippi River Sedimentshed
New Orleans District
78 million cy /
year of maintenance dredge
1 mcy dredge – maintenance
1 mcy dredg – deepening projects
Sedimentshed
Waterbodies

How much mud do we move?

We move a lot of sediment. The US Army Corps of Engineers coordinates the dredging of approximately 210 million cubic yards of material a year for channel maintenance, channel deepening, and other uses.[3]

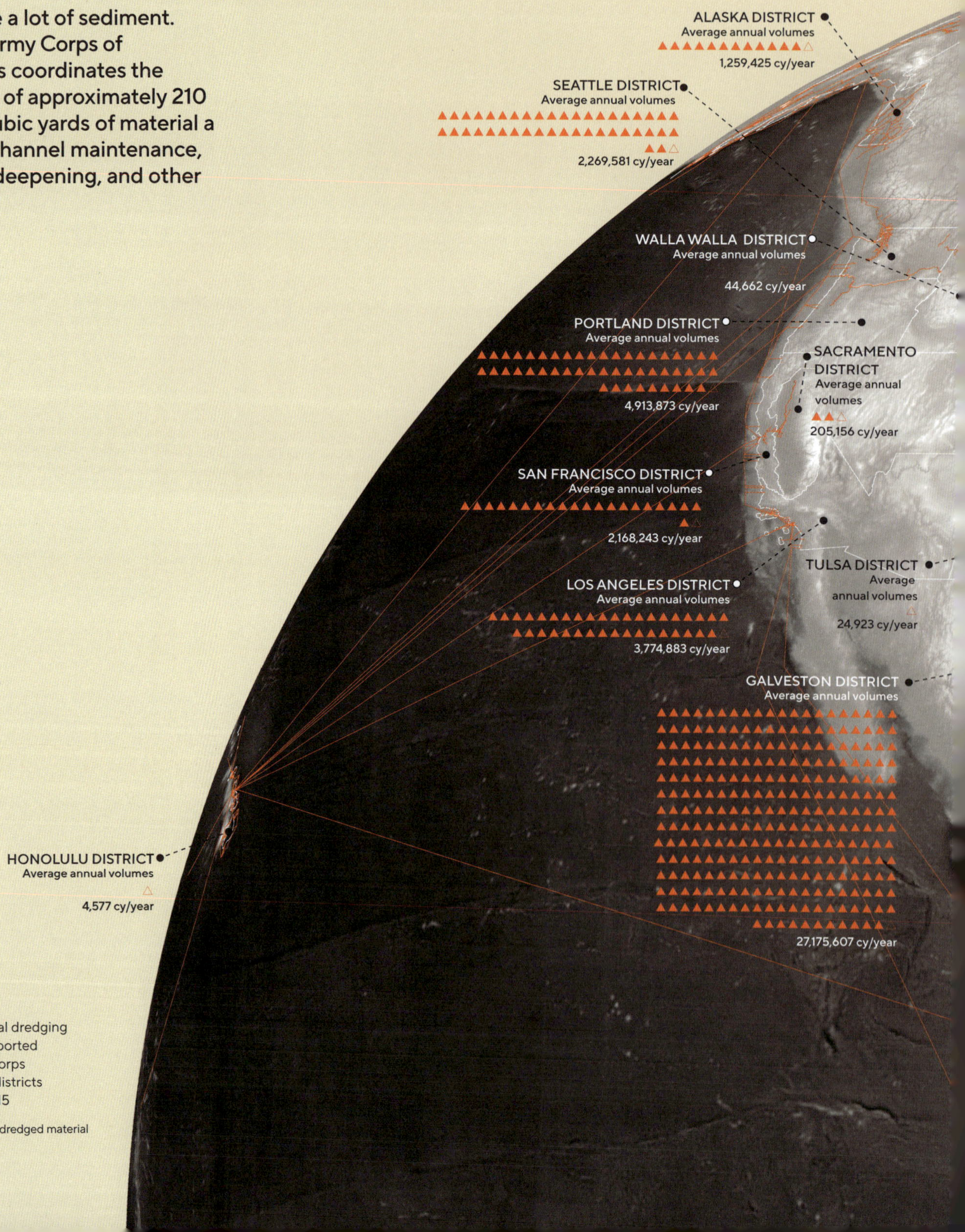

Average annual dredging volumes as reported by US Army Corps of Engineers districts from 1990-2015

▲ = 100,000 cy dredged material

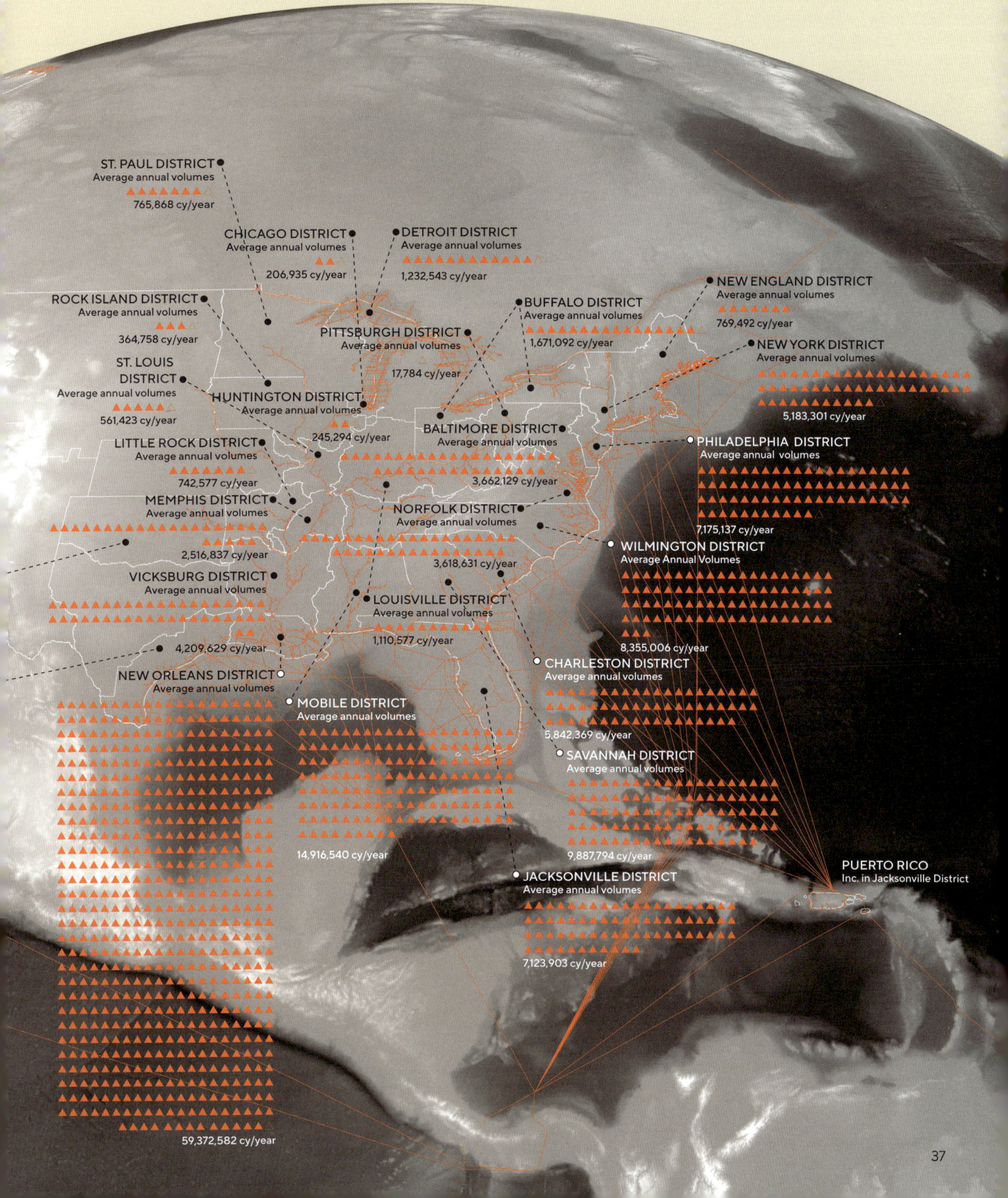
ST. PAUL DISTRICT
Average annual volumes
765,868 cy/year
CHICAGO DISTRICT
Average annual volumes
206,935 cy/year
DETROIT DISTRICT
Average annual volumes
1,232,543 cy/year
ROCK ISLAND DISTRICT
Average annual volumes
364,758 cy/year
PITTSBURGH DISTRICT
Average annual volumes
17,784 cy/year
BUFFALO DISTRICT
Average annual volumes
1,671,092 cy/year
NEW ENGLAND DISTRICT
Average annual volumes
769,492 cy/year
NEW YORK DISTRICT
Average annual volumes
5,183,301 cy/year
ST. LOUIS DISTRICT
Average annual volumes
561,423 cy/year
HUNTINGTON DISTRICT
Average annual volumes
245,294 cy/year
BALTIMORE DISTRICT
Average annual volumes
3,662,129 cy/year
PHILADELPHIA DISTRICT
Average annual volumes
7,175,137 cy/year
LITTLE ROCK DISTRICT
Average annual volumes
742,577 cy/year
MEMPHIS DISTRICT
Average annual volumes
2,516,837 cy/year
NORFOLK DISTRICT
Average annual volumes
3,618,631 cy/year
WILMINGTON DISTRICT
Average Annual Volumes
8,355,006 cy/year
VICKSBURG DISTRICT
Average annual volumes
4,209,629 cy/year
LOUISVILLE DISTRICT
Average annual volumes
1,110,577 cy/year
CHARLESTON DISTRICT
Average annual volumes
5,842,369 cy/year
NEW ORLEANS DISTRICT
Average annual volumes
59,372,582 cy/year
MOBILE DISTRICT
Average annual volumes
14,916,540 cy/year
SAVANNAH DISTRICT
Average annual volumes
9,887,794 cy/year
JACKSONVILLE DISTRICT
Average annual volumes
7,123,903 cy/year
PUERTO RICO
Inc. in Jacksonville District

How does this compare globally?

Global dredge capital projects dwarf dredging activity in the United States.

The Dubai Palm Islands used 94 million cubic meters of sand for land reclamation, dredged from offshore seabeds.[4] A single project in the Netherlands, the Maasvlakte 2 port expansion, required 240 million cubic meters of dredged sand for early phase construction, more than the volume moved for all US projects in a year.[5] At a planetary scale, people are becoming the primary agents of earthmoving—through dredging, agriculture, mining, and other anthropogenic activities—surpassing the natural erosive forces of wind and water.[6]

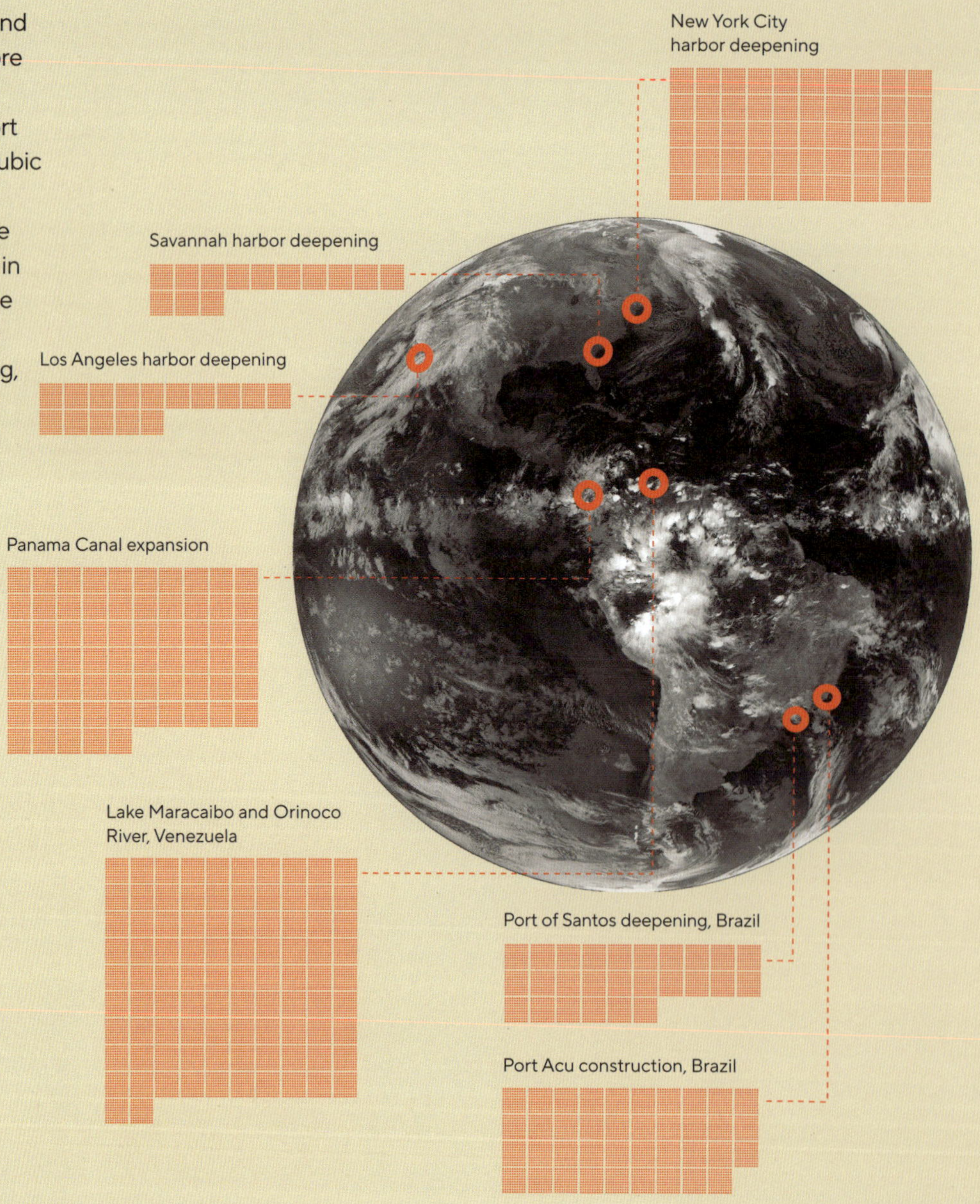

Volumetric comparison of a selection of recent global dredging projects

▪ = 1,000,000 cy dredged material

Maasvlakte land reclamation, Netherlands
Sabetta Port construction, Russia
Denmark tunnel construction
Delfland Sand Engine, Netherlands
Lower Niger River
New Suez Canal construction, Egypt
Pearl Island land reclamation, Qatar
Palm Islands land reclamation, Dubai
Poyang Lake sand mining, China
Hambantota land reclamation, Sri Lanka
Hong Kong airport expansion

Where does dredged material go?

Dredged material has historically been treated as "spoils," a waste product, dumped offshore or contained in nearshore landfills in perpetuity.

Open water disposal or direct land application has been the usual method of dredge disposal for centuries. Efficient and cheap, these practices historically have had negative environmental and social consequences, imposing unjust environmental burdens on underserved communities, and harming underwater ecosystems. Legislation like the Clean Water Act slowed the dumping of dredge and led to the development of alternative containment techniques.[7]

Confined Disposal Facilities (CDFs):
CDFs were developed to contain dredged material within a confined perimeter in perpetuity. While CDFs limit the negative impacts of disposing material in open-water environments, CDFs treat sediment as a waste product to be stabilized and stored, rather than a valuable resource.

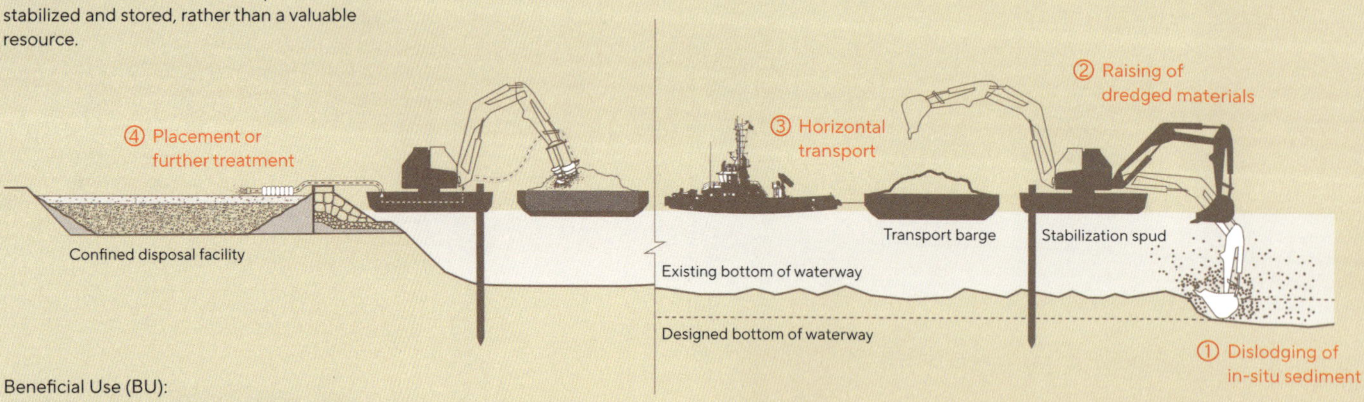

Beneficial Use (BU):
The term "beneficial use" has been coined to describe contemporary practices that treat sediment as a resource. Beneficial use typically only applies to clean sediment, which is used as construction material or for ecological purposes. Common beneficial uses of dredge are landfill cover, roadway base, beach nourishment, and wetland restoration substrate.[8]

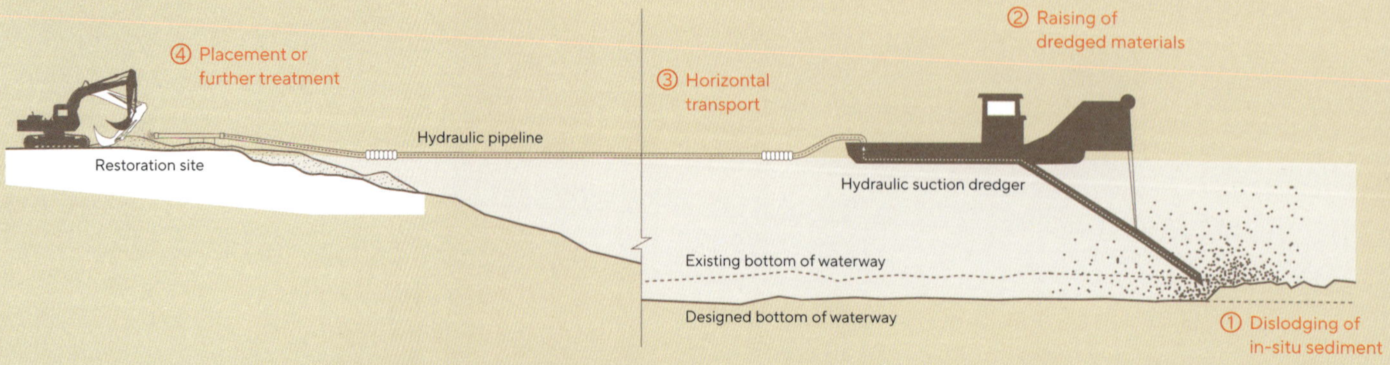

Confined disposal facility (CDF) (Buffalo District, US Army Corps of Engineers, 2011)

Beneficial use of dredged material (BU) for marsh building at Ring Island, NJ (US Army Corps of Engineers, 2014)

Instruments of Dredging

Dredging technology has evolved along with the landscapes it creates.

Narrow constructed water bodies require mechanical dredges, like clamshell or bucket dredge boats that can maneuver in tight spaces and scoop dredged material onto barges (scows) for transport.[9] Open ocean conditions require hydraulic dredge boats, which can withstand large waves while sucking material from the seafloor.[10] Each form of technology moves sediment differently—some are optimized for sandy ocean bottoms, and some are developed specifically for harder, brittle bedrock conditions. Some can work within a precise target area, while others work in a less controlled fashion but move considerably more sediment. Selection of a dredging technology depends on environmental conditions, availability, project budget, environmental law, and sediment type.

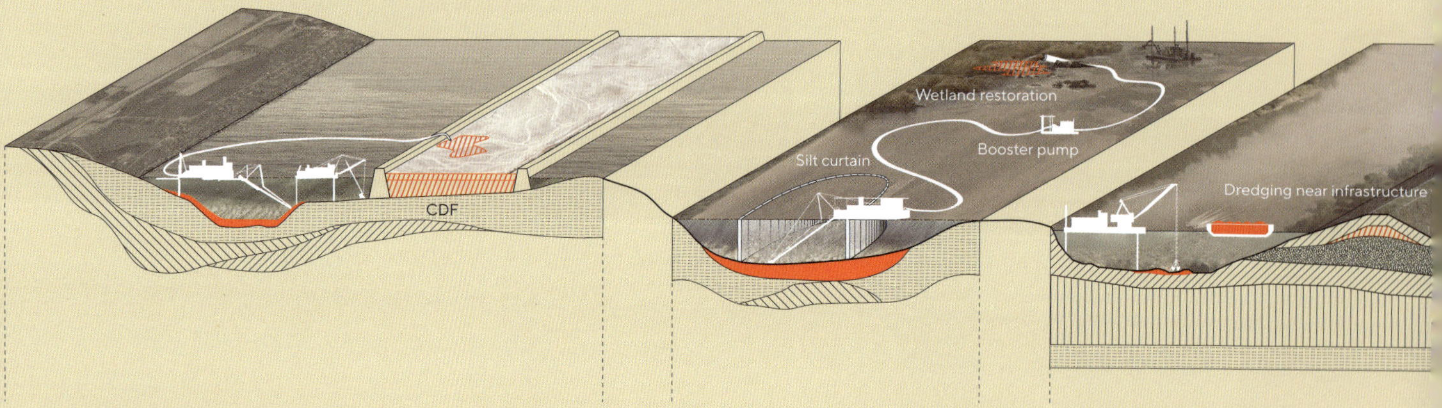

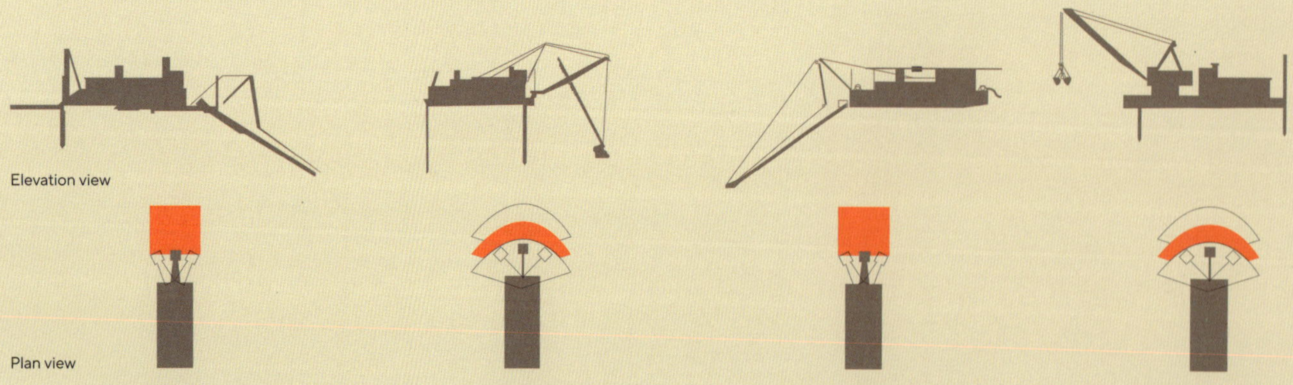

SUCTION DREDGE

Suited for heavy sand and non-cohesive sediment. Used in estuaries, ocean floors, and heavily traffic areas. Dredged material is transported through pipeline to upland areas or directly to site.

DIPPER DREDGE

Suited for rock, clay, hard packed sand, stone, glacial till and other hard, compacted material. Used after blasting, good for small, confined spaces. Material may be transported to scow barge for disposal.

DUSTPAN DREDGE

A hydraulic dredger with a wide, flared suction head that propels water to agitate bedload sediment. Developed for dredging of sand and gravel in uncontrolled rivers with navigation channels.

CLAMSHELL DREDGE

Suited for soft clay, sand, gravel and consolidated material that is not highly cohesive. Used near bridges, piers, docks, harbors with lower traffic and wave action. Dredged material is transported by scow barge to placement area.

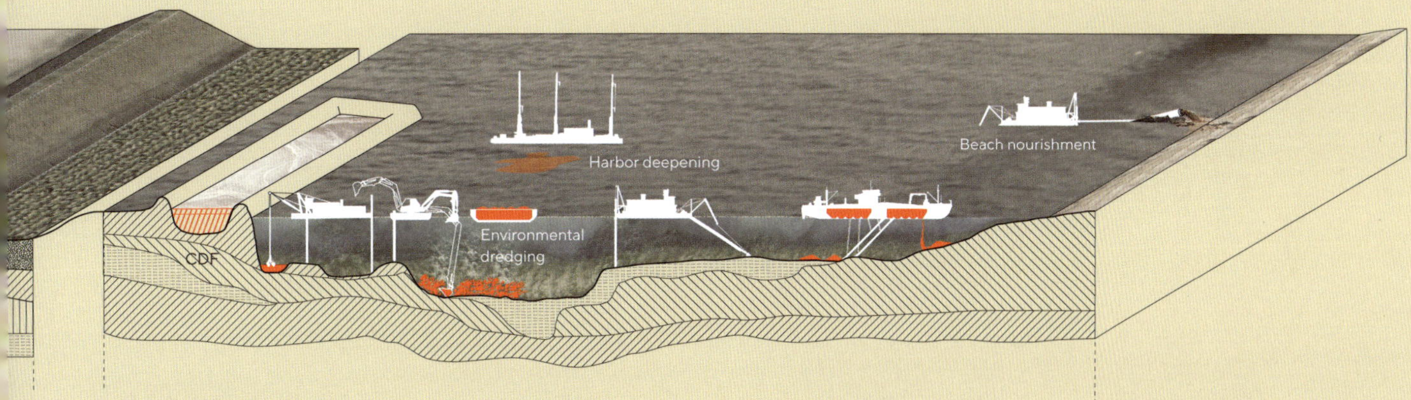

BACKHOE DREDGE

Suited for rock, clay, hard packed sand, stone, glacial till, and hard, compacted material. Used after blasting, good for small, confined spaces. Backhoe dredges are placed on a pontoon and secured with spud poles.

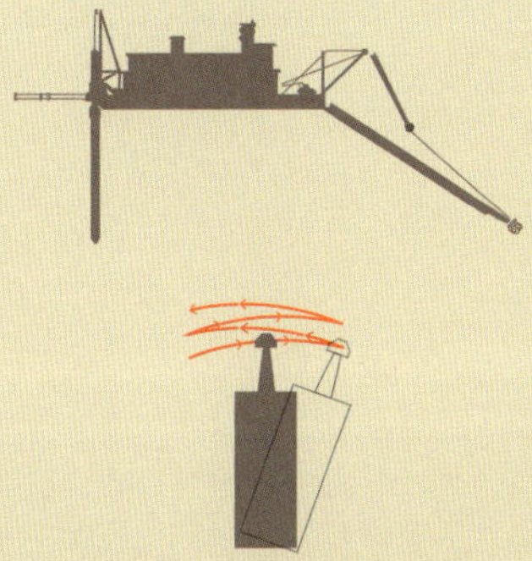

CUTTERHEAD DREDGE

Hydraulic dredger equiped with a rotating cutting apparatus at the intake end of the suction pipe. Suited for clay, silt, sand, gravel, and rock. Used in rivers, harbors, canals where wave action is not excessive. A very common dredge.

MARINE BLASTING EQUIPMENT

Underwater blasting often occurs for operations like harbor deepening that need to remove bedrock or hard materials. Blasted material is often removed with a backhoe or dipper dredge.

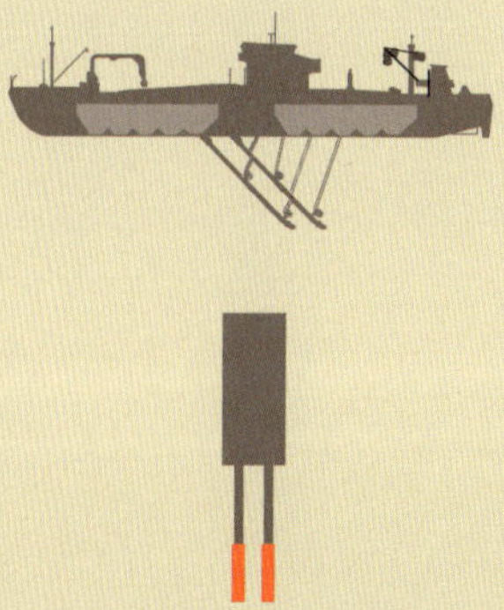

HOPPER DREDGE

Self-propelled dredge ship suited for sand and non-cohesive materials. Used in estuaries and ocean floor sites. Dredged material is stored in the hulls and released directly into open water disposal sites or pumped upland.

What are the impacts of dredging?

Dredging can change ecosystems, impact people, disturb aquatic species, and reshape places.

Dredging affects the surrounding environment, particularly sensitive habitats along the bottoms of water bodies. The extraction of material from seafloors and riverbeds stirs up sediment, which increases turbidity, disturbs benthic (bottom-dwelling) species, and changes the character of underwater habitats.[11] Placement of dredged material can have similar impacts and expose new sites to contamination. Underwater species are often sensitive to water quality and acoustical changes caused by dredging, while larger scale hydraulic dredging and underwater explosions for channel deepening can harm and kill large fish. Because of their potential impact, dredging projects are regulated by federal and state agencies like the National Marine Fisheries Service and the US Fish and Wildlife Service, which limit dredge permits in sensitive underwater habitats and contain dredging events to specific windows of time that avoid fish spawning and migration cycles. Although people rarely notice dredging, they are affected by impacts to the quality of water and

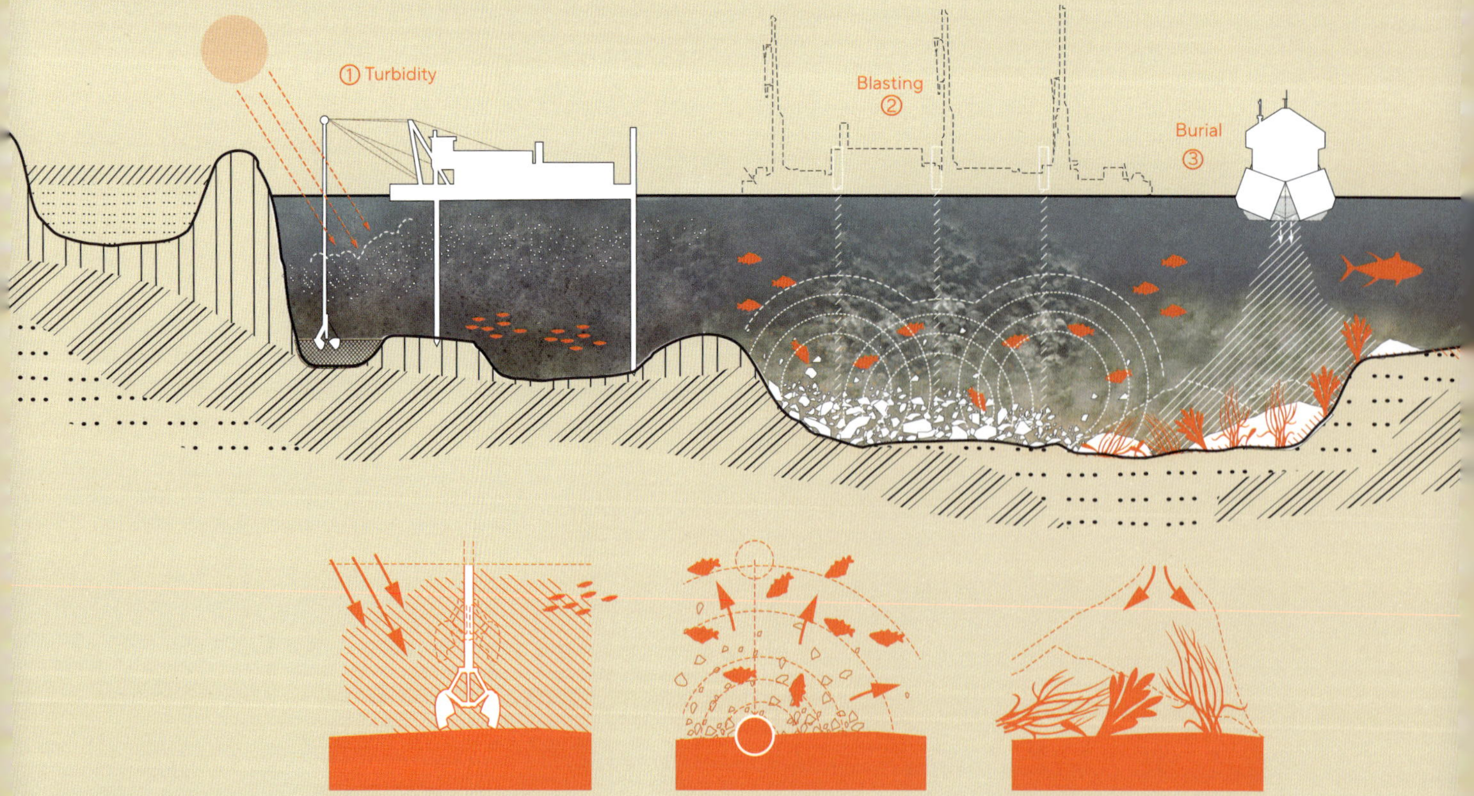

① Turbidity in the water column can block sunlight

② Blasting can kill fish

③ Submerged aquatic vegetation can be buried by open water placement shoreline environments they inhabit. Dredged material has a legacy of environmental injustice, as its treatment as a waste product has led to disposal near disenfranchised communities and marginal, low-cost lands. The cleanup of these sites can take decades when material is contaminated. Yet dredged material has the potential to provide social benefit, as waterfront sites created with clean sediment can be prime spots for recreation, hunting, and fishing.

④ Disturbance of contaminated sediment can release toxins that impact the food chain

⑤ Noise can disturb sensitive species and impact fish migration

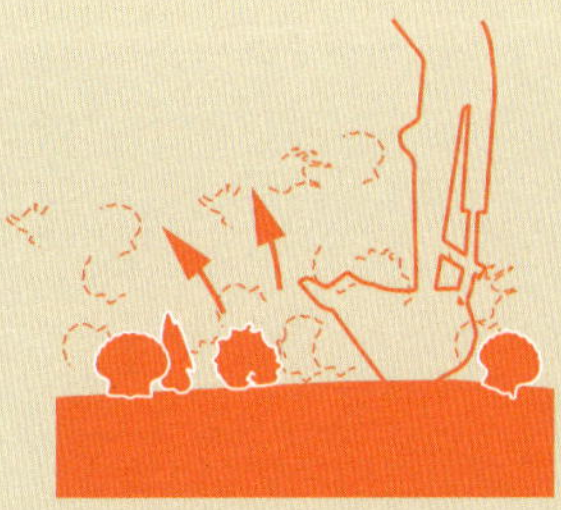

⑥ Dredging can cause direct physical harm

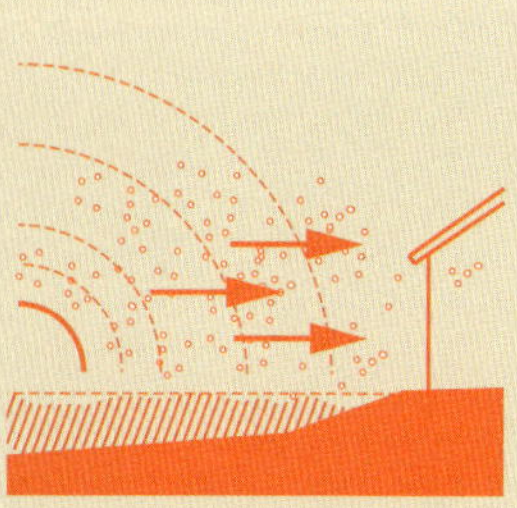

⑦ Potential community exposure to contaminated sediments, odor, and noise

Who dredges?

The US Army Corps of Engineers is the largest manager of dredging activities in the country, commissioning federal dredging projects from private contractors who move dredge vessels from project to project, region to region.

Dredging is a political act, as Congress funds and directs these practices through the authorization of new projects and the definition of operational budgets. Non-federal navigation channels are dredged by private ports and marinas, often funded through local taxes and shipping and docking fees. Dredging for habitat creation or environmental cleanup may be sponsored by federal and state agencies, or local non-profit entities. While the majority of dredge operations in the US are controlled at the federal level, no long-term national plan exists. Dredge decision-making is highly susceptible to political cycles, annual budgets, and changing regional needs.

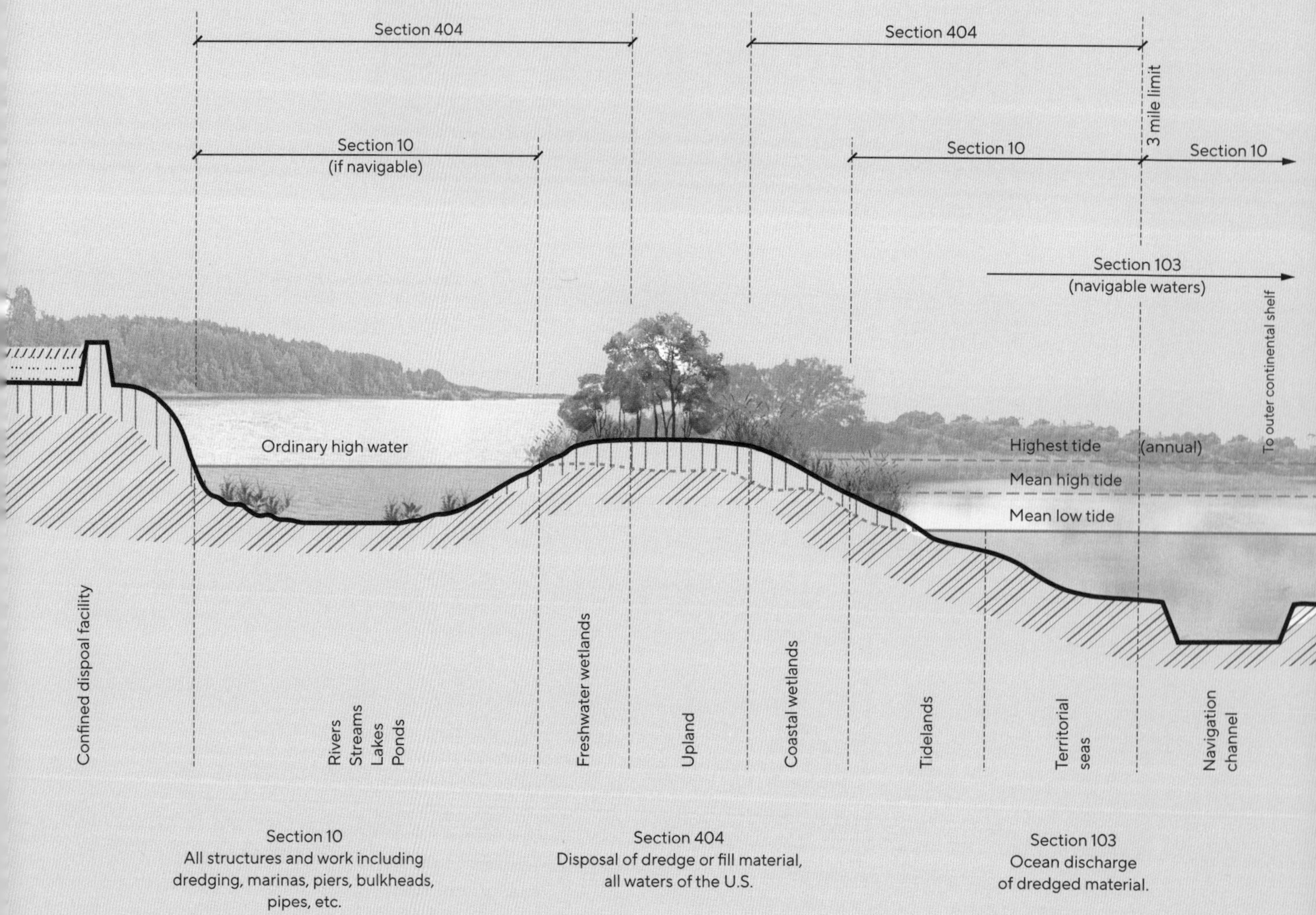

Regulatory jurisdiction of the US Army Corps of Engineers, Clean Water Act

Dredging the Savannah shipping channel (US Army Corps of Engineers, 2010)

Vice President Biden and congressional members visit Delaware River Deepening Project (Tim Boyle, US Army Corps of Engineers, 2014)

The Dredge Cycle

Urbanized landscapes have been completely reconfigured to control the flow of sediment. Dredging is simply one of the more active moments within a larger regime of sediment management.

Upstream, erosion control devices like silt fences and slope stabilization treatments slow the flow of sediments. In larger waterbodies, hard structures—seawalls, armored edges, jetties, and groins—stabilize dynamic edges and fix shorelines in time. Not all sediment controls are intentional. For example, dams trap a staggering amount of sediment behind their walls, creating operational and capacity challenges over time. Deepened dredge channels create faster water, which affects rates of erosion and land loss. Dredging cannot be considered in isolation; it must be contextualized within a larger suite of sediment management operations that alter patterns of sediment transport and often produce a need for more dredging. Dredge policy and funding mechanisms treat dredging as a single, isolated act, but in reality dredging is part of a hybrid interaction between large-scale landscape processes and human modification. We call this the Dredge Cycle.

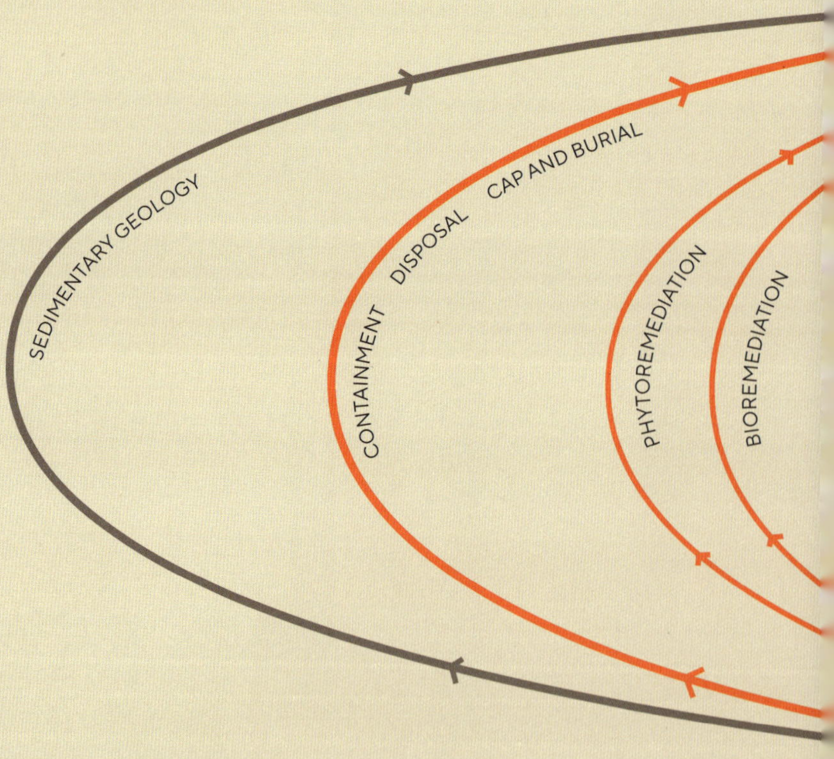

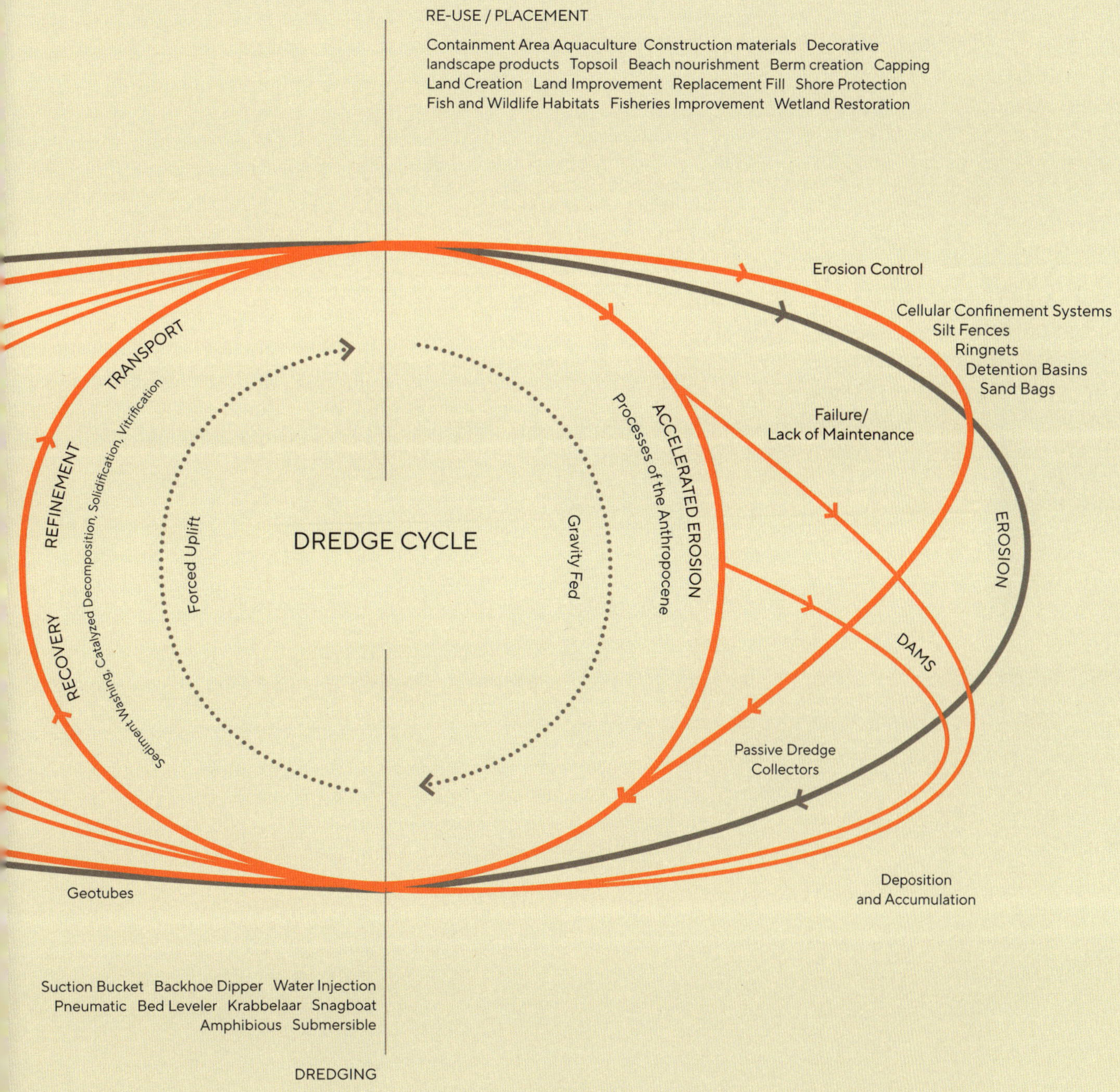
RE-USE / PLACEMENT
Containment Area Aquaculture Construction materials Decorative landscape products Topsoil Beach nourishment Berm creation Capping Land Creation Land Improvement Replacement Fill Shore Protection Fish and Wildlife Habitats Fisheries Improvement Wetland Restoration
Erosion Control
Cellular Confinement Systems
Silt Fences
Ringnets
Detention Basins
Sand Bags
TRANSPORT
REFINEMENT
RECOVERY
Sediment Washing, Catalyzed Decomposition, Solidification, Vitrification
Forced Uplift
DREDGE CYCLE
Gravity Fed
Processes of the Anthropocene
ACCELERATED EROSION
Failure/
Lack of Maintenance
EROSION
DAMS
Passive Dredge
Collectors
Deposition
and Accumulation
Geotubes
Suction Bucket Backhoe Dipper Water Injection Pneumatic Bed Leveler Krabbelaar Snagboat Amphibious Submersible
DREDGING

Four Coasts

The dredging industry is insular and slow to adapt, but practices are gradually changing across the country to treat sediment as a resource, rather than a waste product. This book investigates dredging and sediment management practices within four American sedimentsheds across four coasts, to better understand the impacts on these systems and consider ways that design can improve sediment management.

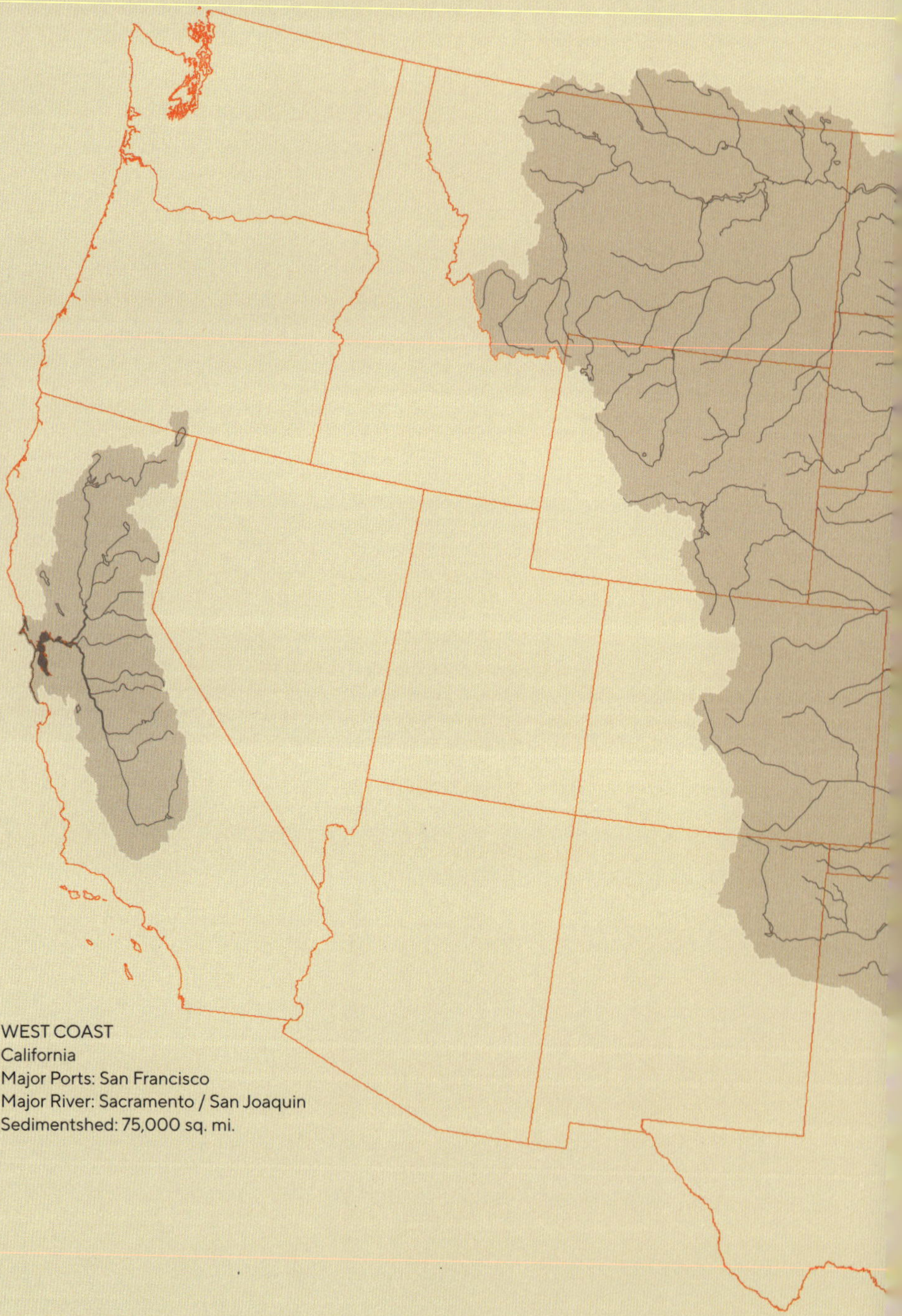

WEST COAST
California
Major Ports: San Francisco
Major River: Sacramento / San Joaquin
Sedimentshed: 75,000 sq. mi.

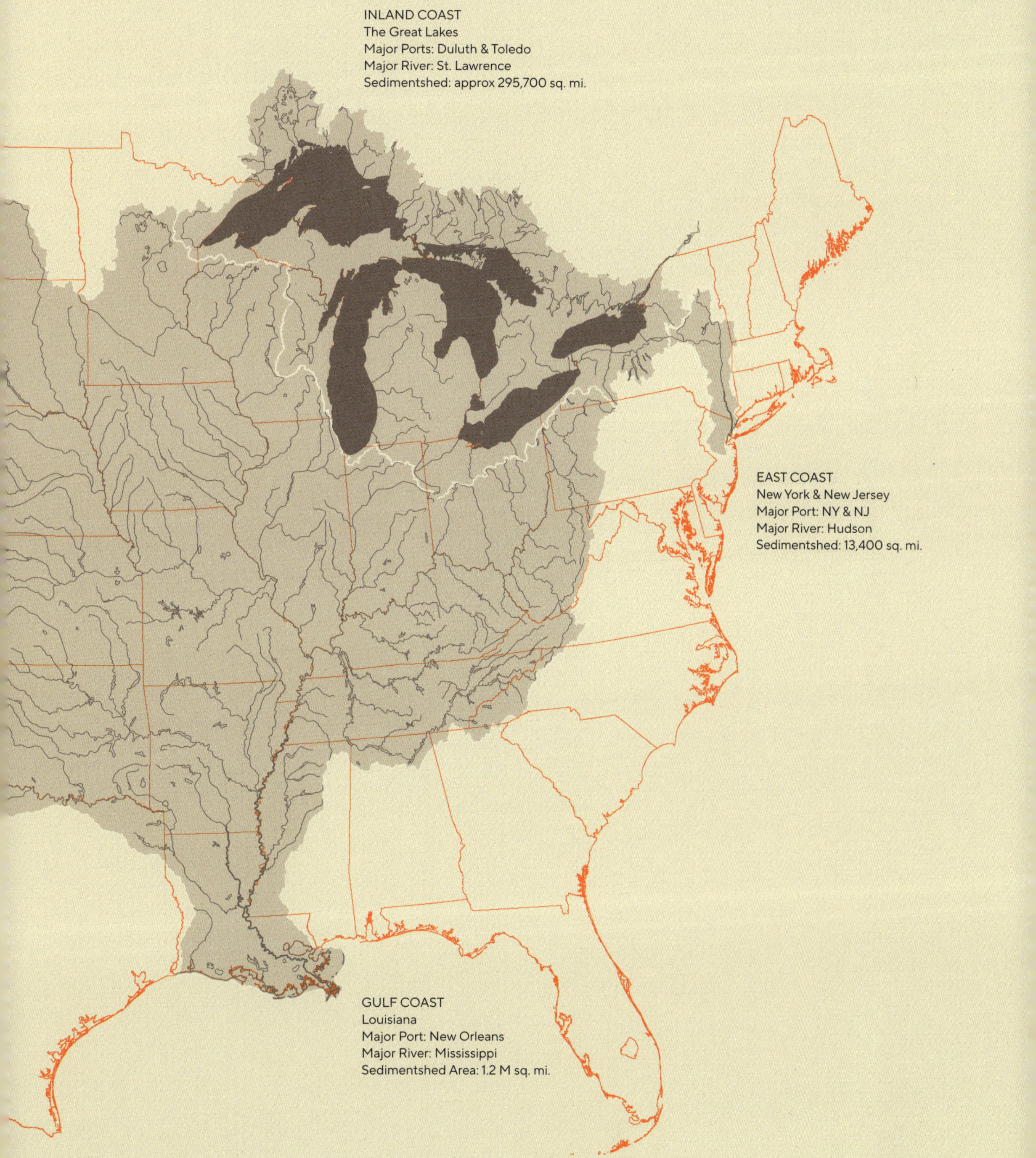
INLAND COAST
The Great Lakes
Major Ports: Duluth & Toledo
Major River: St. Lawrence
Sedimentshed: approx 295,700 sq. mi.
EAST COAST
New York & New Jersey
Major Port: NY & NJ
Major River: Hudson
Sedimentshed: 13,400 sq. mi.
GULF COAST
Louisiana
Major Port: New Orleans
Major River: Mississippi
Sedimentshed Area: 1.2 M sq. mi.

Endnotes:

1. "Dredging Facts." Dredging Contractors of America, October 18, 2019. Accessed April 23, 2022. https://dredgingcontractors.org/dredging-101/.

2. "What Is Dredging?" NOAA's National Ocean Service, US Department of Commerce, September 6, 2013. Accessed April 23, 2022. https://oceanservice.noaa.gov/facts/dredging.html.

3. "Value to the Nation: Navigation." US Army Engineer Institute for Water Resources (IWR). Accessed April 23, 2022. https://www.iwr.usace.army.mil/Missions/Value-to-the-Nation/Navigation/.

4. "Palm Jumeirah: Know about the Artificial 'Islands' Made from Rock and Sand." News9Live, January 28, 2022. https://www.news9live.com/knowledge/palm-jumeirah-know-about-the-artificial-islands-made-from-rock-and-sand-150024. https://www.fircroft.com/blogs/engineering-feat-of-the-month-dubai-s-palm-islands-65019956026

5. "Maasvlakte 2, Creating the Port of the Future." Start Dredging, March 5, 2018. https://startdredging.com/maasvlakte-2-netherlands/. Accessed April 23, 2022.

6. Roger LeB. Hooke. "On the History of Humans as Geomorphic Agents." *Geology* 28, no. 9 (2000): 843–46. https://doi.org/10.1130/0091-7613(2000)028<0843:othoha>2.3.co;2.

7. "Section 404 of the Clean Water Act: Permitting Discharges of Dredge or Fill Material." EPA. Environmental Protection Agency. Accessed April 23, 2022. https://www.epa.gov/cwa-404.

8. Strelzoff, Audrey. "Discover, Learn, and Grow Beneficial Uses of Dredged Sediment." US Army Corps of Engineers Dredging Operations Technical Support. Accessed April 23, 2022. https://budm.el.erdc.dren.mil/.

9. US Army Corps of Engineers. Engineering Manual: Dredging Equipment and Techniques. Chapter 3. EM 1110-2-5025. March 1983. https://www.mvn.usace.army.mil/Portals/56/docs/OPS/BUD/DredgingEquipTech.pdf

10. Ibid.

11. Wenger, Amelia S., Euan Harvey, Shaun Wilson, Chris Rawson, Stephen J. Newman, Douglas Clarke, Benjamin J. Saunders, et al. "A Critical Analysis of the Direct Effects of Dredging on Fish." *Fish and Fisheries* 18, no. 5 (2017): 967–85. https://doi.org/10.1111/faf.12218.

Image Credits:

Spread 1:
George Stringham, US Army Corps of Engineers St. Paul District. "Dredge Goetz." (Photograph), 2014. https://www.mvp.usace.army.mil/Media/Images/igphoto/2001309706/

Spread 2:
(left)
US Army Corps of Engineers (USACE), US Department of Transportation (USDOT)/Bureau of Transportation Statistics (BTS). Navigable Waterway Lines. Shapefile. 2006. https://data-usdot.opendata.arcgis.com/datasets/c91bc22014744f9389488f8ca203b4d4_0/about
US Army Corps of Engineers. National Channel Framework. 2017. Shapefile. https://geospatial-usace.opendata.arcgis.com/maps/9227967a2748410983352b501c0c7b39/about
(right)
Vince Elias, US Army Corps of Engineers. "Deepening Shipping Channels." (Photograph) https://www.usace.army.mil/Media/Images/igphoto/2000745992/

Spread 3:
(left)
US Army Corps of Engineers. National Channel Framework. 2017. Shapefile. https://geospatial-usace.opendata.arcgis.com/maps/9227967a2748410983352b501c0c7b39/about
San Francisco Estuary Institute (SFEI). 2016. "Flood Control 2.0 Sediment Load Data." Accessed January 1, 2022. http://www.sfei.org/projects/flood-control-20
Louisiana Coastal Protection and Restoration Authority (CPRA). 2017 Coastal Masterplan: Restoration Projects. Shapefile. 2017. https://cims.coastal.louisiana.gov/masterplan/GISDownload/
Hugo R. Valverde, Arthur C. Trembanis, Orrin H. Pilkey. "Summary of Beach Nourishment Episodes on the US East Coast Barrier Islands." Journal of Coastal Research. Vol. 15 No. 4, The Coastal Education and Research Foundation, 1999.
(right)
Dredge Research Collaborative. "Dredging sediment from the Mississippi River for marsh creation in Louisiana." (Photograph) 2017.
US Army Corps of Engineers. "Ongoing Long Beach Island Construction." (Photograph). 2015. https://www.flickr.com/photos/usacehq/25021680342

Spread 4:
(left)
Wentworth, Chester K. "A Scale of Grade and Class Terms for Clastic Sediments." The Journal of Geology 30, no. 5 (1922): 377–92. https://doi.org/10.1086/622910.
(right)
Chris Gardner. New York District, US Army Corps of Engineers. "Post-Sandy sand placement activities at Rockaway Beach 8-15-13." (Photograph), 2013. https://www.flickr.com/photos/newyorkdistrict-usace/9521638871/in/photostream/. License: https://creativecommons.org/licenses/by/2.0/

Spread 5:
US Army Corps of Engineers, New York District. Update: Dredged Material Management Plan for the Port of New York & State of New Jersey. 2008. https://www.nan.usace.army.mil/Portals/37/docs/harbor/dredmgmt/Dredged%20Material%20Mgmt%20Plan.pdf
"San Francisco Bay Long Term Management Strategy." Environmental Protection Agency. Accessed April 23, 2022. https://www.epa.gov/ocean-dumping/san-francisco-bay-long-term-management-strategy-dredging.
"Environmental Assessment #567: Louisiana Coastal Area." Beneficial Use of Dredged Material. US Army Corps of Engineers, New Orleans District. Accessed April 23, 2022. https://erdc-library.erdc.dren.mil/jspui/bitstream/11681/34833/1/LCA%20Beneficial%20Use%20Dredged%20Material%20Barataria%20Bay%20Waterway%20EA_2019.pdf.
Great Lakes and Ohio River Division, US Army Corps of Engineers. Great Lakes System Dredged Material Management Long Term Strategic Plan. April 2012. https://www.lre.usace.army.mil/Portals/69/docs/Navigation/Great%20Lakes%20Dredged%20Material%20Management%20Long%20Term%20Strategic%20Plan.pdf

Spread 6:
US Army Corps of Engineers. Navigation Data Center Dredging Information System (DIS). 1990–2015. https://www.iwr.usace.army.mil/About/Technical-Centers/NDC-Navigation-and-Civil-Works-Decision-Support/NDC-Dredges/
US Army Corps of Engineers (USACE), US Department of Transportation (USDOT)/Bureau of Transportation Statistics (BTS). Navigable Waterway Lines. Shapefile. 2006. https://data-usdot.opendata.arcgis.com/datasets/c91bc22014744f9389488f8ca203b4d4_0/about
US Department of Transportation/ Bureau of Transportation Statistics. Major Ports. Shapefile. 2019 https://hifld-geoplatform.opendata.arcgis.com/datasets/geoplatform::major-us-port-facilities/about

Spread 7:
Benecki, Paul. "Digging Deep." The Maritime Executive. Accessed April 23, 2022. https://www.maritime-executive.com/magazine/digging-deep.
"Expanded Panama Canal Open for Business." Dredging Today, June 27, 2016. https://www.dredgingtoday.com/2016/06/27/expanded-panama-canal-open-for-business/.
Lester, Liza. "Building with Nature: The Dutch Sand Engine." Ecological Society of America. Accessed April 23, 2022. https://www.esa.org/esablog/2016/01/27/building-with-nature-the-dutch-sand-engine/.
"Engineering Feat: Dubai's Palm Islands." NES Fircroft. Accessed April 23, 2022. https://www.nesfircroft.com/blog/2020/01/engineering-feat-of-the-month-dubais-palm-islands.
Ohimain, Elijah. "Environmental Impacts of Dredging in the Niger River Delta." International Association of Dredging Companies / Terra et Aqua, December 2004. https://www.iadc-dredging.com/wp-content/uploads/2017/03/terra-et-aqua-97-complete.pdf.
Chernov, Vitaly. "Million-Scale Dredging." Port News. Accessed April 23, 2022. http://en.portnews.ru/comments/1748/.
"Minister Opens Maasvlakte 2 (The Netherlands)." Dredging Today, May 22, 2013. https://www.dredgingtoday.com/2013/05/22/minister-opens-maasvlakte-2-the-netherlands/.

Safi, Michael. “Sri Lanka’s ‘New Dubai’: Will Chinese-Built City Suck the Life out of Colombo?” The Guardian News and Media, August 2, 2018. https://www.theguardian.com/cities/2018/aug/02/sri-lanka-new-dubai-chinese-city-colombo.

“Reclamation of the Pearl-Qatar Island Completed.” Al Bawaba, October 8, 2006. https://www.albawaba.com/business/reclamation-pearl-qatar-island-completed.

Andrew S. Erickson, Kevin Bond. “South China Sea and Beyond: Why China’s Huge Dredging Fleet Matters.” The National Interest. The Center for the National Interest, August 12, 2015. https://nationalinterest.org/blog/the-buzz/south-china-sea-beyond-why-chinas-huge-dredging-fleet-13562.

Mathias Schloesser, J. B. Fontes. “Deepening of the Port of Santos – Brazil Dredging to Results 5,000 km surveyed.” Geology, 2012. https://doi.org/10.3990/2.235.

“Video: Dredging the Açu Port in Brazil.” Dredging Today, November 30, 2017. https://www.dredgingtoday.com/2017/11/30/video-dredging-the-acu-port-in-brazil/.

“New York and New Jersey Harbor Deepening Channel Improvements (NYNJHDCI).” New York District Website Website. US Army Corps of Engineers. Accessed April 14, 2022. https://www.nan.usace.army.mil/Missions/Navigation/New-York-New-Jersey-Harbor/NY-NJ-HDCI/.

Vince Beiser. “Sand Mining: The Global Environmental Crisis You’ve Never Heard Of.” The Guardian. Guardian News and Media, February 27, 2017. https://www.theguardian.com/cities/2017/feb/27/sand-mining-global-environmental-crisis-never-heard.

“Venezuelan Dredging Fleet Builds Economy.” Ellicott Dredges, December 29, 2020. https://www.dredge.com/2018/09/venezuelan-dredging-fleet-builds-economy/.

“Corps of Engineers Completes Deepening of Savannah Harbor’s Entrance Channel.” Savannah District. US Army Corps of Engineers, February 28, 2018. https://www.sas.usace.army.mil/Media/News-Releases/Article/1453646/corps-of-engineers-completes-deepening-of-savannah-harbors-entrance-channel/.

Spread 8:

(right)

Buffalo District, US Army Corps of Engineers. “Dike 10b.” (Photograph). 2011. https://www.flickr.com/photos/buffalousace/5693412723/ License: https://creativecommons.org/licenses/by-nd/2.0/

US Army Corps of Engineers.

“New Jersey Intracoastal Waterway Dredging and Marsh Restoration.” (Photograph), August 8, 2014. https://www.flickr.com/photos/usacehq/14914518250/in/photostream/

Spread 9:

US Army Corps of Engineers. Engineering Manual: Dredging Equipment and Techniques. Chapter 3. EM 1110-2-5025. March 1983. https://www.mvn.usace.army.mil/Portals/56/docs/OPS/BUD/DredgingEquipTech.pdf

Spread 10:

Drawing of negative impacts informed by:

Wenger, Amelia S., Euan Harvey, Shaun Wilson, Chris Rawson, Stephen J. Newman, Douglas Clarke, Benjamin J. Saunders, et al. “A Critical Analysis of the Direct Effects of Dredging on Fish.” Fish and Fisheries 18, no. 5 (2017): 967–85. https://doi.org/10.1111/faf.12218.

“The Facts about Dredging.” The Guardian. Guardian News and Media, December 11, 2013. https://www.theguardian.com/environment/2013/dec/11/the-facts-about-dredging.

Spread 11:

(left)

Adapted from:

“Jurisdictional Determinations.” Savannah District, US Army Corps of Engineers. Accessed April 23, 2022. https://www.sas.usace.army.mil/Missions/Regulatory/Jurisdictional-Determination/.

(right)

US Army Corps of Engineers. “Dredge.” (Photograph). July 23, 2010. https://www.flickr.com/photos/usacehq/14287895306/in/photolist-nLzcuw-fmscV5-9PZE2u-vKAHER-HD4neT-2jt3m7w-ch6Goq-9dC3J5-2jt3m6V-2jt3maT-LxysCE-2jt24kg-nhQbxN-nhQbDj-cD2iZJ-2jsYeXX-nfKDHd-a9CT9B-jCJh-16-2jtCcwU-ggo6v8-cVvLfy-jCJfrK-2jsYf1h-a9FJdJ-bmK9qz-9ZPVsv-bv8iJC-bmK9r2-2jt3mcm-cVvLdm-9U29ha-hyUG8Y-cD2iPf-hyUxNk-2jtC5Df-bxefEK-2jt3mef-2jsYf1T-hyW95H-cD2iTw-bXXHSL-cVvLaw-hyVfBo-bmHLt2-2jtCEug-8RSqT1-nSNozs-oahGXe-2jtGP7H

Tim Boyle, US Army Corps of Engineers. “Vice President Biden, Congressional Members visit Delaware River Deepening Project.” (Photograph). 2014. https://www.flickr.com/photos/usacehq/15559065172/in/photolist-pGUh1m-q2LNhM-nQppcT-nLzfAm-iLxEw1-iLvU1k-fAsA2C-E85DPL-T1gCvC-ciKgQA-EiyjjF-mbSFuM-tKLAwQ-qG8K7x-qYpTD8-ToGtFc-oxvEWw-eXheAn-nbAZTs-bv8iSQ-9CmoHC-WcFWcB-nt8EyU-o8fraj-9fnPnH-nr3UC9-piZkMP-nbB1BS-fAsA5f-TAbBnr-ovG7Go-aJG3bD-nt6isv-Hvrodr-sRgtqx-whHZzd-nt8FCC-LE9i7f-nt6jqH-nbAPFK-nbAPA-p-sR5vCb-nt6iAB-tvw5cs-oxHKXk-nt6juF-sRgqpH-sR697y-SmfvWi-2jtC5xP

Spread 13:

US Government Accountability Office. Report to the Committee on Transportation and Infrastructure, House of Representatives. San Francisco Bay-Delta Watershed. GAO-18-473. August 2018. https://www.gao.gov/assets/gao-18-473.pdf

Ohio Department of Natural Resources, Office of Coastal Management. Ohio Coastal Atlas. Chapter 2, Lake Erie Watershed. Third Edition. 2018. https://ohiodnr.gov/static/documents/coastal/ohio-coastal-atlas/CH02_LakeErieWatershed.pdf

“Hudson River Watershed.” Hudson River Watershed Alliance, April 28, 2021. https://hudsonwatershed.org/hudson-watershed/#:~:text=The%20Hudson%20River%20watershed%20covers,and%20Hudson%20River%20estuary%20watersheds.

“Mississippi River Facts.” National Parks Service. US Department of the Interior. Accessed April 23, 2022. https://www.nps.gov/miss/riverfacts.htm#:~:text=Watershed%20Size&text=The%20Mississippi%20River%20drains%20an,of%20the%20continental%20United%20States.

Choreographing Sediment

1. In a Disappearing Delta

Louisiana is washing out to sea. Since the 1930s, the state has been losing as much as 25,000 acres of land to the Gulf of Mexico every year.[1] This is dramatic, but on the ground, it can seem like a subtle process. It's not as if cliffs are crumbling away, stone and soil tumbling suddenly into the sea. The land being lost is wetlands—liminal terrain where elevation is measured in inches.

1 Lane, Kemp, and Day, "A Brief History," 17.

Tidal wetlands already belong in part to the Gulf. The lowest parts of a marsh or swamp are inundated almost all the time, only occasionally emerging from the water at the lowest of tides. In the highest parts, the tide only comes in during spring tides, which happen about every two weeks when the earth, moon, and sun are aligned, causing solar and lunar tides to reinforce each other. Vegetation like marsh grass arranges itself in accordance with periods of inundation: smooth cordgrass, for instance, is dominant in Louisiana's saline marshes, and the species is split into two phenotypes, one taller and adapted to deeper parts of the marsh, the other shorter and adapted to higher elevations. The Gulf of Mexico has a very limited tidal range, or difference between high tide and low tide. In most parts of the Gulf, including coastal Louisiana, it varies from about one to two feet. This is why the elevation change within the coastal wetlands is so minimal: they lie within that tidal range.

Consequently, what is lost as land washes away is as much a matter of time as of space. A portion of the wetland that was inundated only at high tide is now almost always below water; the fringe of the wetland slips fully beneath the water, becoming a subtidal flat. Bald cypresses, Louisiana-native swamp trees which are well-adapted to frequent inundation, cannot survive complete inundation of their roots, particularly as waters grow deeper. Their "knees" can reach upward toward air from their submerged roots, but the root systems still require oxygen. Marsh plants, like cordgrass, die out as they too are submerged past their tolerance; the tall and short phenotypes of smooth cordgrass migrate inland with the tides. Without anchoring vegetation, sediment erodes more rapidly, and the process of land loss continues.

Despite their subtlety, the impacts of this erosion are clear to people who live on Louisiana's coast. The French-speaking members of the Isle de Jean Charles Band of the Biloxi-Chitimach-Choctaw Tribe are among many communities in rural parts of the Mississippi River Delta who have watched the land around—and beneath—their homes disappear. On Isle de Jean Charles, land has been

2 Yawn, "As Gulf Swallows Louisiana Island."

3 Isle de Jean Charles Biloxi-Chitimacha-Choctaw Tribe, *Preserving Our Place*.

4 Dermansky, "Louisiana Breaks Ground."

5 At the end of August 2021, category 4 Hurricane Ida destroyed more than half of the 15 homes that remained occupied on Isle de Jean Charles.
Baurick, "On Sinking Isle de Jean Charles."

The links between climate change and storm intensity are a matter of on-going scientific investigation, but the preponderance of evidence suggests increased storm intensity is linked to climate change.
Rutherford et al., "Energy and Climate-Global Trends," 81.

6 Muth, "The Once and Future Delta," 10.

7 Lane et al., "A Brief History of Delta Formation," 11–14.

lost at an average rate of about 300 acres a year since 1955. Today, 98 percent of the island is gone. Only a slim high ground remains, a couple dozen buildings and structures along Island Road.

The band, whose members are descendants of indigenous Louisianans, French settlers, and tribes who fled the Trail of Tears, has seen a whole world disappear with that land. Thousands of acres of fertile marshes are gone, and with them the minks and muskrats the band and its ancestors hunted, and the pastures, cropland, and orchards they tended.[2] The tribe now faces an uncertain future. Its predicament has received a great deal of attention. Hundreds of stories have been written about it since 2016, including one in the *New York Times Magazine* that proclaimed the tribe "America's first climate refugees."[3] The tribe has argued that the state of Louisiana has been hijacking its resettlement efforts, forcing it to choose between accepting the state Office of Community Development's resettlement offers and continuing to pursue the tribe's original vision of a recentralized, sustainable tribal community.[4]

So what is driving the land loss that imperils communities like the Isle de Jean Charles Band?

It might seem that climate change is the main cause. It is certainly involved. The Gulf of Mexico is rising, like all oceans, as the planet warms. The storms that slam into its shorelines, like Hurricanes Laura (2020) and Ida (2021), are growing in size and intensity.[5] The funds earmarked to help Isle de Jean Charles are part of a series of federal grants designed to help communities adapt to climate change. And most of the newspaper and magazine stories about the island have centered on sea-level rise. Low-lying coastal wetlands like the swamps, marshes, and bayous of the Mississippi River Delta are experiencing the effects of climate change as rapidly and as intensely as any place on Earth. But the full truth is more complicated.

For most of the past seven thousand years—up until the twentieth century—the Mississippi River was very good at making land.[6] Seasonally, the river swells and recedes in response to rainfall throughout a vast watershed: rain that falls in Iowa, Nebraska, and Kentucky reaches the delta roughly two months later. Because these Midwestern rains are concentrated in the wet spring months, the river regularly rises in late spring. Historically, when the river rose enough, it would overtop some of its banks and the "natural levees" that lie just behind them, carrying sediment-laden water into its floodplains, including wide expanses of the delta. In other places, it would break its banks in violent sprays of water and sediment, called crevasses. And, before Europeans arrived, the river coursed through the delta not only along its main stem, but also along a fine network of distributaries, smaller channels that also flooded, overtopped, and even crevassed. The accretion of the thin layers of annual deposition that result from these processes is the core land-building process in a delta.[7]

There are other forces at work, though, which contribute to land loss. Freshly formed land, like that of a delta, compacts under its own weight, losing elevation even when the amount of sediment in place holds steady. This is a form of subsidence. And, at the edges of a delta, or along channels that run through it (whether formed by geologic processes or shaped by human tools), waves, particularly the strong waves stirred up by storms, erode soil, chipping away at the land.

When the amount of sediment being deposited exceeds the overlapping effects of erosion, subsidence, and sea-level rise, Louisiana grows.[8] But if the sediment's deposition rate slows, Louisiana's delta erodes, subsides, and drowns.[9] That's what's happening now.

Yet there is still more to the rhythm of growth and decline along Louisiana's coast. The river itself moves. This is subtly evidenced by the series of oxbow lakes, narrow backwaters, and accumulating battures that trace the shifting course of the lower Mississippi through Arkansas, Mississippi, and Louisiana. The pattern of these depressions and mounds can be seen from above, in a plane arriving to Baton Rouge, or at a computer, panning satellite view on Google Maps. These are remnants of former river courses, traces of an unruly history. The river, like most any river, has always sought new courses. Along most of its length, it has eroded its "cut banks" (the outside banks, where water moves most rapidly) and grown through deposition on its "point bars" (which form on the inside banks, where water slows, dropping sand and other heavier sediments). Over time, the river wanders in serpentine fashion across its floodplain, carving backwaters and leaving behind oxbow lakes.

As the Mississippi closes in on its outlet to the Gulf, this wandering becomes more dramatic. In the delta, the river's waters have always sought the shortest—most steeply sloped—path to the Gulf. Over decades and centuries, as the river deposited sediment along its course, those courses grew longer and longer, flatter and flatter, until the river, typically in a spring flood that pushed it over its banks, found a new, shorter path to the Gulf. This new path might at first be a distributary, a smaller, weaker branch off the main stem of the Mississippi. But, occasionally, roughly once a millennium for the past seven thousand years, the river has abandoned its prior course and settled fully into a new one.[10] This process, called "avulsion," generates the characteristic many-lobed form of large river deltas.[11]

So, land loss is not a simple math problem. At pretty much any point in the delta's history, some parts of it have been shrinking while others have been growing.[12] After each avulsion, a new delta lobe forms, while older delta lobes gradually begin to decline.[13] Until the twentieth century, the overall growth exceeded the loss. Today, two small deltas are growing, the Wax Lake Delta and the Atchafalaya River Delta, but this is greatly outweighed by a larger pattern of loss.

* * *

What is different now is that the Mississippi River Delta is starved of sediment.

This is because the movements of the river are not compatible with the way that Americans have, since European colonization, settled the Mississippi River Valley.[14] The land use patterns of the United States of America, our rhythms of life, our industries, our legal systems are all dependent on predictability and stability. At an individual level, this is understandable: people who live in the river's floodplain have decided, quite reasonably, that they do not want to risk seeing their homes, communities, and farms wash away. To reduce that risk, Americans have tasked government agencies—most prominently, the US Army Corps of Engineers—with re-engineering the river.[15] Over 2,000 miles of levees hold back spring floods, armored with articulated concrete mattresses.[16] Stand at the foot of a Mississippi River levee, and its height—often twenty or thirty

8 Other contributors to land loss include the carving of channels through the delta's wetlands for oil and gas extraction; the closure of older distributaries; drilling and withdrawal of water, oil, and gas; *Myocastor coypus* (the aquatic rodent known as nutria); and perhaps the inflow of freshwater from existing river diversions. Though there are dissenting voices (the experienced delta scientist Eugene Turner argues, controversially, that canal-digging is the primary driver of delta land loss), most scientists believe that reduced deposition, storm erosion, sea-level rise, and subsidence are the most crucial forces shaping Louisiana's trajectory of land loss.
Muth, "The Once and Future Delta," 16–17.
Sasser et al., "The Nutria in Louisiana."
Turner, "Doubt and the Values of an Ignorance-based Worldview."

9 We generally use the term "delta" to refer to the entire landform produced by Holocene deposition of Mississippi River sediments, inclusive not only of the primary, active Plaquemines-Balize ("Birdsfoot") delta lobe, but also of the other abandoned lobes that form the larger delta plain of south Louisiana.

10 All of the world's present deltas date to roughly the same 7,000–8,000 years BP, "when the rate of sea-level rise slowed markedly following the latest Pleistocene glacial maximum. At this time, all the world's great rivers started building deltas into the marine environment resulting in a seaward advance of Holocene shorelines."
Giosan and Freeman, "How Deltas Work," 31.

11 Barnett, *Beyond Control*.

12 CPRA, *Louisiana's Comprehensive Master Plan* (2017).
Giosan and Freeman, "How Deltas Work," 29–31.
Muth, "The Once and Future Delta," 9–16.

13 Blum and Roberts, "The Mississippi Delta Region."

14 Colten, "Levees and the Making of a Dysfunctional Floodplain."
Mandelman, "Porous Boundaries and Permeable Places."

This stands in stark contrast to indigenous modes of settlement, such as that of the Mississippian culture, whose largest cities were topographic monuments to the long rhythm of life in a floodplain.
Davis, "Land Making Machines."

15 Barry, *Rising Tide*.
Campanella, *Bienville's Dilemma*.
O'Neill, *Rivers by Design*.
Shallat, *Structures in the Stream*.

16 Bentley et al., "Using What We Have," 87.

17 Blum and Roberts, "The Mississippi Delta Region", 664.
Lane, Kemp, and Day, "A Brief History," 15.

18 On Taylorist infrastructure, see "Dredge Matters."

19 McPhee, "Atchafalaya."

20 Jacobson, Blevins, and Bitner, "Sediment Regime Constraints on River Restoration."

21 Bentley et al., "Using What We Have," 87.

22 Muth, "The Once and Future Delta," 14-18.

feet—will impress on you the scale of engineering required to tame the river's floods. These levees have largely accomplished their goal: no major river flood in the Mississippi River Valley has approached the level of destruction produced by the tragic 1927 flood.

In preventing flooding, though, the levees have helped to largely end the deposition of sediments that once followed spring rains.[17] They have also halted the river's meandering: from Cairo, Illinois to the end of the East Bank levees in Bohemia, Louisiana, the course of the river has been fixed, its perpetual migrations halted. Along this course, spillways and floodways—Bonnet Carré, Morganza, West Atchafalaya, Bird's Point-New Madrid—provide intermittent release valves, permitting the Army Corps to regulate maximum flows, working to ensure that the lower reaches of the river never exceed the flow limits articulated in the Project Design Flood. (The rigidly Taylorist Project Design Flood diagram's order stands in stark contrast to the technicolor geomorphic maximalism recorded in Harold Fisk's 1944 maps of the alluvial valley.[18]) And at Old River Control, about two hundred miles upriver from New Orleans, a series of concrete control structures hold the Mississippi River and its largest distributary, the Atchafalaya, in stasis, freezing rivers and delta on the brink of an avulsion—a switch from the current main course along the Mississippi, past New Orleans, to the shorter and steeper course of the Atchafalaya, a switch which now may never arrive.[19]

Far north of Louisiana, dams control the flow of water on the Upper Mississippi and many of its tributaries, like the Ohio, Tennessee, and Illinois Rivers. The largest of those dams were built in the early twentieth century on the Mississippi's biggest tributary, the Missouri. Behind each of the six Missouri River dams is an enormous reservoir. Three of these reservoir lakes—Lake Oahe, Lake Sakakawea, and Fort Peck Lake—are among the twenty largest lakes in the United States. Holding water, though, has also meant trapping sediment, which has been collecting for decades behind each dam, like a series of displaced underwater deltas. Today, the Missouri discharges less than a quarter of its historic sediment load.[20] Collectively, all of the dams, locks, and reservoirs constructed within the Mississippi River Basin have cut the amount of sediment the Mississippi carries to Louisiana in half, from approximately 400 million tons per year to roughly 200 million tons per year.[21]

The unintended effect of all this "control of nature" is that the river no longer builds the delta. So much sediment is trapped behind dams. And the sediment that does make it to the lower Mississippi now shoots down a single through-line. Bypassing the delta's floodplains altogether, sediment either deposits in navigation channels, where it is dredged out, or it shoots out the river's mouth beyond Head of Passes, into the depths of the Gulf of Mexico and toward the edge of the continental shelf.[22] In the delta outside these concrete-hardened banks, the forces of erosion, sea-level rise, salt-water intrusion, and subsidence act in concert to accelerate land loss, and they are no longer checked by the arrival of fresh sediment.

* * *

Spread across the wide, flat delta are many other communities like Isle de Jean Charles. These are places that have always been low and wet—even high ground is often only ten or fifteen feet above sea level. Poised between land and water, they are home to some of North America's most unique cultures: including native descendants of Louisiana's first peoples, Cajun and Creole towns, and more recent immigrants like Vietnamese from the similarly low and wet Mekong. These communities' ways of life are intimately tied to the delta landscape, through the economy the delta sustains, through the food drawn from it, and through the recreation it permits. Loss of land is not merely an abstract or historical loss, but affects and imperils people (the line between safe to live here and no longer safe to live here is not always clear, and shifts suddenly in a storm), communities (individuals can move, but a community is tied to place and pattern), and ways of life (which cannot be reconstituted elsewhere, for there is nowhere enough like the delta).

In other parts of the world, coastal areas are protected by armoring. In the Rhine Delta, Dutch cities like Rotterdam lie behind a centuries-long accumulation of levees, seawalls, and tidal gates.[23] But Louisiana is not the Netherlands. Its area is much larger, its population density radically lower; building a wall around it is not feasible, or even desirable. Even where the land of a given settlement may not be itself eroding—New Orleans, for instance, behind levees—the loss of surrounding wetlands and delta exposes settlement and infrastructures to the full force of hurricanes and storm surge.

Fortunately, Louisiana has an asset that the Netherlands does not: a river that carries a vast supply of sediment, which could build land again, if only it were able to reach the right places.[24]

2. Choreographing Change

Flux is fundamental to sedimentary landscapes.

Sedimentary landscapes are, even more than other landscapes, characterized by ceaseless change. Geologically, marshes, barrier islands, deltas, mudflats, dunes, sandbars, and rivers are all ephemeral landforms. In a single human lifespan, they are more stable, but they still change rapidly enough for that change to be clear and perceptible. A marsh is half the size it was twenty years ago, a barrier island is cut in half in a day by a hurricane, and a river takes a new course in a week-long spring flood, leaving behind a cutoff oxbow lake. These landscapes are not only characterized by change; they depend on it.[25] A delta that does not receive floodwaters carrying fresh sediment will drown as it subsides. A beach isolated from its coast's sand-sharing system will disappear, as the sand that erodes goes unreplaced. A floodplain that never floods quickly loses diversity of both landscapes and species. In the language of contemporary environmental design, their *resilience* depends on transformation.

Resilience can be a difficult concept to understand and use, because it has been deployed by a variety of fields in ways that sometimes coincide and sometimes contradict.[26] And as it has come to dominate discourse about risk management in environmental policy, finance, military defense, and urban infrastructure,[27] it has been frequently oversimplified, as the term sustainability was before it. At the most basic level, resilience often refers to the capacity to experience a damaging event without catastrophic failure. For instance, a levee

23 Pleijster, van der Veeken, and LOLA Landscape Architects, *Dutch Dikes*.

24 The Rhine is a large river relative to others in Europe, but it has a suspended sediment load of about 3 million tons per year, two orders of magnitude less than the Mississippi's.
Middelkoop, Erkens, and van der Perk, "The Rhine Delta."

25 Collier, Webb, and Schmidt, *Dams and Rivers*.
Dean, *Against the Tide*.
Kondolf, Podolak, and Grantham, "Restoring Mediterranean-climate Rivers."
Pilkey and Cooper, *The Last Beach*.
Renaud et al., "Tipping from the Holocene to the Anthropocene."

26 The taxonomy that follows is itself simplified. Brand and Jax, for instance, identify 10 distinct definitions of resilience in contemporary social and ecological discourse alone.
Baggio, Brown, and Hellebrandt, "Boundary Object or Bridging Concept?"
Brand and Jax, "Focusing the Meaning(s) of Resilience."
Davoudi et al., "Resilience."

27 Walker and Cooper, "Genealogies of Resilience," 144.

system is said to be resilient to a storm if it can absorb the impact of that storm without being breached. This use is common in the engineering and planning of infrastructure.[28]

A second meaning of resilience is tied more closely to sociological and psychological discourse. In this usage, resilience refers to the capacity to adapt during or after trauma or adversity.[29] A community, for instance, is said to be resilient if it can experience devastating hurricane damage but nonetheless pull together, organize repairs, and return to healthy patterns of life.[30] Resilience in this sense is often described as "coping capacity." Thus, while in engineering resilience the aim is to prevent an event or minimize its consequences, in social resilience the aim is to cope effectively with consequences. This social definition bridges between the engineering meaning and a more specialized ecological meaning.[31] In the last third of the twentieth century, ecological science underwent a significant paradigm shift.[32] Ecology had long been underlaid by a belief that balance, equilibrium, and stability were the markers of healthy nature. But empirical observations by ecologists undermined those assumptions, and ecosystem theory began to shift toward a recognition that ecological systems are characterized by stochastic disturbance, flux, and change.[33] Thus, in ecological theory, resilience and stability are often functionally opposed: the most resilient ecosystems are often those with significant fluctuations, which are a sign of adaptive capacity.[34]

Ecosystem theorist C. S. Holling, who helped define the concept of ecological resilience, emphasizes the persistence of a given ecosystem over time despite—or even through—great change occurring within that ecosystem.[35] For instance, populations of various insect, plant, and fish species within a lake may change radically across a period of several decades, with one fish species dominating in one decade, and another species dominating in others. The lake as a whole, though, persists: its flora and fauna do not undergo extinction. Similarly, riverine ecosystems are spiky and flashy: flood levels vary from year to year, sediment loads vary by an order or two of magnitude, and species populations will often rise or crash depending on those spikes of water and sediment. Such fluctuation is fundamental to the ecological function of a river, its floodplain, and its delta as a whole. Without it—if the ecosystem becomes too stable—then the ecosystem as a whole becomes vulnerable to collapse. Pursuing stability and balance through design can thus substantially—even catastrophically—undermine the resilience of landscapes.[36] Resilience should not be detached from change.

Moreover, sedimentary landscapes require the opportunity to migrate. River banks and shorelines are never static. They are constantly subject to erosion and accretion. Even armored edges must be reinforced and rebuilt every few decades. Viewed from longer perspectives, shoreline migration is a constant: sea levels even a few tens of thousands of years ago were far lower than they are today. Many of the major estuaries we know today in North America, like Galveston Bay, San Francisco Bay, and Mobile Bay, are, in the abbreviated geologic time of the recent Pleistocene, drowned river valleys. On more immediate timescales of decades and centuries, migration at scales of acres and miles is continuous and, for the grasses, rushes, birds, and snails who populate marshes, no issue for concern: so long as adequate space exists for marshes to migrate upland with rising seas, they will adapt readily, as they have for millions of years.

28 Engineering also has (confusingly) a narrower interpretation of resilience, too: as the capacity of a material or structure to rebound and return to its original shape after being deformed or bent by force. This material resilience obviously lends itself to supporting the larger, infrastructural meaning of engineering resilience: a bridge with a high capacity for flexing and then returning to shape may be more capable of withstanding stresses than one that is perfectly rigid.

29 Masten and Obradović, "Competence and Resilience in Development."

30 C. S. Holling refers to this as "stability," rather than "resilience": "stability, on the other hand, is the ability of a system to return to an equilibrium state after a temporary disturbance."
Holling, "Resilience and Stability of Ecological Systems," 17.

31 For reflections on this bridging, see: Masten and Obradović, "Disaster Preparation and Recovery."

32 Cook, "Do Landscapes Learn?"
Hill, "Shifting Sites."
Wu and Loucks, "From Balance of Nature to Hierarchical Patch Dynamics."

33 Turner, "Disturbance and Landscape Dynamics."

34 Holling and Meffe, "Command and Control."
Pickett and White, *The Ecology of Natural Disturbance and Patch Dynamics*.
Pickett, Cadenasso, and Grove, "Resilient Cities."

For an account of the history of resilience theory in relationship to landscape architecture specifically, see Fleming, "Lost in Translation."

35 Holling, "Resilience and Stability of Ecological Systems."

36 While we find the concept of resilience useful, given its capacity to bridge between disparate communities like landscape architecture, engineering, and policy, the very idea that resilience should be considered a goal is itself contested, particularly because it shares both genealogical and conceptual characteristics with the ideological program of neoliberalism.
Baker, "Anti-resilience."
Walker and Cooper, "Genealogies of Resilience."

The speed of contemporary landscape migration can be challenging for the concept of resilience. Ecological resilience typically implies change around some assumed stationarity: a lake—though stochastically fluctuating in terms of its species composition, sediment loading and temperature—will over time remain that same knowable lake. All three definitions of resilience describe an envelope within which something retains its basic qualities. We live in a time in which landscape conditions are changing faster than they have in millennia, and in ways that appear to be novel and unprecedented. Sedimentary landscapes, given the prominent role of change within them, are at the forefront of these transformations. At such speeds and scales, resilience ultimately needs to be situated within a broader paradigm of adaptation, aimed at the design of landscapes suited to emerging conditions, not the maintenance of historical base states and maladaptive infrastructures.[37]

* * *

To intentionally orchestrate the movement of sediments in time and space between landscapes that are in deficit, such as the Mississippi River Delta, and landscapes that are in surplus, such as the reservoirs of the great upstream dams on the Missouri River, is to *choreograph sediment*. Given the implication of sediment in many of the environmental inflection points that we are collectively accelerating toward, such as exposure to sea-level rise and the loss of crucial coastal habitats like marshes, intelligent choreography is increasingly urgent.

Thinking of design in terms of choreography is uncommon, but not unprecedented. In the middle of the twentieth century, landscape architect Lawrence Halprin and avant-garde dancer and choreographer Anna Halprin famously collaborated to develop creative processes that merged choreography and design.[38] The Halprins developed tools like "motational" scoring systems to visualize, document, and orchestrate movement through time and space.[39] More recently, landscape architects like Anita Berrizbeitia, James Corner, Julia Czerniak, Elizabeth Meyer, Anuradha Mathur, and Dilip da Cunha have persuasively argued that understanding and manipulating landscape processes should be at the core of landscape design.[40] Designing with sediment requires working with flows and movements, with very real and very muddy sediment in motion over time.[41] Dredgers, fixed infrastructures like levees and seawalls, agricultural practices such as contour-plowing, dams, canals, port operators, congressional appropriations and local regulations: these all speed up and slow down the movement of sediment, shifting its distribution and re-organizing the corresponding migration of landscapes.[42] Sediment design is, in many ways, more like orchestrating the patterned movement of bodies in space, as a dance choreographer does, than it is like designing an industrial object such as an iPhone.[43]

Thinking of sediment through the lens of choreography also shifts the terms on which sediment is engaged. While the word choreography is not historically associated with intervention in sediment systems, there are entities, such as the Army Corps' Regional Sediment Management program, whose work is choreographic. The language and underlying concepts of such entities, though, is generally bureaucratic and managerial: budgets, stakeholders, life cycle costs.[44] Economic concepts like budgets (and scientific concepts like resilience)

37 Kristina Hill describes a series of progressive paradigm shifts in planning and design from suitability to sustainability to resilience to adaptation. For Hill, the goal of adaptation is "to accept new forms for cities and new characteristics of ecosystems that are adapted to new conditions, that are resilient to extreme events, and that reduce carbon emissions." Hill, "Climate Change," 109.

We discuss shifting baselines in greater depth in "Technology, Pace and Adaptation."

38 The Halprins were fundamentally concerned with participation and collaboration, not merely between the two of them as creative partners, but as a restructuring of creative practice. Their choreography was not dictation, but an exploration of how many parties could make beauty together, whether on a dance platform or in a bustling city. Anna Halprin sought an understanding of choreography that was more open and inclusive than many other ways of understanding choreography in dance. This orientation is also fundamental to our understanding of sediment choreography. Choreography will not be just if it is a technocratic and exclusive practice. We examine this aspect of sediment design in much more detail in "Aesthetics and Publics."

39 Wasserman, "A World in Motion."

40 Berrizbeitia, "Scales of Undecidability."
Corner, "Terra Fluxus."
Czerniak, "Appearance, Performance."
Meyer, "The Expanded Field of Landscape Architecture."
Mathur et al., "Preparing Ground."

41 In this sense, choreography is not a metaphor. The weakness of what Julian Raxworthy terms the "process discourse" in landscape architecture is that process and movement have often served more as metaphors or the subject of representation (as in James Corner's two influential essays "The Agency of Mapping" and "Representation and Landscape") than as the substance or medium of actual landscape-making.
Raxworthy, *Overgrown*, 3-4, 169-72.

42 If we take seriously J. B. Jackson's definition of landscape as "space deliberately created to speed up or slow down the process of nature," then sediment design is clearly a form of landscape design.
Jackson, "The Word Itself," 8.
Milligan, "Accelerated and Decelerated Landscapes."

43 Note that the difference here is not in the presence or absence of choreographic effects—all design has choreographic effects—but in the mode and practices of design. In choreography, flows are foregrounded, objects recede; pattern, time, and movement are emphasized. Material is reshaped in both sediment choreography and industrial design, but in choreography, other agents have much more direct influence within that process of shaping than they do in the making of an iPhone.

44 USACE, The Corps of Engineers National Regional Sediment Management Demonstration Program. Lillycrop et al., "Advancement of Technologies."

45 Kondolf et al, "Sustainable Sediment Management."

are certainly important to choreography. But, as the word itself reminds us, sediment choreography will always have an element of art to it: we need to deal creatively with sediment, and the value of the landscapes that we make with sediment should never be reduced solely to their performative capacities, even when those functions are infrastructurally vital. Choreography calls as much for qualities like good judgment, patience, and humility as it does for scientific rigor and economic feasibility.

* * *

Choreography begins by taking account of geography. Sediment, like water, moves within basins that have recognizable boundaries, or *sedimentsheds*. Sediment's movement is usually coupled with water's: washed from soil by rain; carried by streams, creeks, and rivers; coming to rest downstream in ponds, lakes, bays, and oceans. Ridges, as a result, typically separate sedimentary basins from one another. Valleys are corridors of movement, as water and sediment together are drawn downward by gravity. The sedimentsheds that result nest in scale, beginning with local creeks and streams and aggregating into larger units spanning whole river systems.

Sediment moves more slowly than water, and drops out when velocities slow: first large gravels, then smaller gravels, then sands, silts, and finally fine clays. Large volumes, known as bedload, move slowly along the bottom of a river, often taking decades to move from headwater to outlet. Other loads, such as organic material and silts washing off exposed farmland in spring, give floodwaters their characteristic browns and oranges, and move much more quickly.

This variance is also the source of differences between how water interacts with infrastructure and how sediment's movement is altered. Many watersheds have been transformed by the construction of canals, aqueducts, and pumps, facilitating inter-basin transfer that extends the reach of watersheds across ridges, hills, and mountains. The Colorado River Aqueduct, for instance, brings water that began its journey in the Rocky Mountains across the desert to Los Angeles. In other cases, such as the six great paired dams and reservoirs of the upper Missouri River, dams hold water in place and regulate its release, backing up enormous reservoirs behind them that can be drawn on in dry seasons. Such water infrastructure, however, rarely accounts for the movement of sediment.[45] Dams regularly release water, but the sediment that flows into the reservoirs behind them stays, accumulating in the deep pools behind dams until it aggregates into muddy masses large enough to clog those reservoirs. Aqueducts may carry water, even carry water uphill driven by pumps, but they do not intentionally bear sediment. Thus, while watersheds are often extended and expanded by infrastructure, *sedimentsheds* are more often constrained, shrunk, and fragmented.

Dams, river canalization, agricultural development, and upstream urbanization have all modified water and sediment flows in tandem. As a result, the design of sediment systems should take into account not only infrastructures and practices that are typically understood as oriented toward sediment management, such as confined disposal facilities, dredging, and silt fences, but also a broad range of other infrastructures and practices that have incidental or unintentional effects on sediment transport: bank armor that limits upstream siltation, levees that prevent conveyance of sediment into floodplains, bulkheads that restrict tidal exchange, pumps that transport water without sediment, or reservoir releases that can, if properly timed, nourish downriver marshes with spring sediment.

Choreography thus requires mapping, understanding, and accounting for the movement and behavior of sediment flows within these basins, as they are modified by infrastructures.[46]

* * *

Sediment flows also vary significantly across time. In many places, the movement of sediment is concentrated in short spans of time within a year, such as spring floods or a new dredging project. Choreographing with the assumption that the average annual load will arrive in twelve equal installments may easily result in over-designing for nine months and massively under-designing for three others.

Annual averages can be similarly deceptive across longer time spans. Two years that happen to have particularly large floods may carry the majority of the sediment that moves across an entire decade, as sediment loads often increase exponentially with flood size. Correspondingly, dredging in a given region varies in the amount (and kinds) of sediment that it provides as the waterborne delivery of sediment to channels varies with climate and weather. But dredging also varies because funding rises and falls with political and economic fortunes. Dredging is a complicated logistical operation that relies on the availability of physically finite resources like dredge boats and operating crews, and each dredging operation is potentially affected by unique events, like new environmental regulations, the discovery of the presence of an endangered species on a placement site, or a lawsuit.

In many regions, the variability of sediment flows is intensifying. Climate change is one driver. For instance, as California's already-arid climate has grown both on average drier and more intensely flashy in recent decades, sediment flows have been both gradually reducing and concentrating within shorter, more intense bursts.[47] Urbanization is another key cause. Heavily urbanized sedimentsheds often have relatively little exposed soil that can erode and become downstream sediment supply, but they are also, as a result of the spread of impermeable asphalt, concrete, and buildings, subject to particularly extreme flash flooding.

The prevailing engineering paradigm for dealing with variability is to overdesign: a flood control channel, for instance, is sized for a very large flood, such as a 100-year or 500-year recurrence interval, and then sees the majority of its capacity unused most of the time. (This happens to make flood control channels excellent sediment traps.) It is reasonable to design flood protection around large, unusual events, as those are precisely the events that have the potential to be most damaging. But, because the prevailing paradigm for urban design deploys infrastructure to control change, this also results in a binary delineation between territories subject to change (inside a flood control channel) and territories where stasis is enforced (communities and landscapes behind protective levees). A more accommodating paradigm would look for opportunities to blur these boundaries, to develop gray zones, to incorporate more flux. This could lead both to ecologically superior outcomes, as landscapes

46 We must also account for the reverberations of sediment design into other systems. Altering dams to enhance sediment transport, for instance, can have a major impact on water supply for agricultural and urban users. Any holistic redesign of infrastructural processes will require weighing a wide array of distinct impacts, which will not fall evenly on all people, landscapes, or systems. The fragmentation of regulatory and planning authority over these impacted realms, including water, sediment, ecology, infrastructure, human settlement, is a significant barrier to holistic design, particularly in the United States, where that fragmentation is often extreme. The following chapters, especially "Aesthetics and Publics" and "Technology, Pace, and Adaptation," deal in more detail with where these barriers are being overcome now and how they might be more comprehensively overcome in the future.

47 Cloern et al., "Projected Evolution."

that are dependent on change are better supported and facilitated across wider areas, and to enhanced resilience, as communities are encouraged to view flood risks more realistically and discouraged from depending on infrastructure that has the potential to fail catastrophically.[48]

Scientists, engineers, designers, and planners increasingly recognize the vital role of change in landscapes. Yet the cities, infrastructures, and policies that we have are, by and large, designed according to earlier models of static control.[49] For instance, the Mississippi River's valleys and delta have been settled under the assumption of increasingly rigid and efficacious control. Farms and homes are built behind levees in former floodplains. The lower Mississippi between Baton Rouge and the mouth of the Mississippi has been developed as a long, linear port on the assumption that the course of the Mississippi will never shift again. The delta has been surveyed, apportioned, and settled according to an intricate web of property boundaries, many of which delineate properties that either are now subsiding or soon will subside into the Gulf of Mexico. The modern city of New Orleans has been developed on the assumption that its extensive network of levees, canals, and water pumps can hold water indefinitely to a clear boundary, with dry in here, wet out there—an assumption that proved catastrophic during Hurricane Katrina.

This is not just a matter of the legacies of the twentieth century. Even much "climate adaptation" and "green infrastructure" defaults to such outmoded paradigms.[50] In the face of sea-level rise and accelerating environmental change, coastal communities are often offered designs that merely build levees higher, add new sea walls, or tack thin ribbons of green onto barely disguised real estate schemes. In the New York City region, for instance, despite a surfeit of projects, like Rebuild by Design or Structures of Coastal Resilience, that have advanced real possibilities for rethinking our collective relationship to the shore, the primary response to coastal storm risk that the New York District of the Army Corps has put forward for the region is the construction of storm surge barriers. These enormous walls and gates would divide the region's waters and reverse hard-won gains in the health of New York Harbor's marine ecosystems.[51] Such proposals are not truly adaptive; they are a form of resistance to adaptation in the face of trends that render them increasingly obsolete. They assume that control is both possible and desirable when it is neither.

Choreography thus requires substantial shifts in the expectations that shape our infrastructures, cities, and landscapes. Functioning macroecological systems require space for the landscapes and ecological communities that comprise them to migrate. Marshes need uplands to move into when sea-level rises. Beaches and barrier islands need to be able to slide laterally along coasts, responding to the movement-in-aggregate of sand-sharing systems. Rivers need access to floodplains, are healthier when they have space to dance their migrations, and must be able to flood if they are to supply their deltas with sediment. This is particularly urgent in the current context of anthropogenically accelerated environmental change, where migration must happen extremely fast if it is to happen at all.

Choreography will not require abandoning the possibility of living in a floodplain, inhabiting a delta, or building on a coast, but it will require rethinking infrastructures, modes of building, and patterns of settlement in these change-prone and change-dependent regions. Choreography should provide space for change, welcome change, and support building with change, not against it.[52]

48 Macdonald et al., "The Significance of Gilbert F. White's 1945 Paper."
Pickett, Cadenasso, and McGrath, *Resilience in Ecology and Urban Design*.
Rossano, *Floodscapes*.
White, "Human Adjustment to Floods."

49 Taylorist ideology plays a significant role in this. Other ideological tendencies shaping the prevalence of static control include the rationalization of society and landscape through high modernist planning, the tools of modernist planning including (technocratic, elitist) master planning, and the "integrated ideal" of Fordist state infrastructure.
Ford, *The Trouble with City Planning*.
Scott, *Seeing Like a State*.
Scott, "High Modernist Social Engineering."
Graham and Marvin, *Splintering Urbanism*.
Weller, "Landscape (Sub)Urbanism in Theory and Practice."

50 This is often a matter of the disconnect between political will—which often emerges suddenly, in the wake of a disaster, rightly demanding immediate change—and the work of design innovation, which requires substantial time for research and development. Unless innovative designs have already been produced, a region looking for answers will understandably tend to default to pre-existing "solutions." In many cases, this means that the projects that advance after a disaster were designed decades earlier, without the benefit of more recent advances in ecological knowledge and design practices.

51 Stringer, to Colonel Thomas D. Asbery and Joseph Seebode, USACE.

52 Holmes, "Design with Change."
Milligan, "Landscape Migration."

3. Redirecting Sediment

The case of New York's Jamaica Bay provides an opportunity to begin to examine choreography in practice.[53]

Jamaica Bay is an isolated waterbody that lies between the barrier island Rockaways and the mainland of Brooklyn. The bay is connected by a narrow gap to the main estuarine complex of the New York Harbor. It has been dredged since the mid-nineteenth century, when private firms such as the Canarsie Railroad, White's Iron Steamboat Company, and the Knickerbocker Steamship Company dug out small channels to facilitate their operations around the rim of the bay.[54] Dredging in the bay accelerated and intensified at the turn of the century, with the 1907 "Report of the Jamaica Bay Improvement Commission," which recommended transforming the bay from a marshy estuary (by then heavily contaminated by the waste of the burgeoning New York metropolitan area) into an enormous port, which would have been larger than the ports of Rotterdam, Hamburg, and Liverpool combined.[55]

The port scheme, which called for the consolidation of the bay's many marshy islands into two massive concrete islands and the construction of a vast array of piers, was never fully realized. Nonetheless, the plan guided decisions about the bay for over two decades, so that by the 1930s a massive transfer of sediment had taken place. Channels, slips, and basins—fifteen, eighteen, thirty feet deep, and hundreds of feet wide—were excavated from the shallow bay bottom. The excavated sediments were piled onto adjacent islands and marshes, most notably being used to construct Floyd Bennett Field, an early airport that is now a component of the National Park Service's Gateway National Recreation Area.[56]

This transfer roughly fixed the current form of the bay: a deep circular channel rings the bay, separating remnant marsh islands in the middle of the tidal estuary from constructed terrain on the mainland, including massive, paved tarmacs at both Floyd Bennett Field and its successor, John F. Kennedy International Airport. The Army Corps maintains this configuration by dredging, battling continual refill by sediment washing in from the bay's sedimentshed.

The channels enable the mix of activities currently found in Jamaica Bay, including recreational boating and commercial shipping. But dredging is also causing problems. It is among the complex of factors responsible for a sharp decline in the quality and quantity of marsh found in the bay.[57] Surveys found that over half the marsh islands' total area disappeared in the twentieth century. A decade ago, scientists predicted alarmingly rapid rates of collapse for the remaining marshes, often anticipating total disappearance by the 2030s.[58]

Marsh loss is significant because of the variety of important roles that the marshes play. Even with the loss, Jamaica Bay is still the largest tidal wetland complex in the New York metro area. It is home to over eighty fish species and is a key stopover for migrating birds. It includes "the only national park in the United States you can reach by subway," Gateway National Recreation Area.[59] During hurricanes and other major storms, the marshes provide critical flood protection for nearby businesses and residences, slowing and attenuating storm surge.[60] Most of NYC's vulnerably low-lying population is located in the Jamaica Bay watershed. And hydrological models predict that neighborhoods around Jamaica Bay will become more, not less, vulnerable to storm surge if a proposed barrier is built across the Verrazano Narrows between Brooklyn

53 Earlier versions of portions of this section were originally published as an essay in the open-access online publication *Scenario Journal*.
Holmes and Milligan, "Feedback."

The environmental history of Jamaica Bay is related at length in Orff, "Cosmopolitan Ecologies."

54 Black, *Jamaica Bay*.

55 Jamaica Bay Improvement Commission, *Report*.

56 Black, *Jamaica Bay*, 71-73.

57 Kolker et al., "Erosion in Jamaica Bay."

58 Jamaica Bay Watershed Protection Plan Advisory Committee, "An Update."

59 Feuer, "Jamaica Bay."

60 Jamaica Bay Watershed Protection Plan Advisory Committee, "An Update."

and Staten Island.[61] Thus, the protective value of the marsh island complex is difficult to overestimate.

Dredging has simultaneously contributed to the loss of marshes and, by deepening the once-shallow entrance, also increased storm surge within Jamaica Bay.[62] The blame for the resulting vulnerabilities cannot be laid entirely on the design of the channels in Jamaica Bay or on the dredging activities that maintain them, though. Reduced sediment inputs due to urbanization have starved the marshes of nourishment. Toxicity has killed marsh plants that would otherwise hold sediments together and stabilize the islands. Nitrogen discharges from the wastewater treatment plants that ring the bay have degraded the ecosystem. Boat traffic produces wakes that have eroded marsh edges. And sea-level rise has played its role, too, just as it has in Louisiana.

Recognizing this degradation, a complicated network of federal, state, and local entities has partnered to rebuild some of the most severely impacted marsh islands utilizing sand dredged from New York Harbor.[63] Between 2006 and 2013, over 1.6 million cubic yards of sand were placed at five such islands, Elders Point East, Elders West, Yellow Bar Hassock, Black Wall, and Rulers Bar, producing 156 acres of new marsh.[64] Ironically, the majority of the sand being used for this marsh-building project is only available because of another large dredging project, a project driven by the same kinds of commercial imperatives that led to the original dredging of Jamaica Bay.

As the 1980s ended, the Port of New York and New Jersey realized that it would need to deepen its channels if it wanted to maintain its central position in global commerce, as it became clear that global trade would increasingly be dominated by economies of scale associated with enormous cargo ships bigger than even "Panamax" vessels.[65] Dredging and blasting for the resulting "New York and New Jersey Harbor Deepening Program," which began in 2004 and finished in 2016, produced a vast sedimentary surplus.[66] Before the deepening began, the Army Corps dredged hundreds of thousands of tons of sediment from the harbor annually; during it, the Army Corps dredged millions of tons of sediment in New York Harbor annually, an increase of an order of magnitude.[67] This dredging surge has been composed of sandier and cleaner sediments than maintenance dredging, which typically involves more polluted silts recently washed in from upland sources. This vast new supply of clean sand has made the marsh rebuilding project feasible.[68]

But Jamaica Bay also demonstrates a potential pitfall for choreography. As the Army Corps has placed dredged material to rebuild the marsh islands, it has done so with a key restriction: the restoration work must never exceed the exact footprints and elevations that the marsh islands had in 1974.[69] Why 1974?

The 1970s were an important time for the environmental movement in the United States. Groundswells of public activism and mass demonstrations, embodied in events like the Earth Day marches, which began in 1970, generated momentum within state and federal governments to address toxicity, pollution, and the degradation of the environment by industrial society. The core institutions of American environmental planning and regulation were established in this period by federal laws like the National Environmental Policy Act (1970), Environmental Pesticide Control Act (1972), the Endangered Species Act (1973), and the Safe Drinking Water Act (1974).[70]

State governments were also acting on the concerns of their citizens, developing and expanding their own environmental regulations. In 1973, the

61 Orton, "Upcoming Presentation."

62 Orton, "Jamaica Bay."

63 "Jamaica Bay Marsh Islands," USACE New York District.

The coalition includes the US Army Corps of Engineers, the Port Authority of New York and New Jersey, National Park Service (Gateway), New York State Department of Environmental Conservation, New York City Department of Environmental Protection, the National Resources Conservation Service, the New York/New Jersey Harbor Estuary Program, the New York State Department of Environmental Conservation, and the New York City Department of Environmental Protection.

64 USACE, New York District, "Fact Sheet." Baron, "Jamaica Bay Restoration."

The restoration efforts on these islands have drawn on lessons learned from an earlier restoration project in 2002–2003 at Big Egg Marsh. The restoration effort there was limited in its extents because it was deemed too expensive to dredge and transport sand from a distant source, forcing the restoration to rely on a small source of sand within the bed of a channel in the marsh. Five more islands were selected by 2017 for future restoration: Elders Point Center, Duck Point, Pumpkin Patch West, Pumpkin Patch East, and Stony Creek.
Frame, Mellander, and Adamo, "Big Egg Marsh Experimental Restoration."
USACE, New York District, "Coastal Restoration Activities within Jamaica Bay."

65 The dimensions of Panamax vessels are set based on the dimensions of the Panama Canal; a Panamax vessel is the largest ship that can pass through that canal. Regarding the broader context and landscape impacts of the global post-Panamax shipping that has reshaped ports like the Port of New York and New Jersey, see Davis, Holmes, and Milligan, "Isthmus."

66 "New York and New Jersey Harbor Deepening Program," Port Authority of New York and New Jersey.

67 "Dredging Information System," USACE NDC.

68 The marsh islands of Jamaica Bay thus demonstrate how sudden shifts in the equilibrium of a sediment system—often owing to exogenous shocks—can be transformative opportunities for choreography. See "Technology, Pace, and Adaptation" for further discussion of the timing and rate of transformative change in locked-in systems.

69 Messaros, Rafferty, and Woolley, "Challenges and Successes of Tidal Wetlands."

70 Merchant, *The Columbia Guide*, 180-1.

New York State legislature passed a law, the Tidal Wetlands Act, which declared that the New York State Department of Environmental Conservation was responsible for preserving and protecting tidal wetlands.[71] The Department conducted the first comprehensive tidal wetlands inventory in New York in 1974, and it is because of that wetland inventory that 1974 is the first date where survey information for Jamaica Bay is detailed enough to set precise dimensions and coordinates for the historic boundaries of the marsh islands. Thus, from the perspective of the state of New York, the 1974 boundaries are the "regulatory footprint": the legal limits of the extent of the marsh islands.[72]

71 Mushacke and Picard, 1974–1999 *Vegetated Tidal Wetland Trends*, 1.

72 USACE, New York District, "Coastal Restoration Activities," 11.

As a result, it would not be too inaccurate to suggest that the reason that the Army Corps' dredgers and bulldozers have been rebuilding the 1974 contours of islands like Elders Point and Yellow Bar is because of the Earth Day marches, or because of the 1962 publication of Rachel Carson's *Silent Spring*, or, if not a single event like those, then the cumulative effect of the many environmental crises and societal responses that birthed American environmentalism. Such historical chains of incident, of course, can be found and reported for any event or action. But this chain highlights the contingent, political, and human nature of the decision to restore the marsh islands' 1974 topography. The 1974 boundaries are not special or unique from the perspectives of water, spartina, and mud. They are only special and unique within human socio-political history.

In every year of their existence, the marsh islands of Jamaica Bay have been both growing and shrinking, accreting sediment in some places and eroding in others. Some years, the gain has outpaced the loss; beginning in the twentieth century, though, until the recent surplus of material from expansion dredging, the erosion far outpaced the accretion. But the fact of change has been continuous: it is simply how marsh islands are. Picking one point in time—even the furthest point back where we can be certain about their shape—and determining that that point represents the "correct" state of the islands, to which they should be restored is, at best, a bizarre decision. Yet the paradigm of restoration, with its implication of recovering a superior past state, frequently favors such choices. They are often the only politically palatable decisions available.

What is missed in modeling the future after the past is not only that the selection of a single past moment requires ignoring the fundamental dynamism of sedimentary landscapes, but also that the many landscape processes that surround and act on such landscapes—like these marsh islands—are not and cannot be the same as they were in a single past moment. The boundaries of marsh islands are not just lines on a map, but are the product of many forces acting simultaneously. Tides bear sediment in and bear it away. Currents do too. Plants are deposited as seeds, root, grow, spread by rhizomes, die, are carried away by tides and currents, wash up as wrack, decay. People dredge out channels that deepen waterways, affecting tides and currents. People burn carbon that changes the global climate. Using old surveyed perimeter lines is fast, convenient, and, given existing law and regulation, legally necessary, but that does not make it adaptive. Choreography must account for current forces, processes, and trajectories. The past may, at times, be a useful proxy, but it is not enough to guide design now—particularly in light of the accelerating intensity and pervasiveness of human influence and landscape change.

4. Choreography Without Control

But what would it look like to design for the future rather than the past?

Efforts to understand and predict the behaviors of sediments, marshes, and islands have grown in tandem with the increasing urgency of landscape restoration and the emerging recognition of landscapes as protective infrastructure.[73] But those efforts have often seductively promised that landscapes and environmental phenomena can be controlled in ways (and at scales) that they cannot. Choreography requires a different approach, grounded in cultivation rather than control, stewarded by designers who recognize that landscapes are as social and political as they are ecological and geomorphological.

* * *

In 1971, ecologist and systems theorist H. T. Odum published his macroscopic work of scientific philosophy, *Environment, Power, and Society.* At the time, Odum was known, together with his brother Eugene, for groundbreaking work in ecology that had helped establish the core principles of the ecosystem concept that lies at the heart of modern ecological science. The pair had become international stars in their field, winning millions of dollars in research funding for studies that measured, quantified, and modeled ecosystems like the underwater meadows of Silver Springs in Florida, the humid (and radiation-haunted) swamps and forests of the Savannah River Site in South Carolina, and the (also radioactive) coral reefs of the Enewetak Atoll in the Marshall Islands.[74]

The ambition of *Environment, Power, and Society* went far beyond those studies. Odum proposed—and began to conduct—a startlingly expansive project: measuring, modeling, and modifying the entirety of the earth, inclusive of all dimensions of ecological phenomena and human activity.[75] At the core of this project was a unifying concept of "power": the translation of all facets of this planetary system into measurable, quantifiable units of energy. Power, according to Odum, lies behind photosynthesis, religious practices, and cities alike. Using power as a "common denominator," Odum was able to consecutively synthesize holistic explanations of the workings of the biosphere, agriculture, modern (fossil fuel-based) industrial economies, politics, and religion.[76] These explanations were codified in "energy diagrams," drawings that described networks of relation and exchange between the components of each system. Accurate and comprehensive energy diagrams, Odum believed, would facilitate the construction of computer models that could, with enough feedback and refinement, predict the behaviors of even the largest, most complex systems, such as a city, region, or the Earth's biosphere itself.

Accurate prediction would, in turn, facilitate effective management. Odum described a process of "environmental control engineering" wherein humanity, recognizing that it is already rapidly altering global earth systems and ecological processes, would take on responsibility for managing them.[77] Through a series of examples, Odum explained why and how this totalized "partnership with nature" would take place. He imagined "cross-continental" transplantations of species, constructed loops between "city wastes" and "producing lands," and the intensified farming of "marine sea bottoms."[78] In each system, ecological

73 On the recognition of landscapes as protective infrastructure, see "Mud is Vital."

74 Kirby, *Mockingbird Song*, 236–56.

75 Odum, *Environment, Power, and Society*, 305.

76 Ibid, 21.

77 Ibid, 247–303.

78 Ibid, 284, 291, 309.

engineers would set up a series of feedback loops between input and output energies, monitor the system's stability, and coax it toward balance, with each step facilitated by energy diagram-based modeling.

Yet Odum was troubled by an unsolved difficulty at the heart of his vision. As he put it: "for many of the world's systems we do not [yet] have enough knowledge to draw the energy diagrams, predict consequences of treatments, and manage nature."[79] One of his main purposes in writing the book was to build a case for the necessity of remediating this lack of knowledge.[80] This was, for Odum, required by the increasingly rapid changes that "flows of concentrated fossil fuels, coal, and oil" permitted humanity to make to environmental systems.[81] Generalized systems science would provide the knowledge needed to engineer solutions, if only humanity could gather enough information for it. Whether society possessed the capacity and will to obtain the vast volume of data demanded was, for Odum, an open question: "as man's system becomes large enough to control and prevail in the flows of the biosphere, will he understand it well enough to prevent disaster?"[82] But Odum was confident that systems science, if it incorporated every aspect of human endeavor, was an adequate means.

This systematizing project was not Odum's alone, nor was Odum's version particularly idiosyncratic. The middle decades of the twentieth century saw a flowering of visions for generalized systems theories and efforts at deploying them to address grand societal, ecological, and environmental problems. A succession of figures from varied disciplinary backgrounds—Jay Wright Forrester, Ludwig von Bertalanffy, C. S. Holling, Walter Isard, Norbert Wiener—converged on systems theory as a unifying science. They developed fields and offshoot disciplines that took root in academia, government, and corporations: general systems theory, regional science, systems dynamics, cybernetics. Forrester, for instance, began his career engineering control systems for military applications in World War II, before broadening his interests to include industrial organization (*Industrial Dynamics*, 1961), cities (*Urban Dynamics*, 1969), and an expansion of his theory to encompass the entire globe (*World Dynamics,* 1971).

Prompted by an invitation from the elite technocrats of the newly formed Club of Rome, the core of *World Dynamics* was a model, titled WORLD2, which Forrester and his students at MIT made in an effort to relate the many "stocks" and processes that they considered central to the collective behavior of the Earth as a system. These included population and birth rates, land area and crowding, food and pollution, capital investment and natural resources. This led directly to the work that the Club of Rome is still best known for, the deeply controversial 1972 publication *The Limits to Growth*.[83] In that work, Forrester's team of students, led by Donella and Dennis Meadows, used a successor model, WORLD3, to demonstrate that "the earth is finite and cannot support exponential physical growth for very long."[84] *The Limits to Growth* evidenced a strong faith in the capacity of systems science to ask and answer questions about the fate of civilization itself. Its reception helps demonstrate the scale and pervasiveness of the influence of systems theory on American and global society in the 1970s: *The Limits of Growth* dropped like a bomb into the fractured and tense environmental politics of the times. It went through five printings in 1972 alone, eventually selling 12 million copies, enough to make it still the best-selling environmental book ever printed.[85] It attracted fervent

79 Ibid, 292.

80 Ibid, 292.

81 Ibid, 6-7; this aspect of Odum's vision is clearly prescient, even if his solutions have proved unworkable and undesirable.

82 Ibid, 7.

83 Lane, "The Power of the Bond."

84 Meadows and Meadows, "The History and Conclusions of *The Limits to Growth*," 194.

85 Parenti, "*The Limits to Growth*."

86 Odum, *Environment, Power, and Society*, 308.
Meadows et al., *The Limits to Growth*.

87 Many more recent landscape architectural theorists have made and been motivated by critiques of control. This includes much of the process discourse. For instance, James Corner's early critique of the role of positivism in landscape architecture cited a presumption of explanation and control as primary failings of positivist landscape architectural theory: "the primacy of logical and objective reasoning in modern positivism has consequently led to the illusion of humankind's infinite capacity to explain, control, and put to work the forces of nature." Corner went on to explicitly pin-point systems theory as exemplifying positivism in landscape architecture and planning. Another angle of critique, more focused on the role of modernist engineering and planning in shaping urban systems through master planning and rationalization, is represented by Pierre Bélanger's work on landscape infrastructure. In practice, work by designers like Claire and Michel Corajoud, Martí Franch, Teresa Galí-Izard, and Gilles Clément could be understood as materializing critiques of control that are tied to the role of process in landscape-making. Another substantial strand of critique in landscape architecture theory descends from the participatory design traditions of the 1970s that the Halprins contributed to. This includes the work of figures like Diane Jones Allen, Jeff Hou, Randy Hester, and Marcia McNally. We discuss issues of participatory design decision-making in more depth in "Aesthetics and Publics."
Bélanger, *Landscape as Infrastructure*.
Corner, "A Discourse on Theory II," 117.
de la Peña et al., *Design as Democracy*.
Galí-Izard, *The Same Landscapes*.

adherents as well as vociferous critics, who found its methods, its conclusions, or both radically misguided.

Systems theorists sought to redesign human societies and the global environment from the top down, using technocratic methods of expert analysis, prediction, and control. This often led them to propose myopic excesses—Odum sought to codify systems science as a new religion organized around the "commandments of the energy ethic," and *The Limits to Growth* argued for the imposition of a global two-child policy.[86] While the willingness of systems theorists to face enormous planetary problems can be admired, our call for a choreography of sediment is based in more humble methods and aspires to a more inclusive, democratic vision for the future.

Choreography, like the open and participatory forms of dance that Anna Halprin developed and Lawrence Halprin modeled landscape practice after, should enable and facilitate, not dominate or control.[87]

* * *

In introducing sediment choreography, we have both made use of and departed from concepts which are linked to systems theory.[88] The most central of these is the idea of systems themselves.[89]

Systems do not exist in a spatial or material sense. Rather, a system is a purposeful abstraction, a conceptual tool designed for scientific and managerial purposes.[90] When you look out across a river valley or an estuary, you don't say, "I see a system." Rather, you see a physical landscape. The diverse assembly of material things that you see is irreducibly more than any effort to systematize it. A system is real in that mentality is real, and in that ideas and thinking have real effects and expression.[91] But a system is fundamentally an abstraction. To systematize is to describe with rules, operations, flow diagrams, and codes, like in Odum's a-spatial diagrams.

Correspondingly, systems thinking has often tended to suppress differences and spatial diversity, which makes it easier to believe that control is possible. *The Limits to Growth* and WORLD3 assumed a singular, knowable, and controllable world. On a planet with many cultures and political systems, proposing a global two-child policy is hubristic and impractical; only within the simplifying assumption of a singular world would it (or many of the other proposals of early systems theory) make sense.

This is a key reason that contemporary systems approaches have diverged from the methods and concepts of early systems theory.[92] Most contemporary systems theorists draw on a newer branch of systems theory known as complex systems theory, which seeks to understand the behavior of systems which are not determinate, not necessarily stable, and not fully knowable.[93] They emphasize phenomena like emergence, self-organization, and state transitions. They typically align themselves with "post-normal science," which considers the effects of conflicting value-judgments and systemic uncertainties embedded within scientific processes.[94]

Choreography can draw on these insights, but it must also go further. As choreography creatively and collaboratively orchestrates sequences of actions and movements, it must be attuned to spatiality and temporality, to contextual difference and diversity. Throughout this book, we intentionally focus on

88 This is not unusual within landscape architecture. Influential thinkers like Pierre Bélanger, Kristina Hill, and Nina-Marie Lister have all drawn directly on branches of systems theory, particularly ecosystem theory. Generally, the contemporary use of systems theory by landscape architects draws on later stages of systems theory, like complex systems theory, which have critiqued and diverged from early systems theory's emphasis on control.
Bélanger, "Sprawl: A Strategy?"
Hill, "Form Follows Flows."
Waltner-Toews, Kay, and Lister, *The Ecosystem Approach*.

89 It is also common for choreographic work to explicitly align itself with systems theory, as the Corps' Regional Sediment Management program does.
USACE, ERDC, *USACE Regional Sediment Management and Engineering With Nature*.

90 Waltner-Toews, Lister, and Bocking, "A Preface."

91 For an explanation of philosopher Alfred North Whiteheads' notion of mentality, see "What a Body Can Do" in Massumi, *Politics of Affect*.

92 Kay, "An Introduction to Systems Thinking," 8–11.

93 Nicolis and Prigogine, *Exploring Complexity*.

94 Funtowicz and Ravetz, "The Emergence of Post-normal Science."
Funtowicz and Ravetz, "Uncertainty, Complexity and Post-normal Science."

landscapes as real, complex, and spatial things. We understand landscapes as shifting assemblages, composed of things that exhibit distributed agency, and capable of forming themselves in a multiplicity of ways that exceed abstraction and totalized understanding.[95] Like Marisol de La Cadena and Mario Blaser, we see *a world of many worlds*, each different and structured by the unique material, ecological, and socio-political attributes of landscapes and regions.[96]

In our work with the Dredge Research Collaborative, we have focused on this geographic, spatial, and contextual specificity; we have foregrounded the malleability and indeterminacy of landscape. Our landscape approach is founded in embodied fieldwork, in maps and drawings that articulate material and spatial qualities, and in engaging the fundamentally social nature of landscape-making. Still, the behaviors of sand particles, dredging corporations, root masses, regulatory agencies, and levees (to name but a few potential agents) can sometimes be usefully assessed as collective sediment systems, which have aggregate tendencies, discernable individual components, and comprehensible networks of relationships, both internally and externally. In those cases, the tools of contemporary systems theory can be useful, particularly when they are deployed alongside the tools of a landscape approach.[97]

Both systems theory and choreography focus on the importance of dynamic, relational processes. For choreography, these processes, like erosion, dredging, Congressional funding, and climate change, shape the formation of landscapes. The networks of relations that link components of a sediment system—including but very much not limited to physical relations—are conduits of exchange, movement, and transformation. This shapes a basic orientation toward intervention that choreography should share with contemporary systems theory: these processes must be engaged, over time, if the larger structure of a system is to be responsibly stewarded and landscape formation is to be guided in beneficial directions.[98]

Thus, intervention requires feedback about the state of a set of landscapes and mechanisms for altering interventions in response. This recalibration involves monitoring, analyzing, and adapting to changes in both individual landscapes and larger bio-physical regions as they are produced by new actions. Sediment is placed in a river, forming a new island; marsh grasses grow on a bank of the island; sediment traveling downstream is trapped by the grasses and begins to deposit; the river's currents erode the edges of the island on the opposite side, carrying that eroded sediment downriver. Designing this new island intelligently requires not only an informed approach to the initial deposition, but also an ongoing process of information-gathering that may lead to deciding to place more sediment in the future, on the basis of the observed behaviors of river, marsh, island, and sediment.

Odum, Forrester, the Meadows, and other systems theorists of their era believed, at the time, that systems and landscapes could be solved. They believed that adequate data could be collected, that accurate predictions could be made using models, and that the systems they were investigating could be controlled. But we now know—and contemporary systems theorists agree—that this kind of knowledge is not actually possible. We know that the control they sought is not achievable within the complex medium of landscape.

* * *

95 DeLanda, *A New Philosophy of Society*. Dovey, *Becoming Places*. Bennett, *Vibrant Matter*.

96 de la Cadena and Blaser, *A World of Many Worlds*.

97 Below, we discuss philosopher Sandra Mitchell's idea of integrative pluralism, which is a model for doing science and making decisions through this kind of back-and-forth between diverse methods and conceptions. Mitchell, *Unsimple Truths*.

98 "Beneficial" here is not assumed to be a known, positivist constant. What is beneficial is contingent and contextual. It should be articulated through transparent objectives arrived at through socio-politically inclusive modes of planning and design. We discuss this in more detail in "Aesthetics and Publics."

The early systems theorists prescribed reliance on mathematical models, which translated diverse physical phenomena into numerical data. Today, geomorphologists, hydrologists, and engineers use digital computer models called *numerical models* to assess and attempt to predict the behaviors of landscapes.[99] Beaches are modeled in order to inform decisions about nourishing beaches with dredged sand or building groins and breakwaters to protect sand from erosion. Rivers are modeled to understand flood risk, to determine where to build levees, to size storage basins and floodways. Marshes are modeled to understand where they are likely to grow or shrink.

Yet models have real limits.[100] When sediment shapes landscapes, they exhibit complexity, unpredictability, and uncertainty arising from nonlinearity.[101] Water moves sediment, and sediment in turn alters the movement of water. The sheer frequency of these interactions—imagine millions of grains of sand, all shifting in response to each wave that strikes a shore, each wave slightly different in direction, period, and energy than the last, over and over and over—makes it very difficult to model the behaviors of sediment across space and time.[102]

With numerical models like those run in HEC-RAS (Hydrologic Engineering Center's River Analysis System, modeling software developed for the US Army Corps of Engineers) or Delft3D (Dutch modeling software), water flow is modeled with one set of equations and sediment transport is modeled with another. These equations are then coupled in each iteration of the model to capture the dynamic feedback between sediment and water, but processing power and time itself place significant limits on how frequently those interactions between fluid dynamics and sediment transport can be calculated.[103] With large landscapes and fine resolution, model runs can take hours, days, or even weeks. This, in turn, makes it difficult to capture the nonlinearity of sediment flows and displacements, because the frequency of those interactions is crucial in allowing small changes to persist and grow in influence over time. (This is the phenomenon often described as the "butterfly effect" in popular descriptions of chaos theory.) Instead of attempting to model every such interaction, models abstract, reduce, and leave out. This makes it possible to form generalizations about aggregate behavioral patterns, to anticipate consequences in less time than they would take to actually unfold. But there is always difference between a model of a system and the messy landscape the model seeks to understand.

Furthermore, precise modeling requires precise data, both as inputs for models and to describe histories that can be used to verify models' validity.[104] But the field collection of data regarding sediment is limited by the practical requirements of sensor placement and budgets for acquiring, monitoring, and maintaining instruments. Even when data for something like the sediment load of waterway is already being collected, it is typically being measured only in one or two locations along that waterway. And historical data is considerably more difficult to procure, often making something as seemingly simple as determining the historical annual average sediment load of a river impossible, much less building an accurate terrain model over a century of geomorphological change. Thus not only is the translation of landscape into data one of key means by which modeling abstracts, but that process is significantly limited by physical constraints.

99 There is also a long history of physical landscape modeling, as either complement or alternative to numerical modeling.
Holzman, "Landscape Models of the Scientific Imaginary."
Holzman, "Hydraulic Modeling as Craft."

100 For a more strident (and controversial) case against current practices for modeling phenomena like sediment transport, see Pilkey and Pilkey-Jarvis, *Useless Arithmetic*, particularly chapter 6: "Beaches in an Expected Universe."

101 In a nonlinear system, the "outputs" are not always proportional to the "inputs," meaning that the system's behavior cannot be predicted simply by understanding the external forces acting on it.
Phillips, "Sources of Nonlinearity and Complexity."

102 Coulthard and Van De Wiel, "Quantifying Fluvial Non-linearity."

103 Coulthard and Van De Wiel, "Modeling River History and Evolution."

104 Coulthard and Van de Wiel, "Modeling River History and Evolution," 2,136.

It is easy to assume that at least the processes represented by equations in a model are actually understood. Yet this is not true either. Geomorphologists readily admit there is a great deal they do not know about sediment transport processes. Those processes are relatively well understood at very small scales (a grain of sand, suspended in water) and macro scales (geomorphology across millenia), but the medium scales of both time (a year) and space (a river reach) that matter most in landscape design are the least understood.[105]

The best modelers are highly attuned to the propensities of their technological medium, particularly the ways that it abstracts, distorts, and reduces. Models can and do offer useful insights into the behavior of sediment. But they cannot predict the futures of the landscapes that sediment shapes with comprehensive certainty.[106] Making wise design decisions for landscapes that are constantly changing, like marshes, beaches, and floodplains, requires confronting the challenges of indeterminacy.[107]

* * *

And sediment systems are not just geomorphological systems. Like the holistic systems studied by early systems scientists, they are fundamentally social and political.[108] Consequently, our attempts to design for their futures must account for social complexity. Designers in the 1960s who sought to use systems theory to make their work "scientific" quickly discovered that technocratic methods could not resolve conflicts over divergent values.[109]

In a 1973 article entitled "Dilemmas in a Theory of General Planning," two professors at the University of California, Berkeley, Horst Rittel and Melvin Webber, famously argued that the problems confronted by "social professionals"—a category that includes both planners and designers—cannot be solved objectively, because the acts of defining them and of solving them are intrinsically linked.[110] Such "wicked problems," so named by Rittel and Webber in contrast to the "tame problems" of science and engineering, have a variety of key characteristics: they can be both framed and resolved in a multitude of ways; their boundaries are porous, because they are embedded in open society; and, because they involve values, their answers are not judged as true or false, but rather as good or bad. The implication was that, while systems theory may provide a tool for understanding conditions in the world, it cannot provide clear direction on how it should be deployed. Planning, they concluded, is always "a component of politics."[111]

An example may make this more clear. Why is the fact that the Mississippi River Delta is losing land a problem? (Or, more precisely: *how* is it a problem?) Is it a problem because the loss of marsh exposes both rural communities and the city of New Orleans to storms? That is certainly one valid answer. But accepting that answer requires accepting that there should be permanent human communities in the delta, and there is no objective, scientific way to decide whether people should live in the delta or not. Nor is there any objective, scientific way to decide how many resources should be expended to protect people who live in the delta. When and where should enhancing protection give way to designing retreat? When is the problem no longer the loss of marsh, but the refusal to leave? The only possible answers to questions like this are answers that are built on value judgments, and because values are situated

105 Ibid, 2,137.

106 C. S. Holling expressed this concern in 1973, as he introduced resilience theory, helping to develop the shift from first-generation systems theory to its contemporary successor, complex systems theory. Holling cautioned against assuming that comprehensive knowledge of ecosystems is possible, arguing that the complexity of ecosystems is too great and that ecosystems are not determinate. He suggested that qualitative modeling and prediction is more useful and plausible than quantitative prediction. First-generation systems theory (including Odum and Forrester) and complex systems theory are in some basic ways diametrically opposed, as the former has a fundamental disposition toward order, balance, and stasis, while the latter embraces chaos theory, resilience as a goal, and change as a desirable quality.
Holling, "Resilience and Stability of Ecological Systems."
Walker and Cooper, "Genealogies of Resilience," 143.

107 This is something that landscape urbanism never accomplished, despite being a body of landscape architectural theory focused on landscape as a dynamic and changing phenomena. It accurately characterized landscapes as fluctuating and indeterminate, but it was content with a primarily representational approach to engaging change, and never deeply engaged the actual tools, like modeling or sensing, by which trajectories of landscape change might be probed, anticipated, or understood. In the past decade, a new body of work has begun to emerge that is engaging these tools, as evidenced, for instance, by the recent "Instruments of Change" conference organized by the EMLab at the University of Pennsylvania and on-going explorations by the landscape architecture faculty at ETH Zurich.
Belesky et al., "A Field in Flux."
Girot, "The Elegance of Topology."
Milligan, "Making Terrains."
Robinson and Davis, "From Solution Space to Interface."

108 Within systems theory, the technical term for this sort of system is a "socio-ecological system."
Berkes, Folke, and Colding, *Linking Social and Ecological Systems*.

within human communities with diverse contexts, they are always potentially subject to debate. When a planning project, like efforts to build new land in the delta, encounters heated public resistance, this is almost always because a conflict in values is being exposed. One public wants to see sediment deployed to build new land in the delta; another public opposes those efforts because a declining delta fuels the economic sustainability of oyster leases, and reversing that decline could undermine their livelihoods. Rittel and Webber pointed out that, in situations like these, deciding what the real problem is necessarily means deciding what kinds of solutions will be accepted.[112] If the problem is that levees cut the delta off from the river, then the solution will be to reconnect the river to its delta. If the problem is that people should not live in a declining delta as sea-level rise accelerates, then the solution will be to design and manage equitable migration to safer uplands. The problem may actually be that there really is no clear, uniformly agreed-upon problem, and thus multiple and sometimes contradicting perspectives need to be reconciled—or enabled to coexist.

For Rittel and Webber, this irreducibly political nature of engagement with wicked problems means that *judgment* has to be foregrounded.[113] Bounds must be set, the key parameters of a problem selected for focused inquiry, explanations for the origin of a problem developed. But they were not interested in the judgment of the lone expert or even a technocratic elite. Rather, they argued for—and Rittel spent the rest of his career exploring—a thoroughly social judgment, judgment produced in community, through deliberation, collective exploration, and open-minded, good-faith argumentation.[114] They saw this "taming" as a process that should not aspire to final completion. New information, new values, or even the consequences of an implemented solution could always necessitate re-engaging with a wicked problem.[115]

Choreographers would do well to learn these lessons, too: the definition of problems and the setting of goals for choreography are inherently value-laden and thus political tasks. Choreographers should not aspire to technocratic neutrality, but look to consciously, clearly, and communally articulate the values and aims that guide a project. And just as wicked problems are never solved, choreography should never be understood as finished: instead, it should be understood as an ongoing process of learning, care, cultivation, and repair.[116]

* * *

None of this is to say that models should not be used. Models remain a crucial tool for choreography. Used properly, they allow us to interrogate the behavior of sediment systems, to probe landscape futures, frame scenarios, and make better decisions. But, just as discovering that all acts of design and planning are contingent and political should lead us to treat problem-setting as an essential act of design, recognizing the potential abuses of modeling requires that we use modeling judiciously, within appropriate decision-making processes.

In 1985, Donella Meadows and colleague Jennifer Robinson published *The Electronic Oracle: Computer Models and Social Decisions*. Reflecting on their own experiences as modelers (Meadows was the lead author of *The Limits to Growth*) as well as a set of then-contemporary case studies, they argued that the way that systems with social components are modeled is deeply flawed. One of the key flaws they identified concerns the definition of problems. Much

109 Designers have been observing, learning from, and attempting to apply systems theory since the 1960s. Much of this work initially took place within the nascent field of environmental design, which, in keeping with the optimistic gestalt of the era of general systems theory, sought to systematize and scientifically-validate design practices, particularly within architecture, industrial design, and landscape architecture. The theorists and practitioners of the "design methods movement" (including Horst Rittel) consciously tied their efforts to systems theory, confident that the design of buildings, cities, and societies could be optimized through the application of both knowledge and methods drawn from ecological and social sciences.
Sachs, *Environmental Design*.
Cross, "A History of Design Methodology."
Protzen and Harris, *The Universe of Design*.

110 Rittel and Webber, "Dilemmas."

Like the ecological engineers who found their neat models of ecosystems flummoxed by the complex and unpredictable realities of dynamic landscapes, though, environmental designers and the design methods movement found their expectations subverted by complexity. Rittel became critical of the early work of the design methods movement (roughly, 1962–1974) and worked toward a "second generation" of design methods. "Dilemmas in a Theory of General Planning" expresses a part of this effort, much as C. S. Holling's work expresses a self-critique within ecosystem theory.

111 Rittel and Webber, "Dilemmas," 169.

Rittel and Webber argue this, notably, in 1973, only a year after the *Limits to Growth* is published and only two after *Environment, Power, and Society* and *World Dynamics* set out their macroscopic (and apolitical) visions for the restructuring of society according to systems theory.

112 Ibid, 161.

113 Ibid, 164–66.

114 Rith and Dubberly, "Why Horst W. J. Rittel Matters."

115 Rittel and Webber, "Dilemmas," 163.

116 Holmes, "The Problem with Solutions."
Jackson, "Rethinking Repair."
Mattern, "Maintenance and Care."

as Rittel and Weber showed that the way a wicked problem is defined ends up determining the way it is "solved," Meadows and Robinson showed that the way modelers (and their clients) define a problem constrains what results the model will produce. The process of problem definition is irreducibly social and contestable, but it is all too often abbreviated, sidelined, or suppressed.[117] This happens in part because modelers, like the designers of Rittel and Weber's era, have often assumed "the trappings of the established physical sciences," such as privileging mathematical expression over verbal communications, without fully adopting the mode of inquiry or the institutions of shared learning that actually constitute science. Similarly, Meadows and Robinson showed that when models are used to inform decision-making, they are often deployed with the inappropriate expectation that the object of study—in their examples, social systems themselves—can be fully known and fully controlled.[118] These twin errors have been the core errors of systems science.

In *Unsimple Truths*, philosopher of science Sandra Mitchell shows how the irreducibly interconnected nature of causation in complex systems requires shifting our expectations of science away from eliminating uncertainty and toward understanding dynamic probabilities over time. In a developed estuary, for instance, marsh evolution is shaped by contaminant levels, storm events, the decisions of individual landowners, the dispersal of seeds, macroeconomic trends, the introduction of new technologies, global climate change, and regional sediment supply—to name only some factors. Any of these factors might in turn shift suddenly as a result of other, external forces, as sediment supply did in Jamaica Bay when the Port Authority decided to deepen its channels. Instead of depending on models to predict the future and then acting on those predictions as if they were certain knowledge, Mitchell recommends a more flexible, pluralistic, and iterative approach, characterized by the probabilistic evaluation of multiple futures, on-going investigation of both situation and scenarios as they unfold, and responsive re-calibration of action as new knowledge is obtained. She ultimately advocates for an *integrative pluralism* in which models are just one way of understanding the world, which should be combined with others, such as local embodied experience and direct observation. Meadows and Robinson arrived at similar conclusions, arguing that models should "be used to search for imprecise policies that are robust against uncertainties, rather than precise policies that try to optimize something that is not understood."[119]

Within many fields that work with ecosystems and landscapes, this way of working has come to be known and practiced as adaptive management.[120] If choreography is to be an ongoing process of care, cultivation, and repair, which works with landscapes that are most resilient when they are dynamic and changing, then, when it uses modeling, it must use modeling as a tool to facilitate adaptive management, not as an instrument of certainty or control.[121]

Systems theorists once assumed that control was possible and desirable. They no longer do, and neither should choreography. We need what Sheila Jasanoff calls "technologies of humility," practices for working without perfect knowledge, in the context of disagreement over goals and values, and within landscapes that are defined by change.[122]

117 Meadows and Robinson, "The Electronic Oracle," 285.

118 Meadows and Robinson charge modelers and their clients with an unwillingness to admit that they confront "problems without knowable and permanent answers", an unwillingness that is built on the belief that "social systems are basically known and can be controlled". They recommend instead "honest, experimental redesign" of social systems, with modelers as knowingly embedded participants in that process.
Meadows and Robinson, "The Electronic Oracle," 299.

119 Ibid, 300.

120 Johnson, "Introduction to the Special Feature."
Lister, "Sustainable Large Parks."

121 Both institutional inertia and the potential for abuse can make and have made adaptive management a difficult process to deploy in practice, but that does not make learning to do so any less necessary.
Regarding the potential for abuse, see for instance Kristin Shrader-Frechette's critique of *Unsimple Truths*' recommendations in her book review, which notes that "flexible management is often seat-of-the-pants management...continued investigation is often merely paralysis-by-analysis regulatory delay... learning-by-doing is often putting-out-fires-bumbling."
Regarding institutional inertia, it is important to recognize that most of the incentive structures and procedures of the bureaucratic organizations responsible for regulating and managing sedimentary landscapes are currently constructed in a fashion that makes adaptive management difficult, if not impossible. Cost-benefit analysis mandates demand certainty and precision that models should not be asked to provide; funding is usually delivered in large but unpredictable increments, rather than the long-term, steady funding for maintenance and monitoring that adaptive management requires; and the very idea of experimentation, which is at the core of adaptive management, is often anathema to political leaders and publics that demand certainty.
Shrader-Frechette, "Unsimple Truths."

122 Jasanoff, "Technologies of Humility."

We discuss this approach to technology in more depth in "Technology, Pace, and Adaptation."

5. Everything All at Once

The images are iconic: the drilling rig, the *Deepwater Horizon*, engulfed in flames, black smoke curling upward in thick clouds, as fire-fighting ships spray it with water cannons. The oil well, which began to spill uncontrollably on that day in April 2010, continued to leak for at least five months, producing the single largest oil spill in history. Oil spread across the Gulf of Mexico, and the environmental damage was enormous.

BP, the corporation that the *Deepwater Horizon* was drilling for, paid over $20 billion to the US government to cover damages from the disaster. Much of that has been distributed to the states whose coasts were directly impacted: Texas, Mississippi, Alabama, Florida, and, particularly, Louisiana. And much of that money has been specifically earmarked for coastal restoration.

For Louisiana, that has meant an infusion of funding at a critical juncture. In the wake of another disastrous period—the 2005 hurricane season, where Hurricanes Katrina and Rita devastated the Louisiana coast, including New Orleans and the delta plain—the state had created a new agency, the Coastal Protection and Restoration Authority. This CPRA was tasked with addressing the land loss and sediment starvation that threatens coastal Louisiana, and it is the work of the CPRA—together with its partners like the Water Institute of the Gulf, the Coastal Sustainability Studio at Louisiana State University, and the US Army Corps of Engineers—that best illustrates what choreography requires in practice.

* * *

The formation and funding of CPRA have been crucial steps in a long process of transformation in the way that Louisiana has managed its delta and coasts.[123] As early as the 1920s, before construction had begun on the Mississippi River and Tributaries Project that would eventually encase the lower Mississippi in earthen levees and concrete armor, contrarian voices recognized that disconnecting the river from its floodplain would harm the region's ecosystems.[124] But through the bulk of the twentieth century, as the levees, spillways, and revetments were constructed, managers treated the river as a problem to be controlled. Evidence grew that the delta was losing ground, though, and the environmental movement increasingly argued for the value of marshes. By the late 1980s, Louisiana and the federal government agreed that coastal wetland restoration should be prioritized. CPRA provided a coordinated mechanism for restoration, and *Deepwater Horizon* provided an infusion of funding to scale the work up.

CPRA's strategy centers on three primary types of landscape infrastructure: barrier island nourishment, marsh creation with dredged material, and the construction of river sediment diversions. Barrier island nourishment and marsh creation are relatively similar. In both, sediment (primarily sand) is mechanically dredged, transported, and placed in a desirable location in order to build land. For nourishment, this typically means mining offshore sand deposits that have been selected for their abundance of appropriately sized sands. Marsh creation is more likely to rely on the opportunistic use of sediment dredged from the Mississippi's navigation channels, which the Corps and its partners refer to as a "beneficial use of dredged material."[125] Both nourishment and marsh creation are fast: sediment placed becomes land built immediately. But they are also

123 Killebrew and Khalil, "An Overview of History of Coastal Restoration."

124 Killebrew and Khalil, 30.

125 The idea of beneficial use exists in contrast to the earlier paradigm for dredged material placement, in which dredged material was treated as a waste product (or "spoils"), which should be disposed of as cheaply as possible, without regard for the potential value of dredged sediment as a resource. We explore these paradigms in more depth in the next chapter, "Mud is Vital."

both energy-intensive (the machines that do this work must be fueled in order to pump sediment-laden slurries against gravity) and expensive (crews must be paid, ships and machines maintained). And land built by marsh creation or beach nourishment is not permanent. It, like any land at water's edge, is subject to erosion, subsidence, and sea-level rise. Thus, it requires further placement of dredged material in order to maintain built land. This sedimentary life-support will become increasingly cost-prohibitive in future decades, as the price of carbon fuels—both financially and in terms of damage to the planet as a whole—rises.[126]

Sediment diversions are different. Rather than relying on dredges, pumps, and bulldozers to do work, sediment diversions put the Mississippi River back to work building land. Two diversions are presently in engineering design, ten more are planned. Each diversion is a massive artificial outlet for the river, built primarily of concrete and steel.[127] Their main purpose is to convey the sand, silt, and clay that the river carries out into basins, sounds, and floodplains beyond the levees. To capture the river's sediment load, each diversion will reach into the riverbed with an armored intake. The flow of diverted water and sediment will be controlled by a series of enormous gates, opened intermittently in rhythm with the heavy loads of sediment that the river carries when it swells in spring. CPRA anticipates that each diversion will supply 2-3 million cubic yards of sediment annually.[128] This sediment will accumulate with each release like a miniature delta-within-a-delta, building new marsh quickly.[129]

The diversions will be large, immobile infrastructures, requiring significant up-front investment in engineering and construction. Once built, though, they will operate relatively cheaply, in terms of both energy and money. As a result, where nourishment and marsh creation are effective quickly, but increasingly expensive over time, sediment diversions will be slow to gain momentum in building land, but will become increasingly efficient and effective over their operative lifespans, which may be a century or more long.

This makes sediment diversions the crucial hinge of the Louisiana Coastal Master Plan, CPRA's iteratively evolving plan guiding efforts to build land in the delta and the adjacent chenier plain. In the long run, only sediment diversions have the capacity to build enough land at a low enough cost to make continuing to live in many parts of the delta feasible.[130] Though funding in Louisiana to date has been concentrated toward marsh creation and beach nourishment, the latest Coastal Master Plan makes the case that the weight of effort must shift toward using sediment diversions.[131]

* * *

The plan has necessarily evolved in the context of imperfect knowledge. Locally, the interactions of the complex of causes that is producing land loss in the delta have been a matter of significant scientific and policy debate. For instance, which has more heavily impacted land loss: the loss of sediment supply due to the confinement of the river, or the fragmentation of the delta through the excavation of canals to facilitate oil and gas extraction? Depending on how you answer that question, you might favor investing in publicly funded sediment diversions or you might want to focus on legal battles with the oil and gas industry aimed at winning damages to support a broader array of physical interventions.[132]

126 Wiegman, "The Costs and Sustainability," 100.

127 To some extent, the sediment diversions are patterned after existing hydrological diversions, like the Caernarvon and Davis Pond Diversions, which were built to increase freshwater flows into the delta, combatting saltwater intrusion.

128 "Mississippi River Mid-Basin Sediment Diversion Program," CPRA.

129 Day et al., "Large Infrequently."

130 Day et al., "Large Infrequently."

131 CPRA, *Louisiana's Comprehensive Master Plan* (2017).

132 For a perspective informed by the latter view, see Turner, "Doubt and the Values of."

The capacity of the tools available to CPRA is another key uncertainty. How much land can a sediment diversion build? How quickly? How is that land-building capacity affected by the location of a diversion along the river course? How frequently should a diversion be operated? Much scientific investigation and modeling effort has gone into attempting to refine answers to such questions, but the empirical data is limited. One key investigation was able to collect eleven examples of structures (like the hydrological diversion at Caernarvon or the Bonnet Carré Spillway) and events (like the 1884 Davis crevasse) that provide precedent for understanding how sediment diversions will function. That is certainly enough examples to provide useful data and to understand key trajectories within the broader uncertainties. The investigation concluded, for instance, that large but infrequently operated diversions would build land more effectively and with less negative impacts than small, frequently operated diversions.[133] But past precedents are not—cannot—provide enough information to be certain precisely how new diversions will perform at other locations, on a river that continues to evolve, in a global climate accelerating toward an unknown future.

Uncertainty in the causes of land loss, the contributions of possible interventions, and the global effects of climate change all make planning for the future of the delta necessarily improvisational.[134] As new scientific information is produced, plans must be adjusted. When an intervention is constructed, its actual performance—as opposed to modeled predictions of its performance—must be studied, so that actual performance can inform both future predictions and further interventions. And there is no time to wait for perfect scientific knowledge of the future before beginning work: land loss is accelerating, and the work must begin before CPRA or any other entity can be certain of the results of their efforts.

The Coastal Master Plan shows that these conditions do not make planning impossible or obsolete (in fact, they make planning crucial), but they do require planning differently. The plan was not released once and considered finished. Implementation is not a second step after planning is completed. Rather, in CPRA's work, planning and implementation are interwoven. Coastal restoration projects in Louisiana began decades before the plan was begun, and CPRA has learned from those projects. An updated plan is released roughly every five years.[135] This work of planning can be expected to continue indefinitely into the future, in parallel to the physical work of building land on the coast. Choreography is always in the making.

* * *

The basic mechanism for responding to Louisiana's wetland loss is straightforward: the delta needs more sediment. This has been clear at least since the first plan for the restoration of the delta was published in 1975[136]—and the idea of reconnecting the river to its floodplains goes even further back, to biologist Percy Viosca Jr.'s proposals in the 1920s for "artificial distributaries."[137]

But the planning and implementation of the Coastal Master Plan is complicated by disagreements that spring from the fact that it is impossible to alter the delta at the scale required without disadvantaging some portion of Louisiana's population. Investments, for instance, in fixed protective

133 Day et al., "Large Infrequently."

134 The trajectory and precise consequences of anthropogenically accelerated climate change is easily the single most studied (and debated) scientific uncertainty of the twenty-first century. While the fact of that change and the general direction of its vector—toward a hotter, wetter planet with higher sea levels—is well-established, predictions for local effects, timings, and intensities necessarily vary widely, given the mathematical complexity and uncertainty of the environmental processes involved.

135 The first plan was released in 2007; another followed in 2012, and the most recent in 2017. The next is planned for 2023.

136 Gagliano and van Beek, "An Approach to Multiuse Management."

137 Killebrew and Khalil, "An Overview of History," 30.

infrastructure like levees—or marshes built with dredged material—require decisions about where funding will be invested and where it will not. The cost-benefit analyses typically used to justify these decisions favor denser urban populations over rural populations and wealthier populations with higher real estate values over marginalized frontline communities.

The necessity for sediment diversions has exposed another, more unique political rift.[138] Many commercially valuable species, including shrimp, oysters, and saltwater fish, prosper in a declining delta, as salinity increases in the absence of freshwater flows from a connected river. Fishers have opposed the sediment diversions due to the belief that the diversions would displace the species they harvest,[139] a concern that the 2017 Coastal Master Plan concurs with when it predicts that implementation would reduce habitat for "adult bay anchovy, spotted seatrout, small juvenile white and brown shrimp, oyster, and green-winged teal."[140] Other species and fishers would benefit from the planned projects—crawfish, bass, and duck are all predicted to gain in a reconnected delta—but as commercial fishers tend to specialize, this hardly allays the concerns of oyster lease-holders.

As with uncertain knowledge, such political challenges have not prevented the work of the CPRA from proceeding, and recognizing that choreography will be entangled with political questions—questions of value that cannot be settled by reference to some objective authority—does not mean that choreography is doomed. But, for the CPRA, it has meant not only making plans, but doing persuasive work, engaging effected communities, and, at times, finding balances between what is technically preferable and what is politically feasible.[141] Choreography cannot be pursued as a technocratic project somehow outside or beyond politics; it is too interwoven with humanity, our ways of life, and our disagreements.

* * *

The Mississippi River Delta exemplifies the dynamic change that characterizes landscapes shaped by sedimentary processes. Though the delta has not always been threatened like it is today, at no point in its 7,000-year history has it been stable. Rather, change—in the form of sediment-laden floods and newly deposited layers of silt—has always been the mechanism by which the delta has prospered and grown. With the possible exception of *homo sapiens*, the plant and animal species that call the delta home are adapted to prosper amidst this change.

The constant revision that the Coastal Master Plan undergoes is one way that it has responded to this context. But the dynamic quality of delta landscapes has also resulted in the use of specific tools that are appropriate to a changing environment. CPRA is shifting away from structural protection projects, like levees, toward measures aimed at building land, like sediment diversions. Since 2007, CPRA has completed $4.3 billion in structural protection projects. This is more than five times as much as has been invested in either of the next two most heavily funded categories of project, barrier island restoration ($780 million) and marsh creation ($585 million).[142] Over the projected 50-year span of the latest Coastal Master Plan update, though, this trajectory will invert. The 2017 plan still allocated $19 billion to structural protection, but also planned

138 Nost, "Climate Services for Whom?"

139 Day et al., "Large Infrequently Operated...," 123.

140 CPRA, *Louisiana's Comprehensive Master Plan* (2017).

141 For some examples of this work in Louisiana, see:
Hemmerling, et al., "Elevating Local Knowledge through Participatory Modeling."
Baustian et al., "Engaging Coastal Community Members."
Peyronnin et al., "Optimizing Sediment Diversion Operations."

142 CPRA, *Fiscal Year 2021 Annual Plan*, 35.

143 CPRA, *Louisiana's Comprehensive Master Plan* (2017), 104.

144 We focus here on modeling, but there is also substantial literature on the use of scenarios in approaching indeterminate environmental situations. In particular, exploratory scenarios hold particular promise as tools for enhancing decision-making in the context of complex systems and wicked problems.
Camillus, "Strategy as a Wicked Problem."
Salewski, *Dutch New Worlds.*
Wilkinson and Eidinow, "Evolving Practices in Environmental Scenarios."

145 Wiegman, Rutherford, and Day, "The Costs and Sustainability of Ongoing Efforts."

146 The 2017 CPRA plan proposes using sediment to maintain and/or build about 800 square miles of land against a projection of around 2,500 square miles of loss.
CPRA, *Louisiana's Comprehensive Master Plan* (2017), 96-97.

a greater $25 billion for land-building projects, including marsh creation and sediment diversions.[143]

Another key dimension of CPRA's toolkit for the delta has been the use of modeling and scenarios. Modeling permits interrogating the behavior of a changing landscape, while scenarios organize and structure expectations about future trajectories in a manner that facilitates decision-making.[144] Both the models themselves and the institutions that produce them have been evolving in tandem with the plans. CPRA has found that the complexity of the physical phenomena it deals with—salinity gradients, sediment flows and accretion rates, marsh plant communities—requires increasingly sophisticated numerical modeling, particularly where those phenomena intersect with political disagreements.

The process that has led to the current plan to build the Mid-Barataria and Mid-Breton sediment diversions is illustrative.[145] In 2007, there were two main scenarios proposed for diversions. Both scenarios located diversions on the Mississippi River below New Orleans, but one scenario located two diversions in the Barataria Basin and Breton Sound, just below New Orleans, while the other scenario located its diversions in lower reaches, closer to the mouth of the river. The Barataria and Breton diversions were anticipated to build more land. But large stretches of the shallow waters where this land would be built are apportioned into long-term oyster leases by the state, and leased to commercial oysterers. Operating the diversions would reduce salinity, likely killing off some of the oyster beds. This is obviously controversial; it could also potentially make the entire project financially unfeasible, as it would give the oyster bed lease-holders grounds to sue the state for their prospective losses. CPRA determined that modeling was needed to understand the nature of these tradeoffs in more detail. Would the diversions work well enough to justify the anticipated costs?

Over the following five years, a new model studying both salinity impacts and land-building was developed. It informed the study of four scenarios for the 2012 plan: a no-action (baseline) future, a scenario with a few large diversions, and scenario with many small diversions, and a scenario with action but no diversions. Only the diversion scenarios were effective in sustaining land-building, while the models showed little variance in overall ecological outcomes between scenarios. These kinds of relative and qualitative comparisons are precisely where modeling is most effective. By 2017, the modeling verdict was clear. The general trajectory of land loss is not reversible. The delta is destined to continue shrinking, even with the massive ongoing efforts. But the overall trend can be reversed locally, if the river can be released to build land again.[146]

The reason that the delta is destined to continue to lose land is that the pace and scale of change in the delta has accelerated rapidly over the past two hundred years and is continuing to accelerate. This acceleration ramped up with the continentally scaled transformation of the Mississippi River Basin in the twentieth century, with the building of dams, levees, and spillways, with the dredging of navigation channels and oil-company canals. Today, the pace is even greater. Continued acceleration is driven by global forces: as temperatures rise and local sea-level rises increasingly rapidly, ever-greater volumes of sediment will be required to build marshland and swamp.

CPRA's scenarios anticipate that trend will continue. Choreography will, for the foreseeable future, take place in environments like coasts, rivers, and deltas that are on the frontlines of climate change. In the context of a dynamic environment, modeling and scenarios do not provide certainty, only opportunities to probe the trajectories of change.[147] Choreography must deploy tools—physically, in planning, and institutionally—that are suited to working with change.

147 In "Technology, Pace, and Adaptation," we discuss prototyping and experimentation, two other methods by which change trajectories can be probed and adjusted. Both of these methods are also fundamental to adaptive management.

* * *

On Louisiana's coast, we can see a future that is arriving there now, but will soon be shared much more widely. The twentieth century has left the United States with a legacy of aging infrastructures intended to control water, constructed without consideration for their impact on the movement of sediment. Sediment systems have been fragmented, disconnected, and broken. Drowning marshes and imperiled communities lie ahead. As climate change accelerates—and, at the same time, many other vectors of human impact also accelerate—the pace of change on coasts, along rivers, and in deltas, places which much of humanity call home, will also accelerate.

Amidst this acceleration, sediment has been growing and will continue to grow increasingly valuable. In Louisiana, wasting dredged sediment—sediment which mere decades ago was described as "spoils"—is already unthinkable. The revaluation of sediment as an economic and ecological resource, rather than a waste product, underlies the desire to choreograph its spatial and temporal distribution. This revaluation is playing out in a diverse range of ways, in waterways and waterbodies across North America. In the next chapter, "Mud is Vital," we consider the new economic and ecological patterns that both are and should be emerging in tandem with sediment's revaluation. If we are wise and opportunistic, then we can shape these patterns to benefit people, places, and ecosystems.

Economic and ecological value are not the only kinds of value that choreography encounters, though. Choreography is also interwoven with social systems and the human experience of landscape. When we say that choreography cannot be a technocratic practice, we mean that the design of sediment systems requires accounting for, including, and designing alongside the people and communities who live, work, and play in sedimentary landscapes. Moreover, these communities are often in disagreement with one another, which makes choreography a political practice—as is evidenced by the struggles of the Isle de Jean Charles band for the future of their community, and the on-going debates over the impacts and design of sediment diversions. We explore the implications of this in our third chapter, "Aesthetics and Publics."

With all these vectors of change overlapping—social, economic, ecological, climatic—technologies and policies must adapt. In Jamaica Bay, dredged sediment is being used to restore long-lost marsh islands. In Louisiana, by contrast, the goal of delta-wide restoration has been tacitly abandoned, replaced by the goal of wisely shaping the novel delta that can emerge given the constraints of sediment supply, climate change, and human desires (even though the language of restoration remains impactful and present). But the institutions that manage sediment systems too often constrain choreography with procedures, regulations, and approaches that make the adaptation and experimentation that novel conditions require impossible. In "Technology, Pace, and Adaptation," we consider why this is and how our policies and machines might evolve to face the challenges our sediment systems confront.

Finally, in this chapter we've briefly alluded to design as a field of practice, particularly in exploring the insights of the systems design thinkers like Horst Rittel, but the protagonists of choreography in Louisiana and Jamaica Bay have largely been engineers, scientists, and planners. In the concluding chapter "Why Design?" we demonstrate the role that design—including, in particular, the discipline of landscape architecture—can play in doing choreography. Through the example of Public Sediment we see how sediment design can operate and what kinds of worlds are possible if we learn to design with sediment.

Bibliography:

Jacopo A. Baggio, Katrina Brown, and Denis Hellebrandt, "Boundary Object or Bridging Concept? A Citation Network Analysis of Resilience," *Ecology and Society* 20, no. 2 (2015).

Shalanda H. Baker, "Anti-resilience: A Roadmap for Transformational Justice within the Energy System," *Harvard Civil Rights-Civil Liberties Law Review* 54 (2019): 1-48.

James Barnett, *Beyond Control: The Mississippi River New Channel to the Gulf of Mexico* (Jackson, MS: University Press of Mississippi, 2017).

Lisa Baron, "Jamaica Bay Restoration," in *Harbor Inspection: Vision of a World Class Harbor Estuary*, USACE New York District, September 16, 2015, https://www.nan.usace.army.mil/Portals/37/docs/harbor/Harbor%20Insp%209%20Sep%2014/2_Partner%20NYNJ%20Harbor%20Ins%2016%20Sept%202015%20.pdf.

John Barry, *Rising Tide: The Great Mississippi Flood of 1927 and How it Changed America* (Simon & Schuster, 1998).

Tristan Baurick, "On Sinking Isle de Jean Charles, Ida Shredded the Few Remaining Homes," *nola.com*, September 4, 2021, https://www.nola.com/news/environment/article_b3ce8e12-0d0f-11ec-8b3e-efcab7617cd6.html.

Melissa M. Baustian, Hoonshin Jung, Harris C. Bienn, Monica Barra, Scott A. Hemmerling, Yushi Wang, Eric White, and Ehab Meselhe, "Engaging Coastal Community Members about Natural and Nature-based Solutions to Assess their Ecosystem Function," *Ecological Engineering* 143 (2020): 100015.

Pierre Bélanger, *Landscape as Infrastructure: A Base Primer* (New York: Routledge, 2017).

Pierre Bélanger, "Sprawl: A Strategy? From Closed System Dynamics to Open Systems Ecologies," Holcim Foundation Green Workshop: Compact City—Sustainable or just sustaining economic law? (2013), https://src.holcimfoundation.org/dnl/60234c0a-f496-4c6f-b495-86817f917398/F13_Workshoppaper_GreenWS_PierreBelanger.pdf.

Philip Belesky, Rosalea Monacella, Mark Burry, and Jane Burry, "A Field in Flux," paper presented at ACADIA 2015: Computational Ecologies: Design in the Anthropocene, Cincinnati, OH (October 2015).

Jane Bennett, *Vibrant Matter: A Political Ecology of Things* (Duke University Press, 2010).

Samuel J. Bentley, Angelina M. Freeman, Clinton S. Willson, Jaye E. Cable, and Liviu Giosan, "Using What We Have: Optimizing Sediment Management in Mississippi River Delta Restoration to Improve the Economic Viability of the Nation," in *Perspectives on the Restoration of the Mississippi Delta*, eds. John W. Day, G. Paul Kemp, Angelina M. Freeman, and David P. Muth (Springer: 2014).

Fikret Berkes, Carl Folke, and Johan Colding, eds., *Linking Social and Ecological Systems: Management Practices and Social Mechanisms for Building Resilience* (Cambridge University Press, 2000).

Anita Berrizbeitia, "Scales of Undecidability," in CASE: *Downsview Park*, ed. Julia Czerniak (Prestel, 2002).

Frederick Black, *Jamaica Bay: A History* (National Park Service, 1981).

Michael D. Blum and Harry H. Roberts, "The Mississippi Delta Region: Past, Present, and Future," *Annual Review of Earth and Planetary Sciences* 40 (2012): 655–83.

Fridolin Simon Brand and Kurt Jax, "Focusing the Meaning(s) of Resilience: Resilience as a Descriptive Concept and a Boundary Object," *Ecology and Society 12*, no. 1 (2007).

John Camillus, "Strategy as a Wicked Problem," *Harvard Business Review* (May 2008), 1–9.

Richard Campanella, *Bienville's Dilemma: A Historical Geography of New Orleans* (Lafayette, LA: UL Press, 2008).

James E. Cloern, Noah Knowles, Larry R. Brown, Daniel Cayan, Michael D. Dettinger, Tara L. Morgan, David H. Schoellhamer, et al., "Projected Evolution of California's San Francisco Bay-Delta-River System in a Century of Climate Change," *PloS one 6*, no. 9 (2011): e24465.

Coastal Protection and Restoration Authority of Louisiana, *Fiscal Year 2021 Annual Plan*.

Coastal Protection and Restoration Authority of Louisiana, *Louisiana's Comprehensive Master Plan for a Sustainable Coast*, 2017.

"Mississippi River Mid-Basin Sediment Diversion Program: Project Benefits," Coastal Protection and Restoration Authority of Louisiana, https://coastal.la.gov/our-work/key-initiatives/diversion-program/project-benefits/.

Michael Collier, Robert H. Webb, and John C. Schmidt, *Dams and Rivers: A Primer on the Downstream Effects of Dams*. USGS Circular 1126. Washington, DC: US Geological Survey, 2000.

Craig Colten, "Levees and the Making of a Dysfunctional Floodplain," in *Perspectives on the Restoration of the Mississippi Delta*, eds. John W. Day, G. Paul Kemp, Angelina M. Freeman, and David P. Muth (Springer: 2014): 29–37.

Robert E. Cook, "Do Landscapes Learn? Ecology's 'New Paradigm' and Design in Landscape Architecture," in *Environmentalism in Landscape Architecture*, ed. Michael Conan (Washington, DC: Dumbarton Oaks, 2000).

James Corner, "A Discourse on Theory II: Three Tyrannies of Contemporary Theory and the Alternative of Hermeneutics," *Landscape Journal* 10, no. 2 (1991).

James Corner, "Terra Fluxus," in *The Landscape Urbanism Reader*, ed. Charles Waldheim (New York: Princeton Architectural Press, 2006), 21–34.

Tom J. Coulthard and Marco J. Van De Wiel, "Modeling River History and Evolution," *Philosophical Transactions of the Royal Society A: Mathematical, Physical and Engineering Sciences* 370, no. 1966 (2012): 2,123–42.

Tom J. Coulthard and Marco J. Van De Wiel, "Quantifying Fluvial Non-linearity and Finding Self-organized Criticality? Insights from Simulations of River Basin Evolution," *Geomorphology* 91, no. 3-4 (2007): 216–35.

Nigel Cross, "A History of Design Methodology," in *Design Methodology and Relationships with Science*, Marc J. de Vries, Nigel Cross, and D. P. Grant, eds., (Springer: 1993).

Julia Czerniak, "Appearance, Performance," in *CASE: Downsview Park*, ed. Julia Czerniak (Prestel, 2002).

Brian Davis, "Land Making Machines," in *Making the Geologic Now*, eds. Elizabeth Ellsworth and Jamie Kruse (New York: Punctum Books, 2012).

Brian Davis, Rob Holmes, and Brett Milligan, "Isthmus: on the Panama Canal Expansion," *Places Journal* (December 2015), https://placesjournal.org/article/isthmus-panama-canal-expansion/.

Simin Davoudi, Keith Shaw, L. Jamila Haider, Allyson E. Quinlan, Garry D. Peterson, Cathy Wilkinson, Hartmut Fünfgeld, Darryn McEvoy, and Libby Porter, "Resilience: A Bridging Concept or a Dead End?," *Planning Theory & Practice* 13, no. 2 (2012): 299–333

John W. Day, Robert R. Lane, Christopher F. D'Elia, Adrian R. H. Wiegman, Jeffrey S. Rutherford, Gary P. Shaffer, Christopher G. Brantley, and G. Paul Kemp, "Large Infrequently Operated River Diversions for Mississippi Delta Restoration," *Estuarine, Coastal and Shelf Science* 183 (2016): 292–303.

Cornelia Dean, *Against the Tide* (New York, NY: Columbia, 1999).

Manuel DeLanda, *A New Philosophy of Society: Assemblage Theory and Social Complexity* (New York: Continuum, 2006).

Julie Dermansky, "Louisiana Breaks Ground on Isle de Jean Charles Resettlement Project Amid Pandemic," DeSmog, May 22, 2020, https://www.desmogblog.com/2020/05/22/isle-de-jean-charles-tribe-louisiana-resettlement-construction.

Marisol de la Cadena and Mario Blaser, eds., *A World of Many Worlds* (Duke University Press, 2018).

David de la Peña, Diane Jones Allen, Randolph T. Hester, Jeffrey Hou, Laura J. Lawson, and Marcia J. McNally, eds., *Design as Democracy: Techniques for Collective Creativity* (Island Press, 2017).

Kim Dovey, *Becoming Places: Urbanism/Architecture/Identity/Power* (Routledge, 2009).

Alan Feuer, "Jamaica Bay: Wilderness on the Edge," *New York Times*, July 29, 2011, accessed February 8, 2013, http://www.nytimes.com/2011/07/31/nyregion/jamaica-bay-a-wild-place-on-the-edge-of-change.html.

Billy Fleming, "Lost in Translation: The Authorship and Argumentation of Resilience Theory," *Landscape Journal* 35, no. 1 (2016): 23–36.

Kristina Ford, *The Trouble with City Planning* (Yale University Press, 2010).

George W. Frame, M. Kathryn Mellander, and Douglas A. Adamo, "Big Egg Marsh Experimental Restoration in Jamaica Bay, New York," in *People, Places, and Parks: Proceedings of the 2005 George Wright Society Conference on Parks, Protected Areas, and Cultural Sites*, ed. David Harmon (Hancock, MI: The George Wright Society, 2006): 123–30.

Silvio O. Funtowicz and Jerome R. Ravetz, "The Emergence of Post-normal Science," in *Science, Politics, and Morality*, ed. Rene von Schomberg (Dordrecht: Springer, 1993): 85–123.

Silvio O. Funtowicz and Jerome R. Ravetz, "Uncertainty, Complexity and Post-normal Science," *Environmental Toxicology and Chemistry: An International Journal* 13, no. 12 (1994): 1,881–85.

S. M. Gagliano and J. L. van Beek, "An Approach to Multiuse Management in the Mississippi Delta System," in *Deltas: Models for Exploration*, ed. M. L. Broussard (Houston, TX: Houston Geological Society, 1975), 223–38.

Teresa Galí-Izard, *The Same Landscapes* (Gustavo Gili, 2005).

Liviu Giosan and Angelina M. Freeman, "How Deltas Work: A Brief Look at Perspectives on the the Mississippi River Delta in a Global Context," in *Perspectives on the Restoration of the Mississippi Delta*, eds. John W. Day, G. Paul Kemp, Angelina M. Freeman, and David P. Muth (Springer: 2014).

Christophe Girot, "The Elegance of Topology," in *Landscript 03: Topology*, eds. Christophe Girot, Anette Freytag, Albert Kirchengast, and Dunja Richter (Berlin: Jovis Verlag, 2013).

Stephen Graham and Simon Marvin, *Splintering Urbanism: Networked Infrastructures, Technological Mobilities, and the Urban Condition* (Routledge, 2001).

Scott A. Hemmerling, Monica Barra, Harris C. Bienn, Melissa M. Baustian, Hoonshin Jung, Ehab Meselhe, Yushi Wang, and Eric White, "Elevating Local Knowledge through Participatory Modeling: Active Community Engagement in Restoration Planning in Coastal Louisiana," *Journal of Geographical Systems* 22, no. 2 (2020): 241–66.

Kristina Hill, "Climate Change: Implications for the Assumptions, Goals and Methods of Urban Environmental Planning," *Urban Planning* 1, no. 4 (2016): 103–13.

Kristina Hill, "Form Follows Flows: Systems, Design, and the Aesthetic Experience of Change," in Nature and Cities: *The Ecological Imperative in Urban Design and Planning*, eds. Frederick R. Steiner, George F. Thompson, and Armando Carbonell (Lincoln Institute of Land Policy, 2016), 345–59.

Kristina Hill, "Shifting Sites," in *Site Matters: Design Concepts, Histories, and Strategies*, eds. Carol J. Burns and Andrea Kahn (New York: Routledge, 2005).

C. S. Holling, "Resilience and Stability of Ecological Systems," *Annual Review of Ecology and Systematics* 4, no. 1 (1973).

C. S. Holling and Gary K. Meffe, "Command and Control and the Pathology of Natural Resource Management," *Conservation Biology* 10, no. 2 (1996): 328–37.

Rob Holmes, "Design with Change," in *Design with Nature Now*, eds. Frederick Steiner, Richard Weller, Karen M'Closkey, and Billy Fleming (Cambridge, MA: Lincoln Institute of Land Policy, 2019).

Rob Holmes, "The Problem with Solutions," *Places Journal* (July 2020), https://placesjournal.org/article/the-problem-with-solutions/.

Rob Holmes and Brett Milligan, "Feedback: Designing the Dredge Cycle," *Scenario Journal* 3 (2013), https://scenariojournal.com/article/feedback-designing-the-dredge-cycle/.

Justine Holzman, "Hydraulic Modeling as Craft," in *104th ACSA Annual Meeting Proceedings, Shaping New Knowledges*, eds. Robert Corser and Sharon Haar (2016), 423–30.

Justine Holzman, "Landscape Models of the Scientific Imaginary: A Visual Textual Précis," in *Codify*, eds. Bradley Cantrell and Adam Mekies (Routledge, 2018), 119–30.

Isle de Jean Charles Biloxi-Chitimacha-Choctaw Tribe, *Preserving Our Place: A Community Field Guide to Engagement, Resilience, and Resettlement*, Isle de Jean Charles: 2019.

J. B. Jackson, "The Word Itself," in *Discovering the Vernacular Landscape* (Yale University Press, 1986).

Steven J. Jackson, "Rethinking Repair," in *Media Technologies: Essays on Communication, Materiality and Society*, eds. Tarleton Gillespie, Pablo Boczkowski, and Kirsten Foot (Cambridge: MIT Press, 2014), 221–39.

Robert B. Jacobson, Dale W. Blevins, and Chance J. Bitner, "Sediment Regime Constraints on River Restoration—An Example From the Lower Missouri River," *Geological Society of America Special Paper* 451 (2009): 1–22.

Jamaica Bay Improvement Commission, *Report of the Jamaica Bay Improvement Commission* (New York: Martin B. Brown Press, 1907).

Jamaica Bay Watershed Protection Plan Advisory Committee, "An Update on the Disappearing Salt Marshes of Jamaica Bay, NY," accessed February 8, 2013, http://nbii-nin.ciesin.columbia.edu/jamaicabay/jbwppac/JBAC_NPS_SaltMarshReport_080207.pdf.

Sheila Jasanoff, "Technologies of Humility," *Nature* 450, no. 7166 (2007): 33.

Barry L. Johnson, "Introduction to the Special Feature: Adaptive Management—Scientifically Sound, Socially Challenged?," *Conservation Ecology* 3, no. 1 (1999).

James J. Kay, "An Introduction to Systems Thinking," in *The Ecosystem Approach*, eds. Waltner-Toews, Kay, and Lister (Columbia University Press, 2008).

Charles J. Killebrew and Syed M. Khalil, "An Overview of History of Coastal Restoration Plans and Programs in Louisiana," *Shore & Beach* 86, no. 1 (2018): 28–37.

Jack Temple Kirby, *Mockingbird Song: Ecological Landscapes of the South* (Chapel Hill, NC: The University of North Carolina Press, 2006).

Alexander S. Kolker et al., "Erosion in Jamaica Bay: Causes, Questions and Sea-level Rise," SUNY Digital Repository, accessed February 8, 2013, http://dspace.sunyconnect.suny.edu/bitstream/handle/1951/48208/Kolker-abst-2001.pdf?sequence=1.

Mathias G. Kondolf, Yongxuan Gao, George W. Annandale, Gregory L. Morris, Enhui Jiang, Junhua Zhang, Yongtao Cao, et al., "Sustainable Sediment Management in Reservoirs and Regulated Rivers: Experiences from Five Continents," *Earth's Future* 2, no. 5 (2014): 256–80.

Mathias G. Kondolf, Kristen Podolak, and Theodore E. Grantham, "Restoring Mediterranean-climate Rivers," *Hydrobiologia* 719, no. 1 (2013): 527–45.

David C. Lane, "The Power of the Bond between Cause and Effect: Jay Wright Forrester and the Field of System Dynamics," *System Dynamics Review* 23, no. 2-3 (2007): 95–118.

Robert R. Lane, G. Paul Kemp, and John W. Day, "A Brief History of Delta Formation and Deterioration" in *Mississippi Delta Restoration*, eds. John W. Day and Jori A. Erdman (Springer: 2018).

Linda Lillycrop, Julie D. Rosati, Jennifer M. Wozencraft, and Rose Dopsovic, "Advancement of Technologies for Practicing Regional Sediment Management," in *The Proceedings of the Coastal Sediments 2011*, 2,061–74.

Nina-Marie Lister, "Sustainable Large Parks: Ecological Design or Designer Ecology?," in *Large Parks*, eds. Julia Czerniak and George Hargreaves (New York: Princeton Architectural Press, 2007).

Neil Macdonald, David Chester, Heather Sangster, Beverly Todd, and Janet Hooke, "The Significance of Gilbert F. White's 1945 Paper 'Human Adjustment to Floods' in the Development of Risk and Hazard Management," *Progress in Physical Geography* 36, no. 1 (2012): 125–33.

Adam Mandelman, "Porous Boundaries and Permeable Places: Ordering Nature in the Mississippi River Delta," PhD diss. (University of Wisconsin: 2016).

Brian Massumi, *Politics of Affect* (John Wiley & Sons, 2015).

Ann S. Masten and Jelena Obradović, "Competence and Resilience in Development," *Annals of the New York Academy of Sciences* 1094, no. 1 (2006): 13–27.

Ann S. Masten and Jelena Obradović, "Disaster Preparation and Recovery: Lessons from Research on Resilience in Human Development," *Ecology and Society* 13, no. 1 (2008).

Anuradha Mathur and Dilip da Cunha in conversation with Nicholas Pevzner and Sanjukta Sen, "Preparing Ground," *Places Journal* (June 2010), https://placesjournal.org/article/preparing-ground-interview/.

Shannon Mattern, "Maintenance and Care," *Places Journal* (November 2018), https://placesjournal.org/article/maintenance-and-care/.

John McPhee, "Atchafalaya," in *The Control of Nature* (New York: Farrar, Straus and Giroux, 1989).

Donella Meadows and Dennis Meadows, "The History and Conclusions of *The Limits to Growth*," *System Dynamics Review* 23, no. 2-3 (2007).

Donella Meadows, Dennis Meadows, Jorgen Randers, and William Behrens III, *The Limits to Growth* (Universe Books: New York, NY, 1972).

Donella Meadows and Jennifer Robinson, "The Electronic Oracle: Computer Models and Social Decisions," excerpted in *Systems Dynamic Review* 18, no. 2 (Summer 2002).

Carolyn Merchant, *The Columbia Guide to American Environmental History* (Columbia University Press, 2005).

Roy C. Messaros, Patricia S. Rafferty, and Gail S. Woolley, "Challenges and Successes of Tidal Wetlands Restoration in Jamaica Bay, New York," in *Watershed Management 2010: Innovations in Watershed Management under Land Use and Climate Change* (2010): 343–63.

Elizabeth Meyer, "The Expanded Field of Landscape Architecture," in *Ecological Design and Planning,* eds. George F. Thompson and Frederick R. Steiner (New York: John Wiley & Sons, 1997): 64–70.

Hans Middelkoop, Gilles Erkens, and Marcel van der Perk, "The Rhine Delta—A Record of Sediment Trapping over Time Scales from Millennia to Decades," *Journal of Soils and Sediments* 10, no. 4 (2010): 628–39.

Brett Milligan, "Landscape Migration," *Places Journal* (2015), https://placesjournal.org/article/landscape-migration/.

Brett Milligan, "Making Terrains: Surveying, Drones and Media Ecology," *Journal of Landscape Architecture* 14, no. 2 (2019): 20–35.

Sandra Mitchell, *Unsimple Truths: Science, Complexity, and Policy* (University of Chicago Press, 2009).

Fred Mushacke and Evan Picard, *1974–1999 Vegetated Tidal Wetland Trends of the New York Metropolitan Area and the Lower Hudson River: Final Report* (1999).

David P. Muth, "The Once and Future Delta," in *Perspectives on the Restoration of the Mississippi Delta*, eds. John W. Day, G. Paul Kemp, Angelina M. Freeman, and David P. Muth (Springer: 2014).

Gregoire Nicolis and Ilya Prigogine, *Exploring Complexity: An Introduction* (W.H. Freeman & Company: 1989).

Eric Nost, "Climate Services for Whom? The Political Economics of Contextualizing Climate Data in Louisiana's Coastal Master Plan," *Climatic Change* 157, no. 1 (2019): 27–42.

H. T. Odum, *Environment, Power, and Society* (John Wiley & Sons, 1971).

Karen O'Neill, *Rivers by Design: State Power and the Origins of U.S. Flood Control* (Durham: Duke University Press, 2006).

Kate Orff, "Cosmopolitan Ecologies: Jamaica Bay as Pilot Project," in *Gateway: Visions for an Urban National Park*, eds. Alexander Brash, Jamie Hand, and Kate Orff (Princeton Architectural Press, 2011).

Philip Orton, "Jamaica Bay: Pollution, Flooding, and Human Vulnerability," *SeaAndSkyNY*, May 22, 2011, accessed February 8, 2013, http://seaandskyny.com/2011/05/22/jamaica-bay-pollution-flooding-and-human-vulnerability/.

Philip Orton, "Upcoming Presentation on Surges, Barriers, and Coastal Restoration," *SeaAndSkyNY*, November 1, 2012, accessed February 8, 2013, http://seaandskyny.com/2012/11/01/upcoming-presentation-on-surges-barriers-and-coastal-restoration/.

Christian Parenti, "*The Limits to Growth*: A Book that Launched a Movement," *The Nation*, December 5, 2012.

Natalie S. Peyronnin, Rex H. Caffey, James H. Cowan, Dubravko Justic, Alexander S. Kolker, Shirley B. Laska, Alex McCorquodale et al., "Optimizing Sediment Diversion Operations: Working Group Recommendations for Integrating Complex Ecological and Social Landscape Interactions," *Water* 9, no. 6 (2017): 368.

Jonathan D. Phillips, "Sources of Nonlinearity and Complexity in Geomorphic Systems," *Progress in Physical Geography* 27, no. 1 (2003): 1–23.

Steward Pickett, Mary Cadenasso, and J. Morgan Grove, "Resilient Cities: Meaning, Models, and Metaphor for Integrating the Ecological, Socio-economic, and Planning Realms," *Landscape and Urban Planning* 69, no. 4 (2004): 369–84.

Steward Pickett, Mary Cadenasso, and Brian McGrath, eds., *Resilience in Ecology and Urban Design: Linking Theory and Practice for Sustainable Cities* (New York: Springer, 2013).

Steward Pickett and Peter White, eds., *The Ecology of Natural Disturbance and Patch Dynamics* (Academic Press, Inc.: 1985).

Orrin Pilkey and Andrew G. Cooper, *The Last Beach* (Durham, NC: Duke University Press, 2014).

Orrin H. Pilkey and Linda Pilkey-Jarvis, *Useless Arithmetic: Why Environmental Scientists Can't Predict the Future* (New York: Columbia University Press, 2007).

Eric-Jan Pleijster, Cees van der Veeken, and LOLA Landscape Architects, *Dutch Dikes* (Rotterdam: nai010 publishers, 2014).

"New York and New Jersey Harbor Deepening Program," Port Authority of New York and New Jersey, accessed February 23, 2019, https://www.panynj.gov/port/pdf/Harbor_Deepening_Fact_Sheet.PDF

Jean-Pierre Protzen and David J. Harris, *The Universe of Design: Horst Rittel's Theories of Design and Planning* (Routledge, 2010).

Julian Raxworthy, *Overgrown: Practices between Landscape Architecture and Gardening* (Cambridge: MIT Press, 2018).

F. G. Renaud, J. P. Syvitski, Z. Sebesvari, S. E. Werners, H. Kremer, C. Kuenzer, and J. Friedrich, "Tipping from the Holocene to the Anthropocene: How Threatened are Major World Deltas?," *Current Opinion in Environmental Sustainability* 5, no. 6 (2013): 644–54.

Chanpory Rith and Hugh Dubberly, "Why Horst W. J. Rittel Matters," *Design Issues* 22, no. 4 (Autumn 2006): 1–3.

Horst W. J. Rittel and Melvin M. Webber, "Dilemmas in a General Theory of Planning," *Policy Sciences* 4, no. 2 (1973): 155–69.

Alexander Robinson and Brian Davis, "From Solution Space to Interface: Six Actions for Landscape Infrastructure Design," In *Codify: Parametric and Computational Design in Landscape Architecture*, Bradley Cantrell and Adam Mekies, eds. (Routledge, 2018), 155–68.

Frederic Rossano, *Floodscapes: Contemporary Landscape Strategies in Times of Climate Change* (Rotterdam: nai010 publishers, 2021).

Jeffrey S. Rutherford, Adrian R. H. Wiegman, John W. Day, and Robert R. Lane, "Energy and Climate–Global Trends and their Implications for Delta Restoration," in *Mississippi Delta Restoration*, eds. John W. Day and Jori A. Erdman (Springer: 2018).

Avigail Sachs, *Environmental Design: Architecture, Politics, and Science in Postwar America* (University of Virginia Press, 2018).

Christian Salewski, *Dutch New Worlds: Scenarios in Physical Planning and Design in the Netherlands, 1970–2000* (nai010 publishers, 2012).

Charles E. Sasser, Guerry O. Holm, Elaine Evers-Hebert, and Gary P. Shaffer, "The Nutria in Louisiana: A Current and Historical Perspective," in *Mississippi Delta Restoration*, eds. John W. Day and Jori A. Erdman (Springer: 2018).

James C. Scott, "High Modernist Social Engineering: The Case of the Tennessee Valley Authority," in *Experiencing the State*, eds., Lloyd Rudolph and John Kurt Jacobsen (Oxford University Press, 2006).

James C. Scott, *Seeing Like a State* (Yale University Press, 1999).

Todd Shallat, *Structures in the Stream: Water, Science, and the Rise of the U.S. Army Corps of Engineers* (Austin: University of Texas Press, 2010).

Kristin Shrader-Frechette, "Unsimple Truths: Science, Complexity, Policy," *The British Journal for the Philosophy of Science* 64, no. 2 (2013): 449–53.

Scott M. Stringer, Office of the Comptroller, the City of New York, to Colonel Thomas D. Asbery and Joseph Seebode, US Army Corps of Engineers, New York District, October 23, 2019, https://comptroller.nyc.gov/wp-content/uploads/2019/11/Army-Corps-Letter-10-24-19.pdf.

Eugene R. Turner, "Doubt and the Values of an Ignorance-based Worldview for Restoration: Coastal Louisiana Wetlands," *Estuaries and Coasts* 32, no. 6 (2009): 1,054–68.

Monica Turner, "Disturbance and Landscape Dynamics in a Changing World," *Ecology* 91, no. 10 (2010): 2,833–49.

US Army Corps of Engineers, Engineer Research and Development Center, *USACE Regional Sediment Management and Engineering With Nature 2013 Workshop Summary*, ERDC TN-EWN-14-3, by Jennifer M. Gerhardt-Smith, and Cynthia J. Banks (Vicksburg, MS: Engineer Research and Development Center, 2014).

US Army Corps of Engineers, *The Corps of Engineers National Regional Sediment Management Demonstration Program*, Engineer Research and Development Center Coastal and Hydraulics Engineering Technical Note XIV-1, by Julie D. Rosati, Bruce D. Carlson, Jack E. Davis, and Thomas D. Smith. (Vicksburg, MS: Engineer Research and Development Center Coastal and Hydraulics Lab, 2004).

"Dredging Information System," US Army Corps of Engineers Navigation Data Center, http://www.ndc.iwr.usace.army.mil//dredge/drgcorps.htm.

US Army Corps of Engineers, New York District, "Coastal Restoration Activities within Jamaica Bay—Planning Ahead," by Peter Weppler, presentation to State of the Bay Symposium 2016, June 15, 2016, http://www.srijb.org/wp-content/uploads/2017/06/WEPPLER.pdf.

US Army Corps of Engineers, New York District, "Fact Sheet—Jamaica Bay Marsh Islands, NY, March 24, 2021," https://www.nan.usace.army.mil/Media/Fact-Sheets/Fact-Sheet-Article-View/Article/487604/fact-sheet-jamaica-bay-marsh-islands-ny/.

"Jamaica Bay Marsh Islands," USACE New York District, accessed February 8, 2013, http://www.nan.usace.army.mil/Missions/CivilWorks/ProjectsinNewYork/EldersPointJamaicaBaySaltMarshIslands.aspx.

US Army Corps of Engineers, New York District, "Coastal Restoration Activities within Jamaica Bay—Planning Ahead."

Jeremy Walker and Melinda Cooper, "Genealogies of Resilience: From Systems Ecology to the Political Economy of Crisis Adaptation," *Security Dialogue* 42, no. 2 (2011).

David Waltner-Toews, James J. Kay, and Nina-Marie E. Lister, eds., *The Ecosystem Approach: Complexity, Uncertainty, and Managing for Sustainability* (Columbia University Press, 2008).

David Waltner-Toews, Nina-Marie E. Lister, and Stephen Bocking, "A Preface," in *The Ecosystem Approach*, eds. Waltner-Toews, Kay, and Lister (Columbia University Press, 2008).

Judith Wasserman, "A World in Motion: The Creative Synergy of Lawrence and Anna Halprin," *Landscape Journal* 31, no. 1/2 (2012): 33–52.

Richard Weller, "Landscape (Sub)Urbanism in Theory and Practice," *Landscape Journal* 27, no. 2 (2008): 248–51.

Gilbert F. White, "Human Adjustment to Floods," *Department of Geography Research Paper* No. 29, Chicago, IL: University of Chicago (1945).

Wiegman, Adrian R. H., Jeffrey S. Rutherford, and John W. Day, "The Costs and Sustainability of Ongoing Efforts to Restore and Protect Louisiana's Coast," in *Mississippi Delta Restoration*, eds. John W. Day and Jori A. Erdman (Springer: 2018).

Angela Wilkinson and Esther Eidinow, "Evolving Practices in Environmental Scenarios: A New Scenario Typology," *Environmental Research Letters* 3 (2008), 1–11.

Jianguo Wu and Orie L. Loucks, "From Balance of Nature to Hierarchical Patch Dynamics: A Paradigm Shift in Ecology," *The Quarterly Review of Biology* 70, no. 4 (1995): 439–66.

Andrew Yawn, "As Gulf Swallows Louisiana Island, Displaced Tribe Fears the Future," *Daily Advertiser*, February 27, 2020, https://www.theadvertiser.com/in-depth/news/2020/02/27/isle-de-jean-charles-louisiana-climate-refugees-resettlement/2448973001/.

New York and New Jersey Harbor (Google Earth)

Dredge City

The New York Metropolitan Area

An invisible tale of dredging, deepening, and disposing in the United States' most populous metropolitan area.

Sediment Flows to the Ocean

This map describes the sedimentshed of the New York–New Jersey Estuary—the land masses and topographic forms that divert sediment-rich water toward the Hudson River and New York City's harbor.

Shaped by prehistoric large-scale glaciation, today's landscape is covered with extensive urban and suburban development that increases erosion and the flow of sediment-rich water downstream. The Hudson moves a half megaton of sediment annually,[1] some of which is dredged from ports as far north as Albany and Troy. Downstream, shipping channels that serve the Port of New York and New Jersey generate the highest volumes of dredged material in the sedimentshed, approximately 1 to 2 million cubic yards per year.[2] Much of this material is moved offshore and out of the harbor to the Historic Area Remediation Site (HARS), a 15.7-square-nautical-mile area of mud-placement.

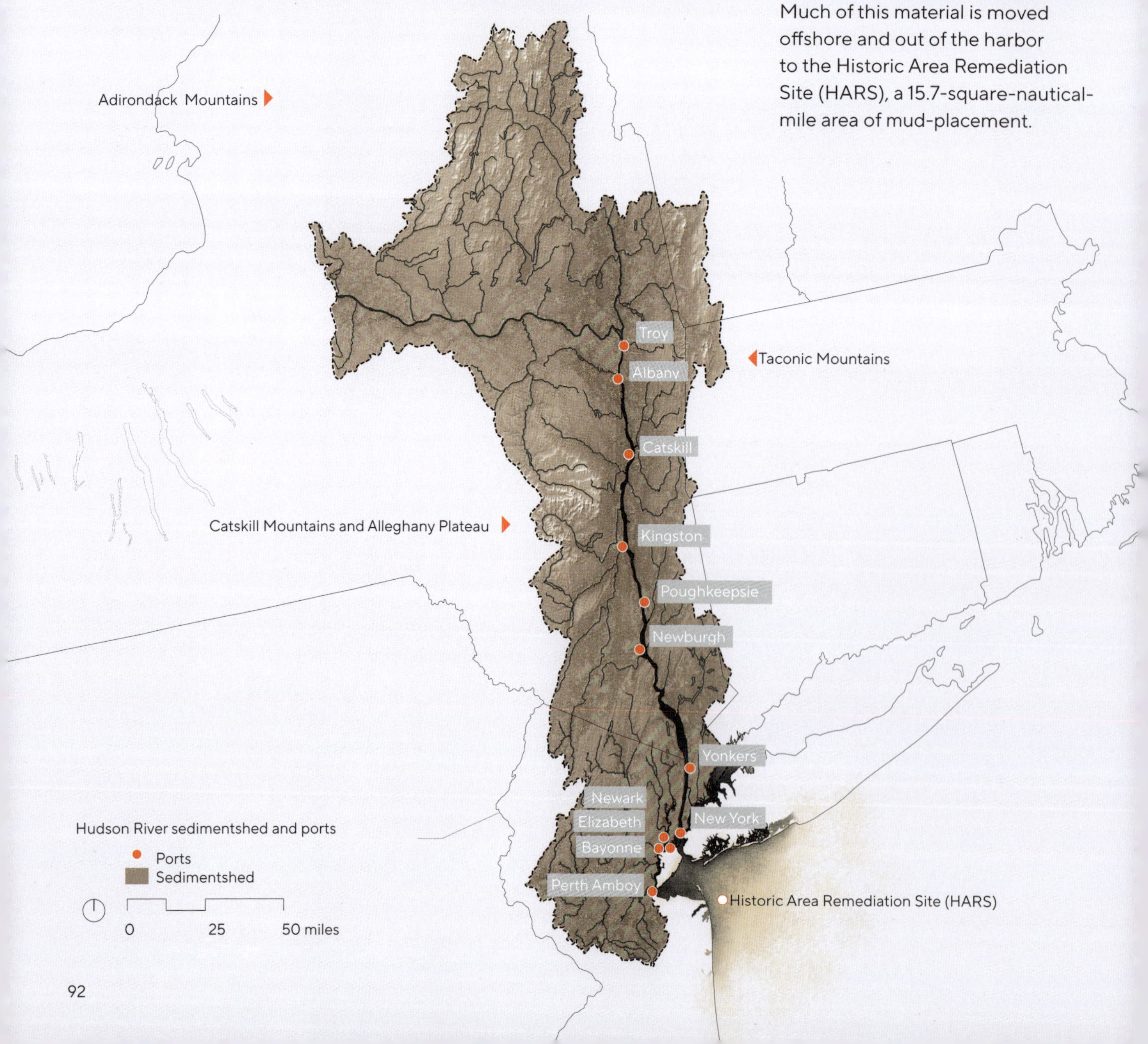

UPSTREAM SEDIMENT SOURCES:
Storm runoff and spring snow melt

Small tributaries channel sediment to larger river systems

Deforestation and developed lands increase erosion and sediment flows downstream

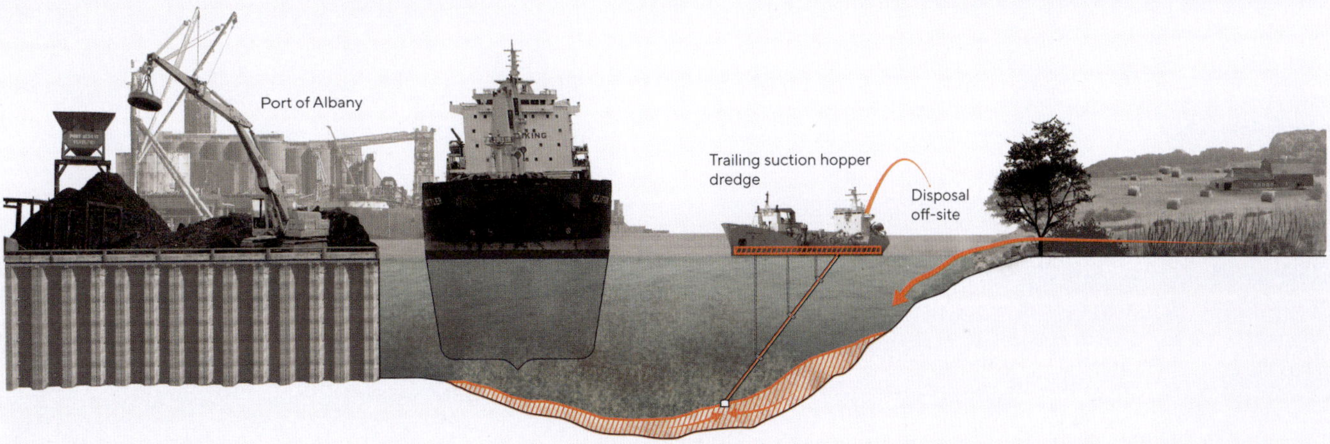

HUDSON RIVER PORTS:
River edges are straightened and hardened for port operation

Sediment accumulates on river bottom, impacting water transport

Dredging occurs throughout the river to maintain navigable waterways

OFFSHORE PLACEMENT:
Much of the dredged material from the New York / New Jersey Harbor is placed offshore

Tug boats push dump scow barges filled with dredge which is then released into the site

Newly placed dredged material caps contaminated spoils below

River Transport

The Hudson's finer-grain sediments remain suspended in the water column and, as far north as Troy, move with the seaward and landward pulsing of the tides. Larger materials, like sands and gravels, settle to the bottom but continue to move as bedload, rolling and skipping across the bed of the river. These shallow-water sediment cores reveal a murky snapshot of this nonlinear movement to the river mouth.

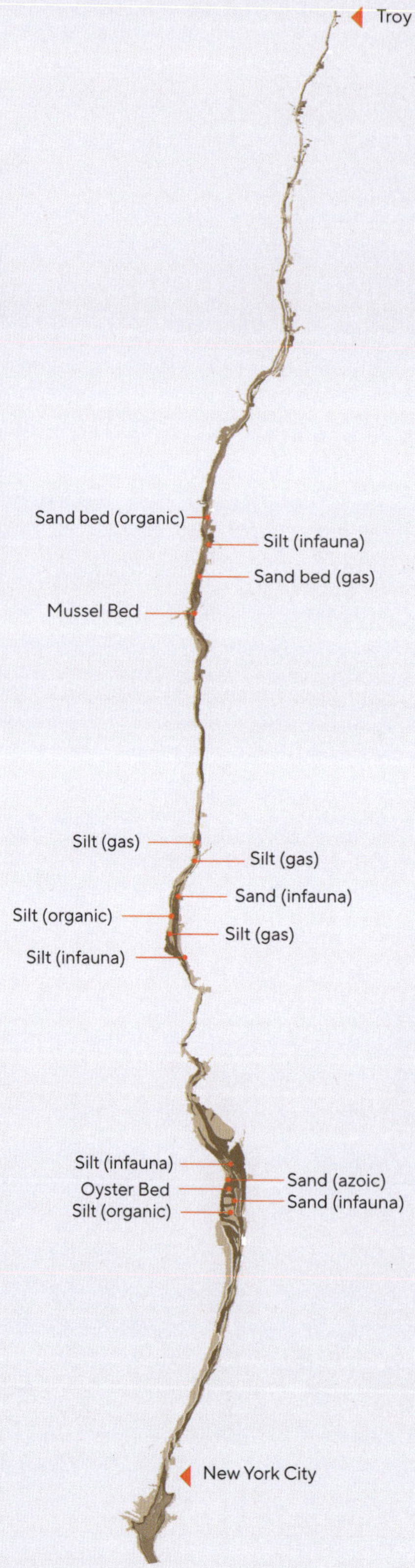

Sediments of the Hudson River, 1998-2004

- Gravel
- Gravelly mud
- Gravelly sand
- Mud
- Muddy gravel
- Muddy sand
- Sand
- Sandy gravel
- Sandy mud

Sand bed (organic)

Silt (infauna)

Sand (azoic)

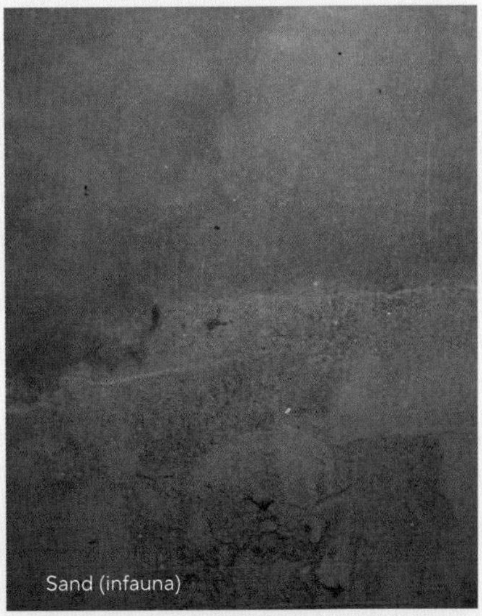

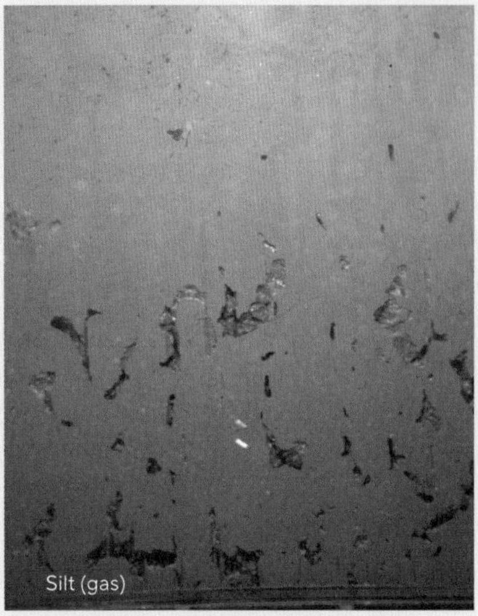

(Hudson River Estuary Sediment Profile Imagery, New York State Department of Environmental Conservation, 2017)

Invisible Superhighway

The region's economy depends upon the movement of mud.

The Hudson's sediment deposits in one of the busiest ports in the United States—the Port of New York and New Jersey. Without human action, this harbor would only be 20 feet deep[3]—virtually impassable to large ships and container cargo. Dredging mediates a conflict between economic interests and the depositional processes of the estuary. The act of sediment removal enables a water-based superhighway system of over two dozen federal shipping channels that transport over 200 billion dollars of critical goods to New Yorkers every year.[4] Annual maintenance dredging preserves these channels as a functional transportation network within the harbor.

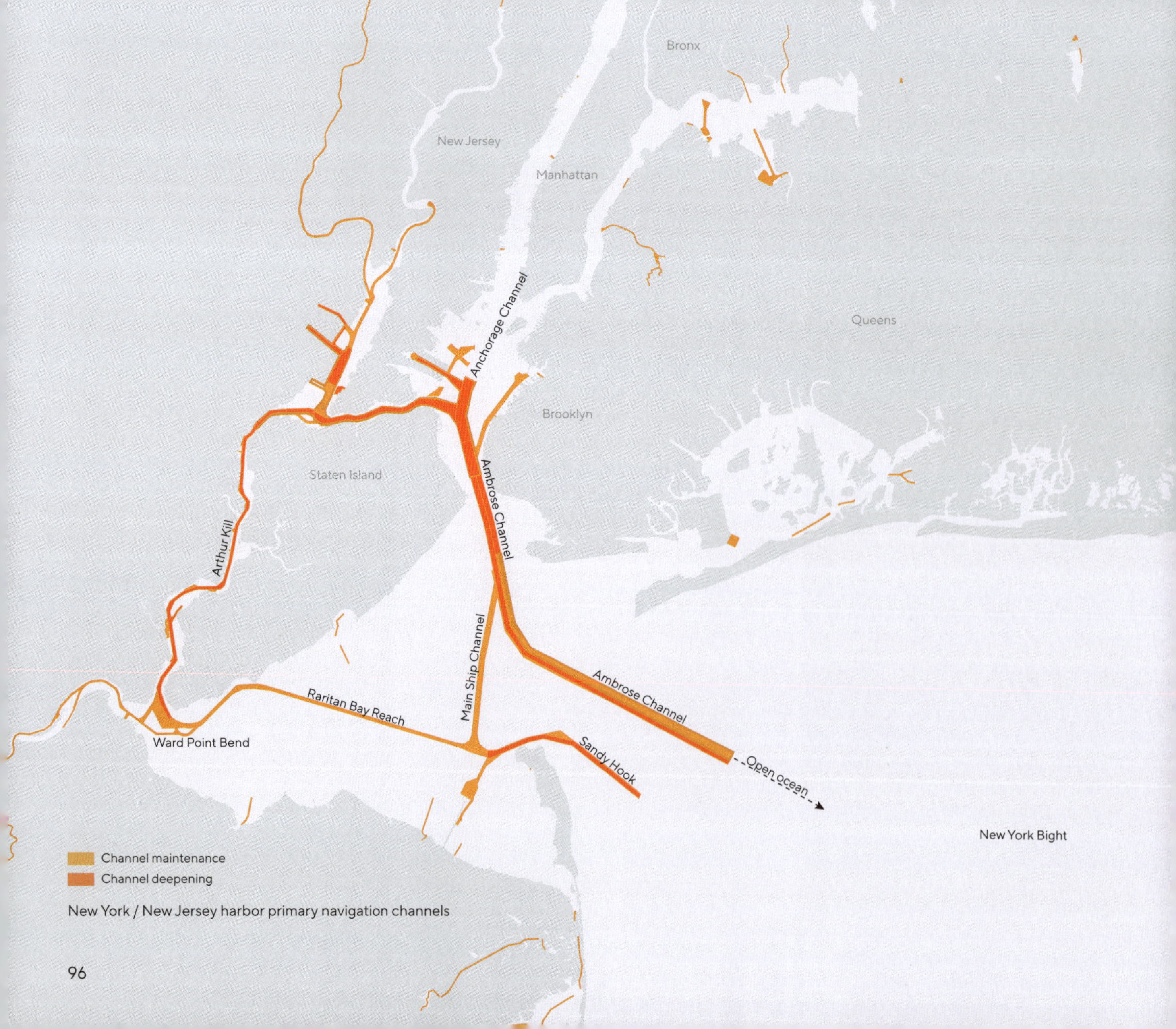

New York / New Jersey harbor primary navigation channels

Arthur Kill Channel Deepening
(US Army Corps of Engineers, 2015)

Race to the Bottom

The upsizing of container ships and the expansion of the Panama Canal has driven port cities nationwide into a competitive dredging frenzy.

Driven by the global push to accommodate ever larger ships, including the post-Panamax and New Panamax container ships, port cities worldwide are competing for the deepest and largest channels.[5] Extensive harbor deepening projects, like the decade-long effort by the Port of New York and New Jersey to deepen channels from 42 to 50 feet,[6] generate millions of cubic yards of sediment to move and manage. This race to the bottom is transforming the underwater landscapes of cities and their marine environments, linking global economic and transportation industry shifts to the continual modification of underwater ecosystems.

Global Container Ship Evolution
TEU= twenty foot equivalent unit

Blasting bedrock to deepen harbor channels for larger container ships

Dredging with Dynamite

Harbor deepening is no easy task—aspirations for deeper channels can literally hit rock bottom. When channel designers encounter bedrock it can be more economical to blast through the stone than to reroute a highly regulated navigation channel. Dredgers detonate the buried rock using underwater explosives and remove the fractured stone with a clamshell or bucket dredge. *The Kraken*, a drill boat pictured here working in the Arthur Kill, blasted through 3.3 million square feet of rock to deepen this heavily trafficked harbor channel to 52 feet.[7]

Stills of the drill boat at work. After blasting, suitable materials were processed and transported to the HARS site, and unsuitable materials were transported to a upland disposal facility. (Alex Chohlas-Wood and Ben Mendelsohn, 2012).

Kraken drill ship, Arthur Kill deepening at Howland Hook (Will Van Dorp, 2012)

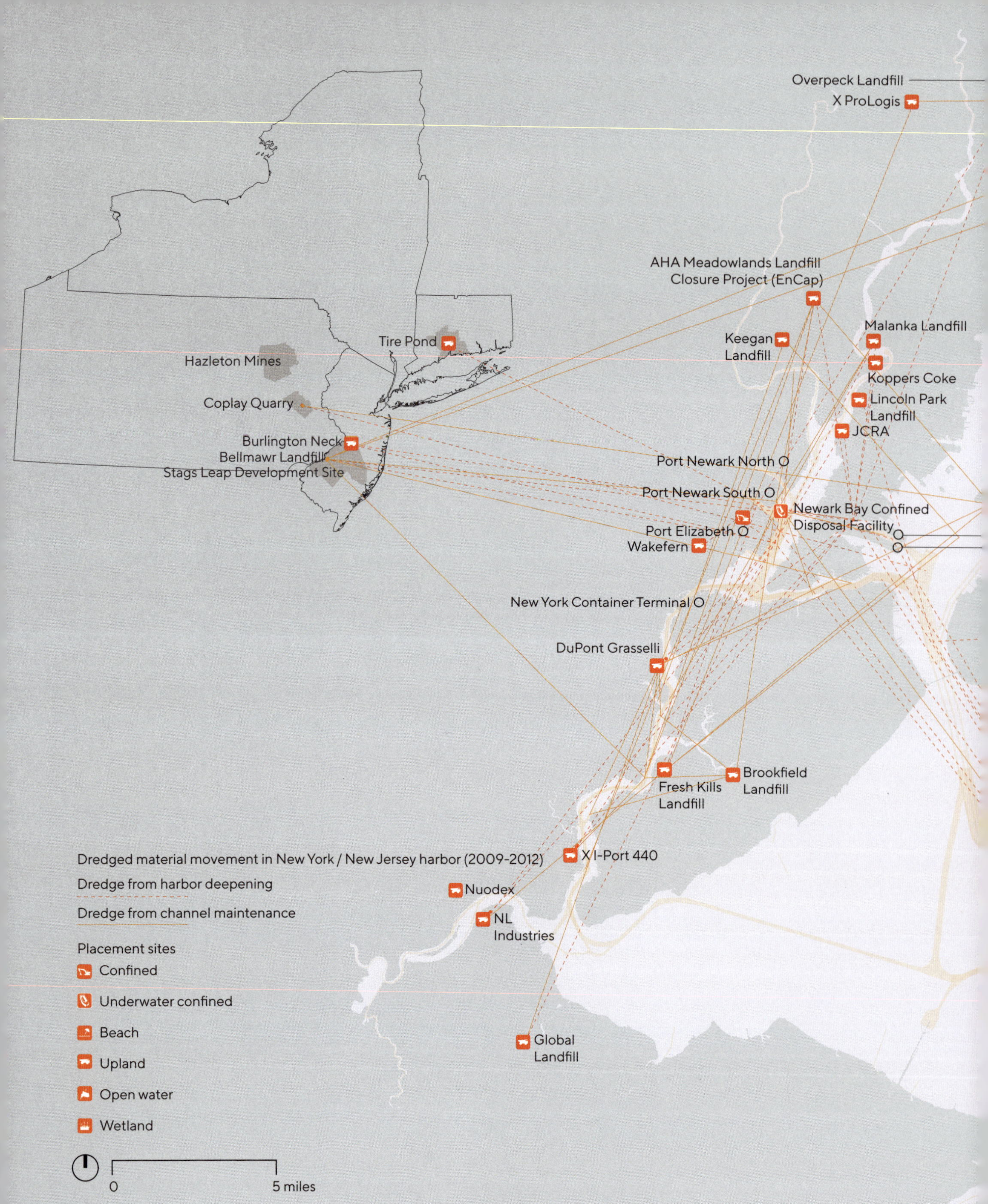
Overpeck Landfill
X ProLogis
AHA Meadowlands Landfill Closure Project (EnCap)
Keegan Landfill
Malanka Landfill
Koppers Coke
Lincoln Park Landfill
JCRA
Tire Pond
Hazleton Mines
Coplay Quarry
Burlington Neck
Bellmawr Landfill
Stags Leap Development Site
Port Newark North
Port Newark South
Newark Bay Confined Disposal Facility
Port Elizabeth
Wakefern
New York Container Terminal
DuPont Grasselli
Fresh Kills Landfill
Brookfield Landfill
X I-Port 440
Nuodex
NL Industries
Global Landfill
Dredged material movement in New York / New Jersey harbor (2009-2012)
Dredge from harbor deepening
Dredge from channel maintenance
Placement sites
Confined
Underwater confined
Beach
Upland
Open water
Wetland
0
5 miles

Choreographing Disposal

Dredging is the simple part. The disposal of dredged material involves a complex web of environmental regulation, logistical decisions, and cost-benefit calculations.

The federal standard adopted by the Corps incentivizes the disposal of dredged material in the most cost-effective manner compliant with environmental regulations,[8] which often limits the creative recycling of this material and leads to more conventional and lower-cost disposal options, like offshore placement and landfill capping.

While sediment is a valuable and critical resource for the coastlines of New York and New Jersey, it is significantly more costly to transport it to shallow-water environments where ecological benefits can be maximized. Below, the costs for moving a cubic yard of dredged material to different environments are highlighted. Even when a sponsor steps in to provide additional funds for beneficial use, factors such as sediment type, seasonal restrictions on dredging and placement, and regulations can complicate sediment matchmaking between source and site.

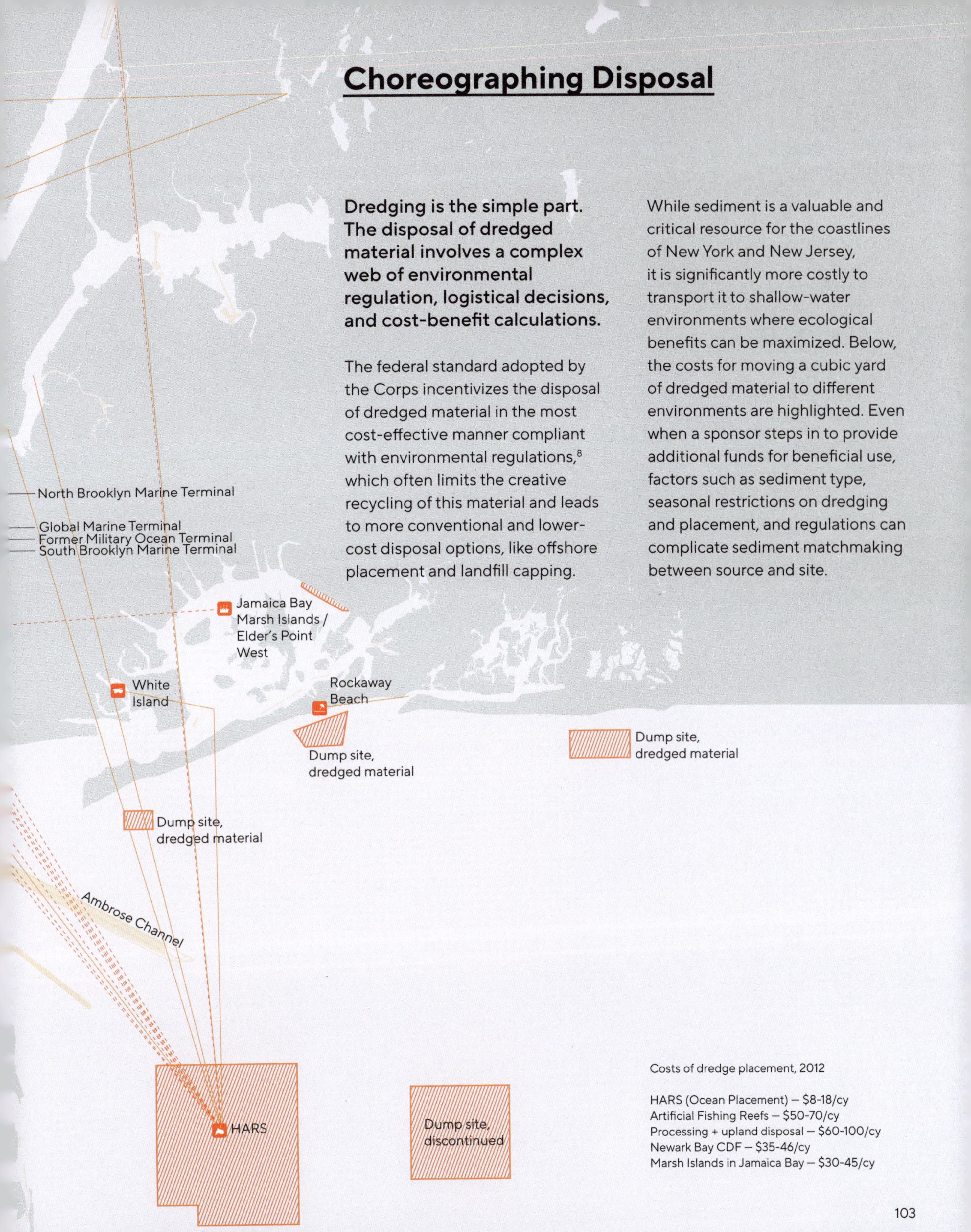

Costs of dredge placement, 2012

HARS (Ocean Placement) – $8-18/cy
Artificial Fishing Reefs – $50-70/cy
Processing + upland disposal – $60-100/cy
Newark Bay CDF – $35-46/cy
Marsh Islands in Jamaica Bay – $30-45/cy

Releasing Dredge

Much of the region's contemporary dredged material is disposed of within the open-ocean Historic Area Remediation Site (HARS), where it is used to cap a historic underwater mud-dumping site. In this regulated but physically unbounded zone, dredge operators place cleaner mud over dirty at the depths of the ocean bottom. While capping this site helps contain historic contaminants offshore, sediment is exported from nearshore environments in the estuary, where it could help sustain intertidal habitats like marshes and mudflats.

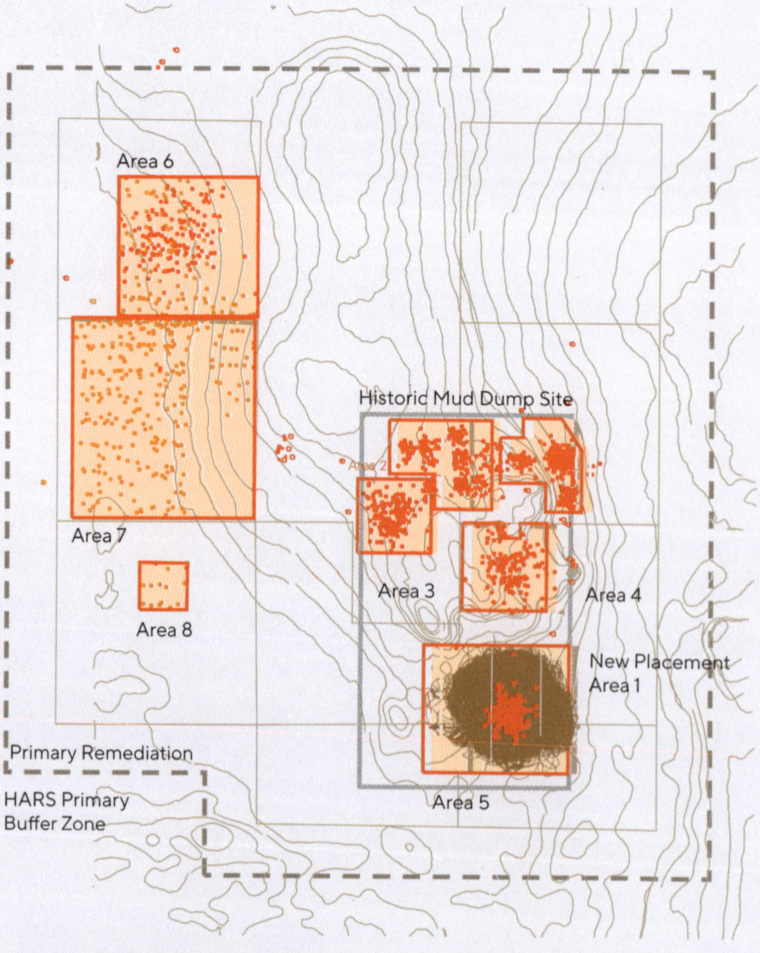

Dredged material placed on the sea floor at HARS between 1996-2000

Clamshell dredging in Anchorage channel (Chris Gardner, US Army Corps of Engineers, 2010)

Deepening the Kill Van Kull with the world's largest backhoe dredge (Chris Gardner, US Army Corps of Engineers, 2010)

A New Use for Dredge Material

Visible in these images is a flat expanse of newly constructed future marshland, composed of 375,000 cubic yards of clean sand dredged from the Ambrose Channel, the main shipping channel leading to the Port of New York and New Jersey.[9]

The economics of dredge can be shifted toward outcomes with greater social and ecological benefit when agencies with different missions work together or choreograph their activities. Marsh restoration with dredged material in Jamaica Bay is one example of this. In 2012, 42 acres of marshlands were created at Yellow Bar Hassock Marsh Island with clean sand from the Harbor Deepening project through collaborations between the National Park Service, the US Army Corps, the Port Authority of New York and New Jersey, the New York State Department of Environmental Conservation, and the New York City Department of Environmental Protection.[10] The resulting new marsh is intended to create habitat for migratory birds, expand nursery grounds for fish, improve water quality, and stabilize the shoreline. It is part of a long-term initiative to restore multiple islands within Jamaica Bay.

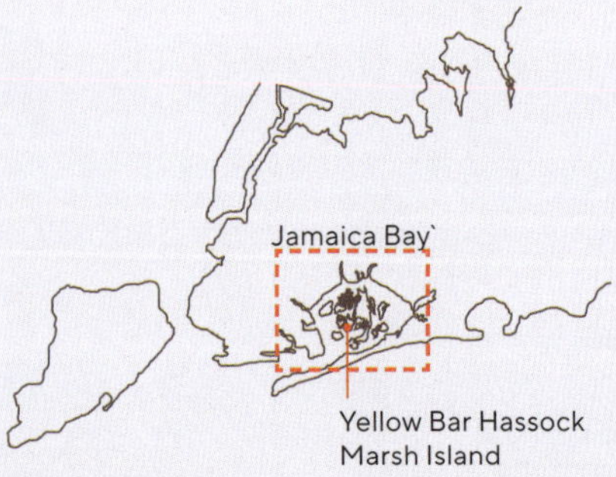

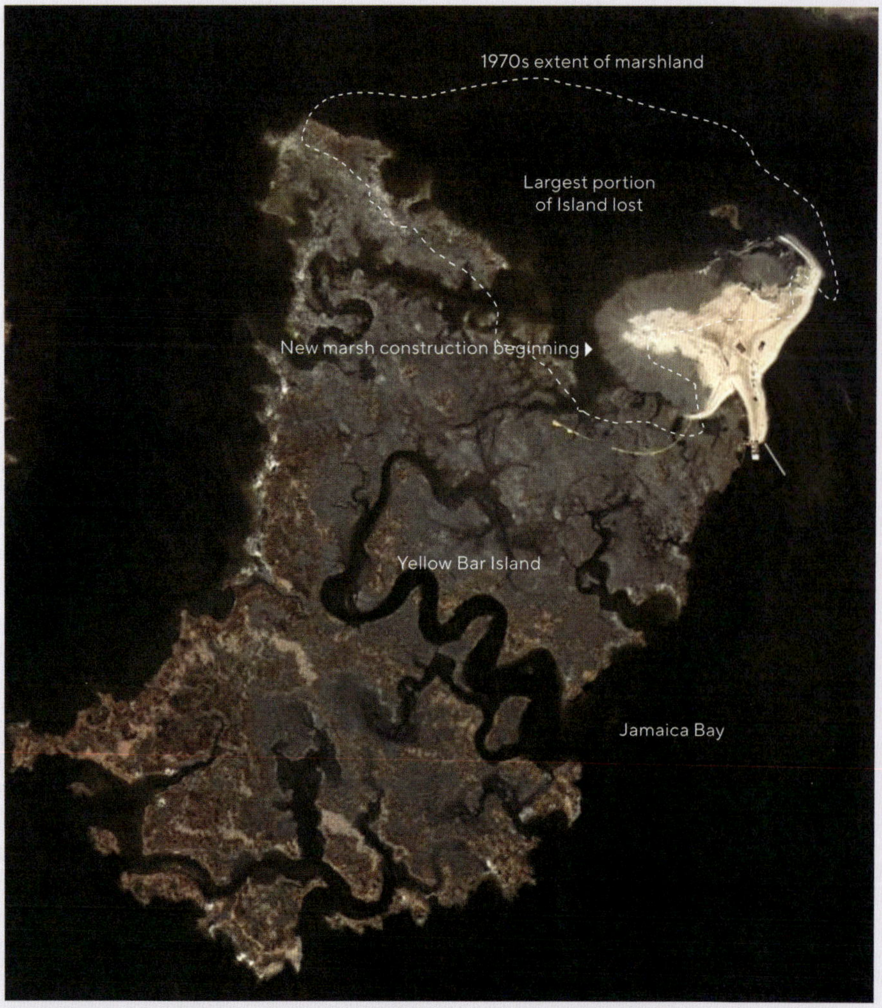

Aerial image of Yellow Bar Hassock at beginning of construction (Google Maps, 2011)

Aerial image of Yellow Bar Hassock under construction
overlaid with construction drawings and balloon photography (2012)

Grassroots Gardening

The newly created marsh islands are dynamic, novel environments established with the help of volunteers.

Saltmarsh cordgrass is visible here as tiny green clumps—miniature living mats of marsh grass and ribbed mussels that were carefully salvaged from the existing island perimeter and spaced out to stabilize the new sandy shore. This unique planting strategy was tested with public funds at Yellow Bar Hassock.[11] Due to the high cost of moving marsh, community nonprofit groups have stepped in to set up future grassroots planting and marsh stewardship initiatives for Black Wall and Ruler's Bar, two other Jamaica Bay islands supplemented with dredged material.[12] Volunteers planted plugs over several seasons and since then have documented plant survival via DIY kite and balloon photography, watching and waiting to see if these constructed ecosystems survive harsh urban waterway conditions and sea level rise.

▲ Construction fence keeps trash out of the marsh and marks the limit of work.

◀ Heavy equipment shapes the dredged sand into specific channels and flats to allow for tidal flushing of the marsh.

Balloon aerial photography of Yellow Bar Hassock during restoration planting

Salt marsh clumps re-establishing

Volunteer planting day on Ruler's Bar Island

Ribbed mussel and marsh grass clumps

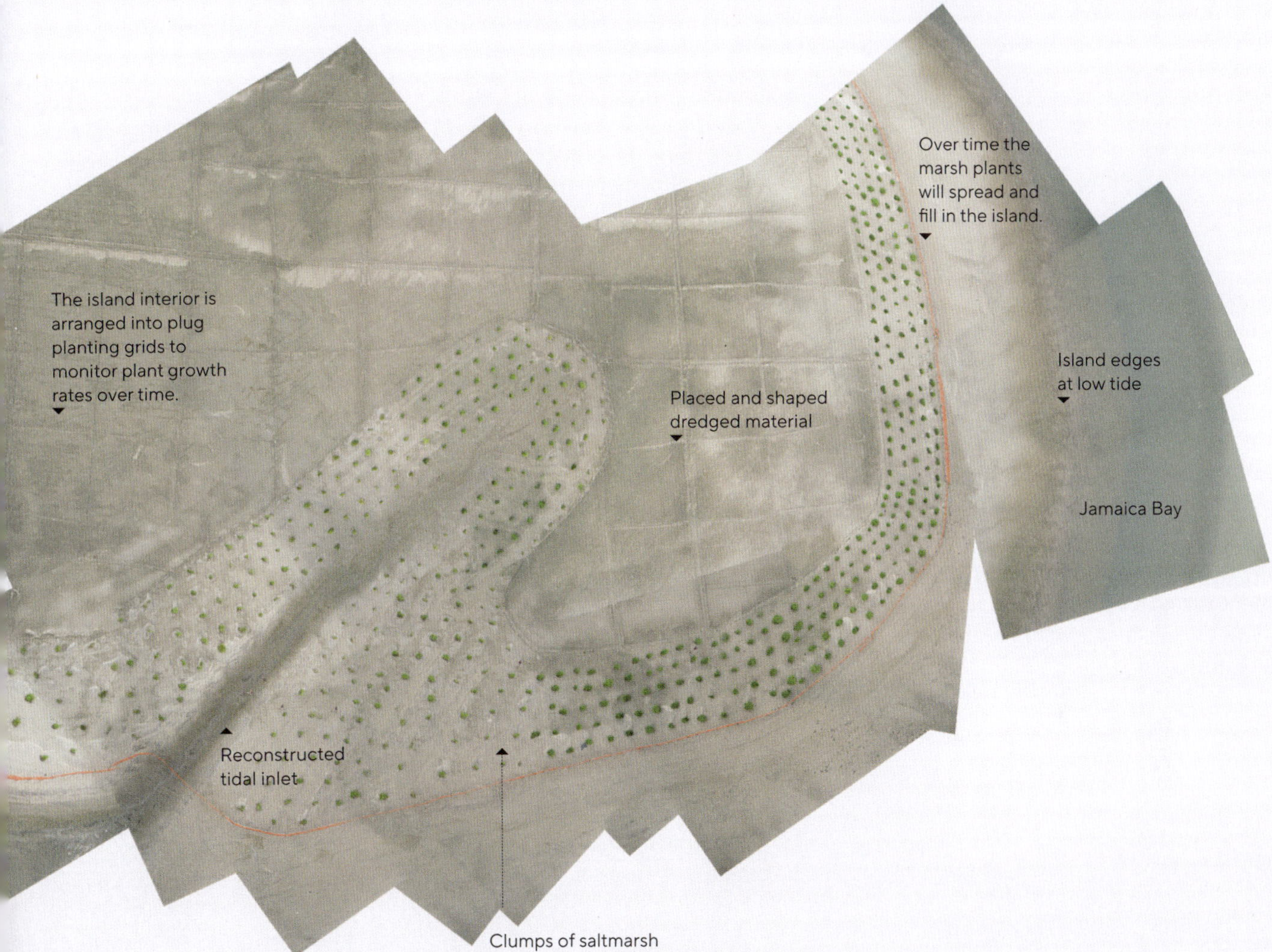

Kite mapping Yellow Bar Hassock

Marsh grass plug planting by volunteers

Juvenile horseshoe crab on Yellow Bar Hassock

Bay Futures

Some 1,400 acres of tidal salt marsh have been lost in Jamaica Bay since 1924 due to anthropogenic impacts, with recent rates estimated at 47 acres per year.[13] Even with projects like Yellow Bar, the scale of marsh loss vastly outpaces the scale of beneficial use for habitat creation. Yet the project remains a model. In the face of sea level rise and global climate change, this intelligent choreography of economic need, ecological value, and social benefit could be scaled up and implemented regionally.

Jamaica Bay Marsh Island Stakeholders
(Ildiko Reisenbigler, US Army Corps of Engineers, 2000)

Endnotes:

1. G. R., E. A. Nystrom Wall, and S. Litten, "Suspended Sediment Transport in the Freshwater Reach of the Hudson River Estuary in Eastern New York," *Estuaries and Coasts* 31, no. 3 (2008): 542–53. https://doi.org/10.1007/s12237-008-9050-y. https://pubs.er.usgs.gov/publication/70033241

2. US Army Corps of Engineers, New York District, Update: Dredged Material Management Plan for the Port of New York & State of New Jersey, 2008. https://www.nan.usace.army.mil/Portals/37/docs/harbor/dredmgmt/Dredged%20Material%20Mgmt%20Plan.pdf

3. Lara Elmayan, "Cities 101: How Does Harbor Dredging Work in NYC?" Untapped New York (blog). Accessed April 14, 2022. https://untappedcities.com/2013/08/12/cities-101-how-does-harbor-dredging-work-nyc/.

4. "Our Port - History Information: Port Authority of New York and New Jersey," Our Port - History Information | Port Authority of New York and New Jersey. Port Authority of New York and New Jersey. Accessed April 14, 2022. https://www.panynj.gov/port/en/our-port/history.html.

5. "Ports Primer: 6.2 Port Factors Impacting the Regional Economy," Environmental Protection Agency (EPA), accessed April 14, 2022. https://www.epa.gov/community-port-collaboration/ports-primer-62-port-factors-impacting-regional-economy.

6. "New York and New Jersey Harbor Deepening Channel Improvements (NYNJHDCI)," New York District Website, US Army Corps of Engineers, accessed April 14, 2022. https://www.nan.usace.army.mil/Missions/Navigation/New-York-New-Jersey-Harbor/NY-NJ-HDCI/.

7. "Arthur Kill Channel, Navigation Improvement 50-Foot Project (S-AK-2)," Cashman Dredging, accessed April 14, 2022. https://www.cashmandredging.com/projects/arthur-kill-channel-navigation-improvement-50-foot-project-s-ak-2/.

8. US Environmental Protection Agency/ US Army Corps of Engineers, The Role of the Federal Standard of the Beneficial Use of Dredge Material from the US Army Corps of Engineers New and Maintenance Navigation Projects, Washington, DC, 2007. https://www.epa.gov/sites/default/files/2015-08/documents/role_of_the_federal_standard_in_the_beneficial_use_of_dredged_material.pdf

9. "NPS, Partners to Restore Yellow Bar Salt Marsh in Jamaica Bay," National Parks Service, US Department of the Interior, accessed February 16, 2005. https://www.nps.gov/gate/learn/news/restore-yellow-bar.htm

10. Ibid.

11. US Army Corps of Engineers, Y- Ambrose Channel (S-AM-3B Contract 14) Navigation Improvement Project, New York Harbor Deepening. // File name-SAMS3B PlansFinal.pdf. 2011.

12. Jason Diaz and Abigail Carney, "Community Comes Together to Restore the Jamaica Bay Salt Marsh," City Atlas: New York, April 5, 2013. https://newyork.thecityatlas.org/lifestyle/community-restore-jamaica-bay-salt-marsh/.

13. Vince Elias, "Army Corps Restores Marsh Islands in Jamaica Bay N.Y," New York District Website, US Army Corps of Engineers, September 13, 2012. https://www.nan.usace.army.mil/Media/News-Stories/Story-Article-View/Article/488022/army-corps-restores-marsh-islands-in-jamaica-bay-ny/#:~:text=From%201994%20and%201999%2C%20an,rapidly%20increasing%20in%20recent%20years.

Image Credits:

Spread 1:
Map showing greater New York and New Jersey Harbor area. Google Earth, earth.google.com/web/

Spread 2:
(left)
Hudson River Watershed Alliance. Hudson River Watershed Maps. pdf. 2015. https://hudsonwatershed.org/hudson-river-watershed-maps/
New York State. New York State Elevation Data. Shapefile. 2015. https://gis.ny.gov/elevation/
(right)
Composite imagery developed through site observation. Not to scale.
Spread 3: (map on left)
Lamont-Doherty Earth Observatory, Columbia University. Sediment Type, Hudson River estuary. 2004. https://cugir.library.cornell.edu/catalog/cugir-007884
(images)
New York State Department of Environmental Conservation. Hudson River Estuary Sediment Profile Imagery (SPI). 2017. https://www.dec.ny.gov/imsmaps/benthic/webpages/spi.html

Spread 4:
(left)
US Army Corps of Engineers, New York District. Update: Dredged Material Management Plan for the Port of New York & State of New Jersey. 2008. https://www.nan.usace.army.mil/Portals/37/docs/harbor/dredmgmt/Dredged%20Material%20Mgmt%20Plan.pdf
(right)
Elias/NAN/US Army Corps of Engineers. Dredge 54 Bucket Arthur Kill Channel Deepening. August 6, 2015. Accessed 2022. https://www.flickr.com/photos/usacehq/21654025230/in/photolist-yZuxhA-EiyjjF-KLkGHk

Spread 5:
(left)
Container ship diagram adapted from:
Jean-Paul Rodrigue. The Geography of Transport Systems. Abingdon, Oxon: Routledge, 2020.

Spread 6:
(left)
Alex Chohlas-Wood, Ben Mendelsohn, "The Kraken," (still images from video footage), 2012.
(right)
Will Van Dorp. Kraken drill ship, Arthur Kill deepening at Howland Hook. 2012. Creative commons, non commercial https://tugster. wordpress.com/tag/kraken/. Used with permission.

Spread 7:
(map)
Map is based on data recording dredging contracts and disposal sites from 2009 through early 2012.
US Army Corps of Engineers, New York District, Update: Dredged Material Management Plan for the Port of New York & State of New Jersey; NY/NJ Harbor Dredge Material Management Dredging Activities, Volumes, and Placement Sites (As of August 2012), accessed 2012. https://www.nan.usace.army.mil/Portals/37/docs/harbor/dredmgmt/TAB1%20AUG%202012.pdf
Dredging sites and dredging dump sites located from:
National Oceanic and Atmospheric Administration Office of Coasts Survey, NOAA Nautical Chart New York Harbor, 12327 and US4nY1AM, accessed 2012. https://www.charts.noaa.gov/OnLineViewer/12327.shtml

Spread 8:
(left)
Mapping adapted from:
US Geological Survey, Mapping the Sea Floor of the Historic Area Remediation Site offshore of New York City, Bradford Butman, Fact Sheet 076-01, Woods Hole, MA. https://pubs.usgs.gov/fs/old.2002/fs001-02/fs001-02.html
(right)
Chris Gardner, US Army Corps of Engineers, Dredge New York in the Kill Van Kull (photograph), October 25, 2010. Public domain. https://www.flickr.com/photos/usacehq/5158224555/in/photolist-8RPfs8-bzqMSu-bEAzy6-bEAzED-bEAzZP-brFEWE-fLcGmY-fKV6yk-jCJrJB-MnusLC-jCMokY-Nh1Ltw-bzqMTJ/
Chris Gardner, US Army Corps of Engineers, "Clamshell Dredging in New York New Jersey Harbor (photograph)," October 26, 2010. Public domain. https://www.flickr.com/photos/usacehq/5158844974/in/dateposted/

Spread 9:
(left)
Google Maps. Yellow Bar Island, Jamaica Bay, NY. 2011 Imagery. https://earth.google.com/web/
(right)
Google Maps. Yellow Bar Island, Jamaica Bay, NY. 2012 Imagery. https://earth.google.com/web/
US Army Corps of Engineers, Y-Ambrose Channel (S-AM-3B Contract 14) Navigation Improvement Project, New York Harbor Deepening. // File name-SAMS3B PlansFinal.pdf. 2011.
Dredge Research Collaborative, "Balloon mapping images of Yellow Bar Island," Public Laboratory, 2012.

Spread 10:
Dredge Research Collaborative, "Balloon mapping images of Yellow Bar Island," Public Laboratory, 2012.
Dredge Research Collaborative, "Image collection of Yellow Bar and Black Wall Islands," 2012–2018.

Spread 11:
Ildiko Reisenbigler, US Army Corps of Engineers, "USACE New York District, Partners at Marsh Islands restoration project, Jamaica Bay (photograph)," January 1, 2000. Public domain. https://www.flickr.com/photos/usacehq/7087847271/in/photolist-bEAzy6-bEAzZP-bNk54i-bEAzED-brFEWE-MSXt9u/

Lisa Baron and Peter Weppler: Building Marsh Islands and Making the Most of Dredging

Lisa Baron has been a Project Manager with the New York District Civil Works and Harbor Programs Branches for almost 15 years. She has more than 30 years of experience managing large-scale ecosystem restoration projects in addition to environmental dredging, ecological risk assessment and site investigations with the USACE, state (NJDOT), Department of Energy, and private sector.

Peter Weppler has been the Chief of the Environmental Analysis Branch, Planning Division, New York District since 2014. With over 30 years of experience in the field of coastal storm damage risk management, watershed management and ecosystem restoration, Peter is the New York District's Senior Subject Matter Expert on Civil Works environmental policy and compliance and is responsible for providing guidance to the project teams on several business lines including navigation, flood and coastal risk management, and aquatic ecosystem restoration.

DRC: Can you describe your job and what you enjoy most about it?

PW: My primary functions are environmental compliance for all of the Civil Works projects that we do and providing the ecosystem restoration expertise to design the projects that we plan and build. The best part of my job is that it's not just theory. We actually accomplish things on a practical basis. Our general mission is to provide solutions to our nation's water resource problems, and I get to actually see what we write about. We get our feet wet and we build something that we spoke about or planned 5 or 10 years earlier.

LB: I get the opportunity to lead a project delivery team of all different disciplines to plan, design, and eventually construct these projects. Leading the effort from start to finish is very rewarding. I'm responsible for the delivery of the project, making sure that the work products are of high quality, and that they're done within budget and on time. I love working with the external partners.

DRC: What is most challenging about your work?

LB: I think that the most challenging part or our work is implementing our studies and ensuring they can be completed in 3 years. Expectations—especially for large studies—are unrealistic. The first milestone that we have when we start a new study is the Alternative Milestone; it has to be scheduled within the first ninety days. For it, we need to formulate and present alternatives that solve water resource problems, and they have to be coordinated with the resource agencies and with public outreach. These milestones are so rigid and the coordination with our North Atlantic Division and Headquarters is tightly monitored. The stress that goes with keeping a study on time is tremendous.

PW: Near-term, definitely the new risk-informed paradigm that Lisa just spoke about. We need to make quicker decisions with less information. [Editor's Note: In 2012-13, USACE initiated a "Civil Works Transformation," an effort to rethink how the Corps plans, manages, and operates civil infrastructure projects.] I understand why it was needed, but the days of the 10-year study that the Corps used to do before it made a decision are gone. So now when stakeholders review our reports, we basically have a target on our back because we're making decisions without the large [quantities] of data we used to collect as part of a study.

As Lisa described, managing deadlines is incredibly challenging. You know you could have the right answer 6 months from now, but you have to give an answer in 3 months. That's tough. Another challenge is our ability to pair up Federal and non-federal money. When the Federal government has money to do something, our local partners—usually the states—often don't have money to do things. You need a perfect storm where everybody has money and can study, design, and construct at the same time.

DRC: You mentioned how a "perfect storm" has to come together to make a project happen. Can you talk a little bit about what it has been taking to restore Black Wall, Rulers Bars, Stony Creek, and other islands with dredged material?

PW: "The perfect storm" is a funny descriptor since most of our funding has been coming after the recent large storms that affected our area. When hurricanes like Ida, Katrina, and Sandy happen, money is usually appropriated to deal with [the aftermath of the] storm. While that money traditionally used to be focused on projects that protect people, now we have a lot of opportunity to do things that we didn't do before. Projects that care more about a multi-purpose response, where you're not just building for people and safety, but you're also considering the ecology of the area that you're building in —and whether that natural ecology and habitat also provides the answer to flooding or coastal storm risk management.

LB: As a Project Manager, I have to deal with the budget, the budgeting cycle, the projection of funding, and coordinating with the sponsors so that funding is aligned. We spend a lot of time convincing leadership that our projects should be funded. It's really challenging to orchestrate when projects will be initiated since we are working on the budgets three years in advance.

DRC: Can you talk about aligning sediment sources for this work?

LB: In the past, some of the sediment for the marsh islands came from maintenance dredging in Jamaica Bay's federal channel. However, the primary sources were Ambrose and Anchorage Channels from the Harbor Deepening project. Right now, I'm working with the Asset Manager for the Operation and Maintenance [dredging] of the Jamaica Bay Channel which would be the sediment source for Stony Creek. The O&M of the Jamaica Bay channel is usually done every three years. Usually, the volume ranges between 300,000 and 500,000 cubic yards. The District hasn't dredged the channel since 2013. Fortunately for the coordination with Stony Creek Marsh Island, none of the users are demanding immediate dredging. Once the infrastructure bill was passed, we now have all the money for construction of the Stony Creek Marsh Island. We now just have to marry up the schedules. I said, "We're going to be ready to construct in 2025. Can you advertise your O&M dredging contract in 2025?" And, luckily, he can.

We're going to have significant commitment from the leadership at the District for this project since we are beneficially using dredged material, keeping the material out of the Historical Area Remediation Site, and utilizing the sand for restoration projects. If the sediment from the O&M project can't be used, the contingency plan is to use dredged material from the Harbor Deepening Channel Improvements Project that will be deepening the port. This project should dredge about 9 million more cubic yards of clean sand for beneficial use. If our schedules line up, sand from Ambrose Channel could be used for the next marsh island. However, coordinating with the HDCI construction may be more challenging because they've got deadlines, they've got money to spend, and it's hard to ask them to wait. We'll be coordinating to see which sand source makes the most sense.

PW: The HDCI project is estimated to dredge a total of 28 to 32 million cubic yards of material. There is an opportunity, like the Corps did as part of the original NY/NJ Harbor Deepening Project, to use the material beneficially for many different types of projects, like capping landfills. We also would have some rock material that we could use to build some more shallow marine habitat, rather than traditional wetlands. The HDCI Study's Chief's Report was signed by the Chief of Engineers in June 2022. One of the big tasks will be to find a home for all the dredged material. However, the overall project will be constructed over a 17-year period, so placement sites will be determined contract by contract... Stakeholders are going to want to use a lot of that 9 million cubic yards of material that Lisa highlighted for beach nourishment within the region, including Raritan Bay beaches, resulting in a big competition for the material. Restoration of the marsh islands not only provides critical wetland habitat, it also results in the creation of early and late successional habitat for threatened and endangered species like least terns, piping plovers, red knots, saltmarsh sparrows, and seabeach amaranth.

DRC: Are there any lessons that you've taken from those built islands that you're applying forward in the future?

PW: The biggest lesson learned on the first set of islands was that the historical peat composition of the island affected the amount of sand required to build them. We originally estimated subsidence at 20-30%. It ended up being close to 50% subsidence, so we needed 50% more sand in order to create the design that we wanted.
The other lesson learned was that we originally thought we would encircle the islands with geotextile tubes, filled with sand. The geotubes would have been big and heavy and would have stayed where they were to build the perimeter. But the National Park Service felt that geotubes were not a natural feature and they didn't want them within their jurisdiction. Therefore, the design changed to coconut matted coir logs, which were much lighter and were supposed to do the same thing. But as construction progressed, the coir logs ended up breaking loose from the bay's large fetch and wave activity, and became hazards to navigation. So, lesson learned there was, don't use coir logs within Jamaica Bay.

LB: The Jamaica Bay Eco-Watchers recently suggested to incorporate more upland habitat in our designs similar to Elders East Marsh Island. The team will be considering this—however, the value of more upland habitat has to be balanced with the concern for air strikes and our FAA coordination [because of proximity to JFK]. Our designs already integrate more high marsh and transitional areas compared to low marsh in order to keep up with sea level rise, and they already consider marsh migration in order to realize the benefits that we predicted which justified these projects. I also want to bring in some additional habitats to the marsh design. Instead of using coir logs, I would like to consider incorporating shellfish habitat along the edge of the island. Whether it's mussels or oyster gabions, it will need to serve as containment and minimize subsidence, while also including another type of habitat and advancing another target goal for the Comprehensive Restoration Plan.

DRC: We understand that some of the islands were planted by the Corps and some were planted by volunteer community groups. Have you seen differences in how these have evolved?

LB: There are two really important aspects of community-based planting. One is cost. I think it could get more done for a cheaper price. But I think the second, probably more important aspect, is that the community, NGOs, and stakeholders have a vested interest in the success of that project. They helped build it and they get to say that they've been a part of the island's restoration. Those types of activities result in the tremendous commitment by the public.

PW: When we are building a project, there's not as much public interaction other than project updates. With community involvement, you can make it part of educational programs, you can make it part of telling people that putting sand in Jamaica Bay is not a bad thing. This is important because there is a constituency out there that feels putting sand in Jamaica Bay is a bad thing because it's contaminated material. You have to remind the public that we're using clean sand and bringing back former marsh islands. Perhaps in the future the Corps could work with local partners in placing the material as the foundation for marsh restoration and have the local citizens follow-up with the plantings—that would showcase successful federal and local relationships. But to me, the biggest thing is letting go of the public demand for immediate gratification. You need to see it more over time with community-based planting.

River Gods, Return

Brian Davis

Weirdly, erosion is the most powerful tool to combat some of the looming effects of rapid climate change. There are two reasons for this. The first has to do with the long carbon cycle. While trees and plants get most of the glory as carbon capturers, rocks do some heavy lifting over time. Plants take in tons (gigatons, actually) of carbon each year, but they don't sequester it for long. Most of it is released back into the atmosphere within the century as the plant decomposes, or burns. But rocks weather. And as they do they capture carbon by freeing tiny little silicates to move to the ocean. There they convert into carbonates that sink to the bottom and are cemented by time into limestone or something similar. The second reason that erosion is a saving grace has to do with rivers.

In an earlier age, in a place called Lydia, in what we now call Turkey, there was a god named Pactolus. He was a river, and the sediments of this river were full of something called *ephestus*, a naturally occurring alloy of gold and silver. It was copious and easy to gather in the river's sediments, a fact that gave rise to a strange practice with profound implications for civilization. According to Heroditus, these Lydians formed it into little lumps, stamped it, and assigned it an arbitrary value. In other words, they made what we now call money. The golden sands of Pactolus owed their provenance to King Midas, who bathed in the river to rid himself of his eponymous curse, thereby turning all the sands of the river into gold for a time. But now, in a time of rapid sea level rise and habitat loss, with so much human population and activity huddling along the coasts, all rivers are Pactolus, and all sediment is gold.

In the bureaucratic systems that regulate so much of our coastal and riverine landscapes now, the simple realization of the value of sediment often goes by more anodyne terms like "beneficial use" that betray the fact that sediment has long been a precious resource and cultural artifact. This realization is slowly making its way into the ways we think about rivers. During the engineering age rivers have been understood mostly as water, with anything that is not-water conceived of as aberrant, bad, a problem to solve. This was not a bad idea, because a lot of the stuff in rivers over the last two hundred years was washed there in great amounts from urban surfaces, agricultural fields, factories, mines, and wastewater ponds. It was disgusting, a danger to public health. Policies, technologies, and projects were adopted and undertaken to correct this—separated sewers and treatment plants, buffers, zoning, low impact development guidelines, the Clean Water Act.

But this reasonable approach to non-water materials in rivers ossified, and we adopted a similar position regarding clastic sediments, the little silts and clays and sands and gravels that shoaled around bends or behind obstructions, that caused streams to braid or anastamose, that filled floodplains and gave them their fertility and elevation. These sediments represented impediments to certain types of navigation, and a great deal of ingenuity has gone into controlling or removing them and disposing of them as waste. This is no small undertaking.

Rivers move massive amounts of sediment, removing land from high areas, and depositing it in lower places. The Army Corps of Engineers dredges more than a hundred million cubic yards per year from the 23,000 miles of navigable rivers and channels within the United States. But this is a pitiful amount if you compare it to the rivers themselves, and it is a pitiful approach to engineering, this solving of single problems. In their seminal paper from 1983, John Milliman and Robert Meade estimated the total amount of sediment moved by rivers to be some fifteen billion tons annually.[1] Rivers can build land. And like the geographer Richard Campanella has pointed out, that has never mattered more than now.[2]

Can we imagine a future where river sediment is taken as a public good, a cause and effect of stewardship, an aim of good policy and practice, and a key method for improving habitability along our nations' watery landscapes? The stakes are high. The 2021 Inland Waterways report from the American Society of Civil Engineers states ominously:

Additionally, the growing severity and frequency of flooding is pushing more sediment into the navigation channels, meaning more frequent dredging is necessary to keep the Mississippi River and its tributaries navigable.

In our society's most common formulation, sediment is still a threat, a problem to solve. Our ongoing processes of urban development and rural living tend to increase the rates of erosion and sedimentation. Sediment becomes an indicator of many of the excesses of modern life, and a tax on the creatures and communities that developed with a different sedimentary regime. Whatever we call it, sediment is not usually considered a means toward making a richer, more habitable world along the rivers. And this is too bad. Most of our great cities and civilizations for a long time have sprung up in the alluvium deposits along their banks, until global commerce became the primary organizer of human effort and landscape-making in the last few hundred years.[3] Sediment has been so foundational for so long that we have completely taken it for granted. But there may be another way.

The fate of that other major constituent of rivers—water itself—offers clues. For the last two hundred years or so, the variable flow of river water has been understood as a problem to solve according to the scientific norms of hydraulics and the values of efficiency and control, despite water's importance in our modern way of life. This sensibility is changing in some places. And a recent turn in landscape architecture and engineering from the last forty years offers a few examples.

Do you know the Guadalupe River Park? It is undoubtedly a high-water mark in the history of modern American landscape architecture. At Guadalupe excess water is returned to its rightful place as a driver of landscape form. The park itself was 2.6 acres of riverland within downtown San Jose, CA, designed by Hargreaves Associates and the Army Corps of Engineers. The project had to satisfy the Corps' flood control plan for the full 14 miles of flood control channel, and the City of San Jose's desire for a public park along the river. Karen M'Closkey explained this tension in her book on the firm's work. She notes that the USACE originally "proposed encasing this portion of the river with vertical and trapezoidal concrete channels, fenced to prevent access, in order to move the water through downtown as quickly as possible" while the City of San Jose wanted a park.[4] The contrasting values and goals of these two constituencies, one technical and one recreational, became drivers for innovation in remaking this landscape.

Speaking of the interrelated roles of engineering and landscape architecture in reimagining this river landscape, Karen M'Closkey says it with clinical accuracy:

designers do not simply calculate forces and then make artifacts to match those forces; instead, designers propose structures and then determine the forces that those structures create and are subjected to. By construing the project first and foremost as a public landscape, rather than flood control infrastructure, Hargreaves was able to create a system that provides a varied experience along an urban river.

But a bit more is needed. Guadalupe is more than a varied experience along an urban river that sometimes floods, just as it is more than an artifact to match forces. It is a truly remarkable thing for this reason: floodwaters are taken as a *feature* of the landscape and a *driver of its form*. While there are other examples of this approach now, most river projects still take the level and velocity of high flows as an aberrant condition to resolve. Guadalupe was pioneering. The lessons are all over the landscape, but a couple of places illustrate the point best.

First, the iconic berms located in the floodplain between Coleman Avenue and the 880 freeway. They were developed through a sophisticated application of a very basic method—clay modeling—and tested by the Army Corps of Engineers using an eighty-foot physical hydraulic model.[5] The berms interpret the braiding gravel dunes common to a western river like the Guadalupe. Here they are a metaphor, echoing this quality of the river to reshape a large floodplain field. Their form minimizes shear stress and erosion during high flows, keeping the berms stable. But they work because they create a varied topography that rescales the paths and spaces as you walk along, while creating small moisture and exposure gradients through variation in slope. These adjustments set the stage for different plants and interactions that you see and smell and hear as you move through the berms. The metaphor is lost, appropriately, and the experience is both basic and elevating—just a park walk, plants and berms, but not just that.

Or consider the overlook behind the Children's Discovery Museum. From above you look over a humble, shaded terrace into a tangle of cottonwoods and ivy. The terrace is made with a tidy wall, and a small tree that bumps off the main path. It is just off-axis from the Museum, about eight feet. But if you venture to the stairs on either side you descend into a chamber of sound and shade. All boulders and weeds and low-flow silty channel and point bar. From below, the monolith of the overlook appears unapologetic, intrusive. Its angles are sharp and its surface smooth and heavy to resist the shear force of the swift floodwaters. Thor was its mason. It is unnecessary, and it is not even clear that it is well-used.

Insights from these moments help structure the park. The paths and vegetation and terraces and railings were designed to withstand the shear force of flood water, the landform meant to reference the fluvial geomorphology, to guide and shape the flow. The program is simple—to walk, to look, to play, to rest. This simple insight—that the high water should have a role, a presence—is becoming a paradigm today and is helping us revalue rivers and their waters. The approach is not yet artfully practiced everywhere, but pale copies of it have proliferated. Tulsa, Oklahoma, and Houston, Texas, have built vast park systems that hold stormwater or floodwater and can withstand massive flooding. River initiatives from Greenville, South Carolina, to Buffalo, New York, show how high water can enrich a public landscape over time. Major projects on the Trinity River in Dallas and the Los Angeles River hold out more hope, promising to use the floodwater to create aesthetic, ecological, and programmatic richness and excess. What we are beginning to do with floodwater, we might also do with sediment.

To do this we need to cultivate a different mindset and approach, pushed along by the demands of accelerated climate change to build better riverlands. This means good public health and planning policy resulting in public access to rivers, often vegetated, flowing with clean water. But picturesque aesthetics does not go far enough.[6] Ecological research stands alongside eons of observation by many cultures to show that the healthiest rivers are those that can migrate within their alluvial plain, evolving to constantly create and destroy landforms on small scales through annual high-water events.[7] Modern human interventions are already trying to work more with these concepts by creating more space for the rivers through policy and planning. Some fields, such as landscape architecture and environmental engineering, are developing a healthy obsession with forms and production techniques that do not rely so heavily on materials such as concrete and boulders, instead working with sediments and vegetation, and other erodible or degradable features. All of this points to a renewed emphasis on rivers moving sediments, including the sands and gravels that often stay stuck upstream behind dams.

The Renaturation of the River Aire by the Superpositions team points a way. In that project, the restoration of the river was achieved not through innovative policy measures that reclaimed a broader alluvial plain for the river only, but also by the creation of striking landforms meant to catalyze a new, complex normal in the morphology. This would encourage ecological diversity and enhance the simple beauty of the river channel. That is the theory, at least. The team worked with sediment and time to create complexity that would have been wildly expensive to build or even imagine by the methods used in Guadalupe River Park, where each mound was made in bespoke fashion. At the River Aire, a portion of adjacent farmland was famously shaped into diamonds by cutting channels into the ground. The shape is intended to optimize the evolution of the river as it found its own form.[8] That is, though the river would have produced a meandering channel on its own eventually once freed from the canal, "the grid of channels serves to accelerate the evolution of the channel, and provides a striking visual pattern for the public."[9]

In some ways, this is an accelerated "Stage 0" stream restoration[10] and indeed, the river's annual spring flows have eroded these simple landforms into complex, braided forms in just a few years. The stakes were perhaps lower here than at Guadalupe. The surrounding landscape was farmland and an old canal. While the place has cultural significance, and flood management did inform the project, the river doesn't course through the downtown of a major city like Guadalupe does. But the stunning beauty and richness of the project suggests there is great potential to position sediment and water alongside other elemental concerns as active agents and cultural drivers in making better riverlands.

Alexander Robinson's work with the Los Angeles River shows how this integration might work to reimagine rivers in more difficult circumstances. His approach takes the taut boundaries of the engineered system seriously and represents them in physical models that make other qualities present, allowing for a more complex picture of the results to emerge. In physical model studies of the Bowtie reach of the LA River, sediment is simulated using ground walnut shells. The particles form a mobile part of the model that slowly flush through the area as gravel bars that would host plants and attract human recreation.[11] Gravel bars already do this, of course, but the art lies in enabling this agency of sediment that can be so difficult to predict, while ensuring the flow and flood control parameters necessitated in the infrastructure design. Here, too, sediment plays a key role, somehow holding its own alongside a cinematic concrete section, devastating floodwaters, and human communities in opening new possibilities to connect habitat, offer recreation, and inspire imagination.

Because it is so humble, paradoxical, and complicated, sediment also offers a radical way of working with rivers. Positioning sediment as an agent, even a means of remaking worlds, is an escape hatch from the dull, technocratic world of problem-solving and policy-making that we have foisted on rivers. By slowly-but-constantly remaking form and reshaping relationships, sediment creates an opening that allows rivers themselves to help us live again. It helps us heed James Hillman's advice from his thoughtful essay "The Practice of Beauty"—it invites the gods to return to our world.

We might still design structures, but smaller. Trapping sediments, or slowing them for a time; using currents to pile them and sort them; allowing them to settle out over low-lying plains or within basins; creating simple erodible landforms whose complexity will increase through erosion and time, bedforms that might support vegetation or spawning—these are some ways we might act in the future.

River edges are desirable economically, aesthetically, and spiritually, and history suggests that if they can be inhabited by humans, they will be. Reimagining them as places for living, not just risk to reduce or problems to solve, has already happened along the Guadalupe and Aire rivers, and very obviously works, but it is rare, and very hard to do in the context of our cities and the legal structures that help define them. Most of our rules were created in the context of rivers that were threatening and bays that were shrinking through filling and reclamation. But today, bays are growing, and rivers are not as commonly seen as enemies. Special projects like the River Aire Renaturation, Robinson's work on the LA River, and the Guadalupe River Park point the way forward. The challenge is how to re-traditionalize these approaches so they are no longer exceptional projects, but merely early indicators of a common condition in a future world. A focus on the sediment in our rivers offers a chance to make a more common ground.

1 J. D. Milliman and R. H. Meade, "World-Wide Delivery of River Sediment to the Oceans," *Journal of Geology*, 91 (1983): 1–21.

2 Richard Campanella, "Beneficial Use: Balancing America's (Sediment) Budget," *Places Journal*, January 2013.

3 Mon Vanderostyne and Marsha Cohen, "From Hand-drag to Jumbo: A Millenium of Dredging," *Terra et Aqua* 77 (1999): 5.

4 Karen M'Closkey, *Unearthed: The Landscapes of Hargreaves Associates* (University of Pennsylvania Press, 2013), 118.

5 Ibid., 118.

6 Kristen Podolak et al., "Designing with Nature? The Persistence of Capability Brown's 18th Century Water Features," *Landscape Journal* 32, no. 1 (January 2013): 51–64.

7 G. Mathias Kondolf, Georges Descombes, and Aude Zingraff-Hamed, "Restoring Dynamic Fluvial Processes in Urban Rivers: Learning from the Aire and Isar Rivers," *Landscape Architecture Frontiers* 9, no. 4 (August 2021): 11–12.

8 Elissa Rosenberg, "Before and After. Both. The Revitalization of the Aire River, Switzerland," *Landscape Architecture Magazine*, June 2019, 130.

9 Kondolf, Descombes, Zingraff-Hamed, "Restoring Dynamic Fluvial Processes," 10+.

10 Brian Cluer and Colin Thorne, "A Stream Evolution Model Integrating Habitat and Ecosystem Benefits," *River Research and Applications*, 30 no. 2, (February 2014): 140–41.

11 Alex Robinson and Brian Davis, "From Solution Space to Interface: Six Actions for Landscape Infrastructure Design," in *Codify: Parametric and Computational Design in Landscape Architecture*, eds. Bradley Cantrell and Adam Mekies (New York: Routledge, 2018) 155–68.

Mud is Vital

1. Sorting Sediment

The largest American port on the Great Lakes is not in Detroit, Buffalo, or Chicago, but rather hundreds of miles away, at the far western end of the basin, where the St. Louis River drains into Lake Superior.

Straddling a state border between Duluth, Minnesota, and Superior, Wisconsin, the port owes its scale to grain fields of the upper Midwest, timber-rich forests, coal mines as far west as Montana and Wyoming, and, in particular, the iron ranges of the Lake Superior region.[1] Among the many striking relics of this industrial heritage are Ore Docks #5 and #6—elevated steel structures jutting roughly a mile into the harbor so that rail cars could drop their ore directly into ships' cargo holds. On both sides of the channel, black coal stockpiles, storage silos, belt conveyors, and shiploaders sit on land that faces the water like rows of enormous rectangular teeth. This orthogonal geometry is a telltale sign of made land: like most ports, Duluth-Superior's facilities are built on fill dredged up from nearby channels and now held in place by bulkheads that demarcate land from water.

One tooth stands out. On the Duluth waterfront, between the Bong Bridge and the long lines of the ore docks, one large square of land is occupied not by steel machinery but by ponds, grasses, stretches of bare dirt, dikes, and access roads. If you could view that square in time-lapse over the past decade, it would be perpetual whir of motion: sands and silts piling up and then being removed by excavators and dump trucks, barges docking, pipes disgorging slurries of water and sediment into the ponds, boundaries between water and earth fluctuating. This is the Erie Pier, and it is a landscape machine.[2]

Like most navigable waterways, the shipping channels that link the facilities of Duluth-Superior to other ports around the Great Lakes (and, through the St. Lawrence Seaway, to the rest of the world) require frequent dredging. The mouth of a river is a depositional environment. Silt, sand, and clay is continually conveyed down the St. Louis from the sedimentshed, and much of it drops out of the water column as the river loses velocity where it meets Lake Superior. Every year, the US Army Corps of Engineers dredges about a hundred thousand cubic yards of sediment from the harbor's 17 miles of navigation channels.[3]

When sediment is dredged from Duluth's harbor, it is often placed at Erie Pier, as it has been since this "Confined Disposal Facility" was first constructed in 1978.[4] Today, the process works like this. A dredger excavates the harbor

1 "Port History," Duluth Seaway Port Authority.

2 Roncken, Stremke, and Paulissen, "Landscape Machines."

3 Brossart, "Erie Pier Briefing."
Clark, "Duluth-Superior Harbor Erie Pier."

4 Erie Pier, for instance, can also be seen as a microcosm of the broader condition of the Great Lakes Basin. Many rivers—Nipigon, Menominee, Fox, Milwaukee, Kalamazoo, Muskegon, Huron, Maumee, Sandusky, Cuyahoga, Black, Saugeen, to name a few—feed into the five Great Lakes, just as the Saint Louis feeds into Superior at Duluth. Though each is relatively small, by comparison with the larger rivers of North America, each does bring sediment to its mouth, where it accumulates. The many harbors and ports of the basin are largely found in these river mouths, and as sediment accumulates, harbors, ports, and their channels fill in.
Over the course of the twentieth century, these Great Lakes ports, like Duluth, generally coped with that accumulation first by dredging sediment and dumping it in open water to the sides of estuaries, harbors, and river mouths, then by building diked confinement facilities alongside navigation channels.
Burkholder et al., "Dredge Dump Dike."
Burkholder, "Designing Dredge."

bottom and places sediment in a hopper barge. When the hopper barge is full, it is pushed by a tug boat to the southern corner of Erie Pier, where it is tied up alongside metal bulkheads. An excavator on the pier reaches into the hopper and scoops out the sediment, piling it on the pier. A bulldozer pushes the sediment from that pile into the sluiceway, a long channel carved into the ground that slopes from a high point at the southern corner of the pier down to a set of ponds. Water is pumped up from the ponds and into the sluiceway, where it mixes with the sediment and runs back downhill as a slurry of water, sand, silt, and clay.

Because the sluiceway is so gently sloped, the heavier sediments—first, the sands—fall out of suspension even before the slurry reaches the ponds. Meanwhile, the fines, which are composed of the smallest sediment particles, settle into the pond at the base of the sluiceway, building a miniature delta. Medium-sized particles settle in between the top of the sluice and this growing delta. As the sluiceway, slurry, and gravity do their work, the dredged sediment is sorted into a gradient of sizes, from large to small, stretched out across hundreds of feet.[5]

Once the sediment is sorted, other excavators and dump trucks reposition it in stockpiles elsewhere on the pier. The sand is easiest to work with; it has been recycled at Erie Pier since 1988.[6] The hydraulic separation (and eventual reuse) of fines is a more recent innovation, emerging here only within the last two decades. This is in part because fines are more demanding. When dry, the fines are light enough that they blow easily in the wind. When wet, the fines retain water much more readily than quick-draining sand, and can become quite heavy.[7] Reusing fines, though, has been a crucial step in the recent transformation of Erie Pier from one of dozens of Confined Disposal Facilities operated around the Great Lakes Basin to the first (and, so far, only) Processing and Reuse Facility: the great bulk, approximately 80%, of the material dredged from the harbor and placed at Erie Pier is fines.[8]

Soon, sorted sediment is excavated again, this time from the stockpiles, and trucked offsite. It has a wide range of destinations. Sand, being relatively valuable, is offered at no cost to contractors, who come and excavate it at their own expense. Most sand is taken away in this fashion; it is destined for use as concrete aggregate, as backfill on construction sites, as bed layer in road construction.

Reuse of the fines requires more complicated linkages between economic interests. Ore mined in the Iron Range northwest of Duluth arrives at the docks in train cars, which typically make the return journey back to the mines empty. Some of the Pier's dredged material is placed in these otherwise empty cars and backhauled to the Iron Range, where it becomes substrate for new wetlands being constructed on closed mines. Since rail lines are already in place and rail cars already making these trips, transport costs are low enough to make sourcing soil from the port feasible.[9]

Aqueous habitat creation within Duluth-Superior Harbor itself has also emerged as significant in the past decade.[10] As in most other ports around the United States, nineteenth- and twentieth-century land reclamation radically altered the ecosystems at the mouth of the St. Louis River. The pre-reclamation riverlands were characterized by shallow, sloping gradients with littoral wetlands full of rushes, reeds, grasses, and arrowheads; beaches; and mudflats. Beds

5 Duluth-Superior Metropolitan Interstate Council, *Erie Pier Management Plan*, 2021.

6 Brossart, "Erie Pier Briefing."

7 Brannan, Skurla, and Castleberry, "Erie Pier."

8 Clark, "Duluth-Superior Harbor Erie Pier."

9 Metropolitan Interstate Council, *Erie Pier Management Plan*.

10 Metropolitan Interstate Council, *Erie Pier Management Plan*, 15.

of aquatic vegetation like wild celery, pondweeds, and lilies were so thick that early settlers described them as obscuring the main path of the river from view.[11] Bulkheaded piers and deep dredged channels have taken their place. Where there was once ecological complexity and diversity, there are now stark divisions between dry land and open water.[12] Moreover, decades of industrial activity and outflows from the creeks and sewers that drain the urbanized watersheds of Duluth have deposited contaminated sediment in many parts of the harbor. Now, dredged material is being used to seal off contaminated sediment and build new habitat at sites around the harbor. This has involved both recycling material through Erie Pier and directly transporting dredged material to destination sites.

The most significant of these sites is an open bay tucked into the northeast corner of Duluth's Inner Harbor, where it is framed by the railroad tracks of a switching yard, the concrete fly-overs of a freeway interchange, and the low, squat tanks of the Western Lake Superior Sanitary District's wastewater treatment facility. The bay's waters are fed by two small creeks, Miller and Coffee Creek, which drain highly urbanized watersheds in Duluth, bringing with them large sediment influxes (and surface contaminants). The wastewater treatment facility's sewer outfall adds an average 43 million additional gallons of treated wastewater to these flows each day. While the bay did have some existing habitat, it was, by the turn of the twenty-first century, of poor quality—vegetation density was about half what would be expected. To the south, the edge of the restoration site is marked by Interstate Island, a small island created by the earlier placement of dredged material. Interstate is regularly managed by the Minnesota Department of Natural Resources to prevent woody vegetation from establishing, protecting its capacity to serve as a nesting site for an endangered bird species, the common tern.

The 21st Avenue West Habitat Restoration project, which is coordinated by the Minnesota Pollution Control Agency, has been transforming this corner of the estuary into healthy, varied habitat.[13] Dredged material is being used to construct a complex, dimpled topography, both underwater and emergent. New fingers of earth are extending from the existing shorelines, dotted by islands (also new), creating a cove at the mouth of Miller and Coffee Creeks, which will be protected from wind fetch and wave energy. Relatively deep, high-energy zones are being transformed into shallower, lower-energy habitat, with much greater diversity and heterogeneity. Vegetation will be encouraged (and planted) along this new depth gradient, from emergent marsh vegetation, like marsh grasses and rushes, to floating-leaf aquatic beds, to submerged aquatic beds, such as wild celery. Diversity of vegetation should in turn foster diverse populations of small fish and insects. With time, scientists expect larger game fish, like walleye, bluegill, and lake sturgeon, to spawn, develop, and forage in these new shallows.[14]

While this work is consistently described as "restoration," which implies the re-emergence of some lost past condition, it might more accurately be described as remediation and habitat creation.[15] The proposed design for the new habitat was based on an evaluation of present conditions and modeling of likely vegetation responses to various options for the placement of dredged material.[16] It bore no precise relationship to any historic configuration of habitat, except in the fact that preindustrial conditions, like the proposed design, were

11 Angradi et al., "Predicting Submerged Aquatic Vegetation."

12 St. Louis River Estuary, "21st Avenue West Restoration Site."

13 Minnesota Pollution Control Agency, *Environmental Assessment Worksheet*. Breneman, "Habitat Restoration Effectiveness."

14 St. Louis River Estuary, "21st Avenue West Restoration Site."

15 We discuss the idea of restoration and the related concept of shifting baselines at further length in "Technology, Pace, and Adaptation."

16 USFWS, *An Ecological Design for the 21st Avenue West Remediation-to-Restoration Project*.
MPCA, *Environmental Assessment Worksheet*.

generally shallower, more diverse in topography, and more heterogenous in plant and animal assemblages. A sediment flow is being choreographed to support the emergence of a new ecological system.

The transformation of Erie Pier illustrates how reconfiguring a web of economic relationships—the local logistical network of prices, supply and demand, contractors, regulators, and earth-moving machinery, tied to the regional and even global networks of commerce that originally called the port into being—can facilitate the making of new landscapes, new landscapes whose value is understood in ecological terms. This is the internalization of ecology, and stands in stark contrast to the historic externalization of ecological value in modern economic systems.[17]

* * *

But what has made this possible? And why has it not happened until now?

In the United States, decisions about where dredged sediments go are governed by a byzantine web of regulations and rules, which interact with local officials, federal agencies, and state bureaucracies. At the heart of this web are decisions about values—particularly how to assess economic value and compare it with ecological value. In the transformation of Erie Pier into a sediment-recycling landscape machine we see how old ways of making those assessments are being altered by new valuations. First, though, we have to understand something of those byzantine workings.

When Erie Pier was constructed as a confined disposal facility, it was projected to have a ten-year lifespan: 1978 to 1988. Confined disposal facilities have finite lifespans determined by the volumetric space between their perimeter dikes. When they fill up with sediment, operators have three options: the dikes can be enlarged to increase capacity, stored sediment can be removed to make room for new deposition, or the facility can be closed. At Erie Pier the dikes have been enlarged several times, and they are now running up against geotechnical limits: if earthen dikes get too tall, they may fail. Even with the sand recycling operations, by 1998 the Pier was close to finally running out of space.[18] That was a problem for the Corps, which is responsible for dredging federally authorized navigation channels like those in Duluth Harbor, and the Duluth Seaway Port Authority, which, as the "non-federal sponsor" for those channels, is responsible for contributing a portion of the costs of dredging.

It might seem that the obvious solution was to construct a new confined disposal facility. With more capacity, these parties could continue to operate exactly as they had for the past three decades. But they have agreed that this is not feasible, given the current political economy of sediment. The last new CDF on the shores of the Great Lakes was built in 1983, in Monroe, Michigan. Nearly half a century later, it's clear that the region's heyday of CDF construction was barely two decades long, and high capital costs together with restricted port budgets mean the era is unlikely to return.[19] Construction costs for a new CDF at Duluth are estimated at $30 million, which greatly exceeds the annual dredging budget of about $650,000.[20] Moreover, there is no space within the Port's existing boundaries, so land would have to be purchased at further expense. Then there are the anticipated objections of environmental groups, who would prefer to see land use within the estuary shift toward habitat, not new industrial facilities.[21]

17 On the shared tendency to produce environmental problems in modern economic systems, see Guha, *Environmentalism*, 69–97 and 125–37.

18 Metropolitan Interstate Council, *Erie Pier Management Plan*, 6–7.

19 The Great Lakes CDF building boom of the seventies was facilitated by Section 123 of the Rivers and Harbors Act of 1970, which "established a funding source for the creation of CDFs and operated under the assumption that after ten years of disposal, the material being dredged would be clean enough to no longer need confinement," permitting the resumption of open-water disposal (Burkholder 2016, 8). What has actually happened is that ports have generally been unable to find the capital funding and physical space necessary to construct new CDFs, preferring instead to extend the lifespans of existing CDFs by raising their dikes, dewatering, and diverting material to alternative uses.
This is not to say that confined disposal and open-water dumping are becoming obsolete at the same pace in all places, or even that they will necessarily be completely abandoned. Other regions and other Army Corps districts, like the Galveston District on the Texas Coast or the Jacksonville District that includes Florida's Atlantic Coast, have either substantial capacity remaining in existing CDFs, typically constructed more recently than Great Lakes CDFs, or are in the process of building new CDFs. But even in those regions, capacity is not unlimited. A national review in 2010 found that two-thirds of Corps districts cited CDF capacity shortage as a crucial issue.
USACE ERDC, "Sustainable Confined Disposal Facilities."
USACE, Great Lakes & Ohio River Division, "Confined Disposal Facilities on the Great Lakes."
USACE and EPA, "Great Lakes Confined Disposal Facilities."

20 A sum of $30 million, of course, only seems insurmountable relative to the funding currently directed toward sediment management for the port. That level of funding is the product of political choices, not a hard fact of physical reality, and it is readily imaginable that under another political economy, constructing CDFs (or, ideally, more PRFs) might seem financially feasible again, just as it is possible to imagine a great deal more investment in American infrastructure, generally. After all, a 2018 study by Martin Associates ("Economic Impacts of the Port of Duluth-Superior") concluded that shipping through the harbor generates over $200 million annually in state and federal taxes alone.

21 Brannan et al., "Erie Pier."

In selecting among potential placement methods—for any dredging project, not just those in Duluth harbor—the Army Corps of Engineers is bound by federal rules to use the *least costly alternatives* that conform to sound engineering practices and meet current environmental regulations.[22] These three criteria together have historically determined, for each and every Corps navigational dredging project, what is called the *federal standard*, a baseline method of dredging and placement that will be employed unless other forces, interests, or funding come into play. Because the Corps is responsible for the great majority of the dredging that happens in the United States, the federal standard has become a tremendously influential policy, shaping decisions across the nation's bays, estuaries, coasts, and waterways, and, in turn, fashioning landscapes in accordance with its cold logistical calculus.

In most cases, the least costly approach is to dredge sediment from the navigation channel in question and dump it back into the water nearby, somewhere out of the way of ship and boat traffic. This is called open water disposal. Historically, most dredged sediment was dumped in this way, unless it was being used as fill for harbor construction.[23] But there are two major problems with open water disposal that prompted the shift toward CDF construction in the 1960s in the Great Lakes.[24] First, much of the sediment that settled into navigation channels in industrial ports in the mid-twentieth century was highly contaminated, laced with toxic chemicals like PCBs and mercury. Dredging can stir those contaminants up into the water column, where it enters the ecosystem, and open-water dumping exacerbates this, as it gives the lightest—and often most contaminated—particles time to float gradually down through the water column and spread horizontally away from the dump site. The second problem is that river and lake bottom habitat—which open water dumping can smother—is quite ecologically valuable. Rich assemblies of species like tubeworms and caddisfly inhabit its muddy layers, and they are foundational to food webs that include larger fish and waterfowl. This fundamental value was enshrined in law and regulation in the latter portions of the twentieth century, and many states, like Minnesota and Wisconsin, have become increasingly protective of it.

In the late nineties, with Erie Pier seemingly approaching full capacity, management agencies worked to establish alternatives for the placement of the harbor's dredged material. They identified two main options: placement of sand on nearby beaches as beach nourishment, and the open-water filling of six "deep holes" that had been excavated in the harbor decades ago. Both plans had serious flaws. The great bulk of the material dredged from the harbor is fines, not suitable for beach placement. And because the holes now functioned as benthic habitat with important fisheries benefits, Minnesota and Wisconsin opposed filling them.[25]

The Corps determined that filling the deep holes would be considerably cheaper than transporting dredged material to Erie Pier by barge. State officials, however, threatened to block that proposal by denying the Corps its Clean Water Act certification.[26] This situation, in which the Corps, guided by its mandate for economic efficiency, finds one plan optimal, while a local or state entity prefers a more expensive option that has other benefits or reduced impacts, is not at all unusual. (In fact, it is common enough that the Corps felt compelled to issue clarifying guidance in 2015.[27]) Typically such situations are resolved by the Corps using the "least costly alternative" to calculate the amount it will pay (the

22 Stockton, "Federal Standard Clarification." Code of Federal Regulations, Title 33, Parts 335 and 336.
Note also the exception in Section 207 of the 1996 Water Resources Development Act: "[this] is a separate authority, which authorizes for navigation projects, subject to certain requirements, the use of a disposal method that is not the least cost option if the incremental costs are reasonable in relation to the environmental benefits."

23 Even as CDFs came online, open water dumping never ceased: from 1999–2008, for instance, roughly a third of dredged material placement by the USACE in the Great Lakes basin continued to be in the form of open water dumping. While Pennsylvania, Minnesota, and Wisconsin banned the practice, Illinois, Indiana, Michigan, New York, and Ohio continue to allow it, provided the dredged material meets certain standards of cleanliness.
Great Lakes Dredging Team, "Beneficial Use."

24 Regarding the role of contamination in the federal development of CDFs, see McCallister and Kavalar, "Confined Disposal Program."

25 Brannan et al., "Erie Pier."

26 Ibid.

27 Stockton, "Federal Standard Clarification."

28 Similarly, with the rebuilding of Jamaica Bay marsh islands described in "Choreographing Sediment," "Aesthetics and Publics," and "Dredge City," the New York State Department of Environmental Conservation and the New York City Department of Environmental Protection served as non-federal sponsors. USACE, "Army Corps Awards Contract for Restoring Black Wall and Rulers Bar."

29 Brannan et al., "Erie Pier," 124–25.

30 Metropolitan Interstate Council, Erie Pier Management Plan, 6-15.

The habitat creation project has also been eligible for a special form of additional federal funding, authorized under Section 204 of the Water Resources Development Act of 1992 and keyed to the beneficial use of dredged material. Section 204 was thus an early encoding in federal law of the value of sediment as a resource, recognizing that it can be worth paying more to place it if that placement can create other benefits, like habitat creation.

31 Transition design is an emerging domain of scholarship and activist practices dedicated to examining and making this shift. It approaches design as ontological and world-making, arguing that the contemporary world has been thoroughly designed and, in turn, designs us. Humans and societies have created many crises through acts of design. In order to change the status quo of a degraded and inequitable world, we need to design fundamentally different ways of designing. We need to figure out how to holistically transition, as a species, and to approach the current era as one of overt, intentional change and transformation.
Fry and Nocek, eds., *Design in Crisis*.
Escobar, "Transiciones."
Tonkinwise, "Design for Transitions."

federal standard), while being willing to support alternatives if local entities or other sources of funding make up the difference. This functions to both limit the capacity of Corps districts to pursue creative, multi-benefit choreography and shifts the burden of internalizing ecological value onto state and local entities, who typically have fewer financial resources than the federal government.

This kind of deal is exactly what happened in Duluth.[28] After state officials threatened to block permitting of the Corps' proposed least cost alternative (open water placement), the states, the Corps, the Duluth Seaway Port Authority, and other local entities, represented by the Duluth-Superior Metropolitan Interstate Council and the Harbor Technical Advisory Committee, reached a compromise in which the Corps agreed to use the least cost alternative only to determine the amount of funding the Corps would contribute, not the actual method of placement.[29] Under this agreement, continued use of Erie Pier has been extended in time by the transformation of the Pier into a Processing and Reuse Facility. Beach nourishment can be employed where dredged material is sandy enough. Recycled fine material can be used in efforts like the 21st Avenue West Channel habitat creation project. Over 400,000 cubic yards of sediment have been placed at Erie Pier since the designation of the PRF, and the current management plan anticipates future opportunities to use Erie Pier to sort, store, and supply sediment for ecological landscape-making. The agreement has also facilitated the direct placement of dredged material for habitat creation, including four years in which the harbor's dredged material was diverted directly to the 21st Avenue West site.[30] Both the Corps and its local partners have agreed that finding additional ways to reuse material dredged from the harbor is a priority, because they recognize that sediment has value as a resource.

And so, well below the radar of most Americans, a complex chain of negotiations in this far corner of the Great Lakes has enabled an estuary to escape the bureaucratic dominance of the least costly alternative, which has so often obstructed efforts to manage sediment intelligently. As the original 2007 Erie Pier Management Plan stated:

> *This plan follows from the premise that the least cost alternatives to dredged material management are not necessarily the most cost effective methods. The currently identified least cost method of disposing of dredged materials, open water dumping, does not consider the environmental and social costs and does not consider the economic benefits of reuse.*

Nationwide, ports, bays, estuaries, rivers, lakes, other waterways, and the myriad sedimentsheds connected to them are all grappling with the legacies of both practices and infrastructures that have this paradigm of economic externalization encoded into them. These legacies are vast, and entrenched. They will not be easily unmade. But they are interwoven with economic structures, values, and ways of designing that have been—that continue to be—tremendously destructive. (And, no less crucially, are becoming increasingly difficult to sustain.) Thus, actively transitioning to new structures, to other ways of being, and to other modes of design is deeply necessary.[31] Efficiency needs to give way to a broader spectrum of values, including ecological fecundity, intergenerational sustainability, the redressment of historic injustices, and the provision of landscapes as public commons. Economic value needs to be defined more broadly, to include these other forms of value.[32] And forms of value that cannot be translated into dollar-equivalents—like the experience of enjoying a warm summer evening amid rushes and reeds, with shorebirds and frogs calling—must be recognized as sediment is choreographed to shape new landscapes.[33]

In this chapter, we largely examine dredging, its particular legacies, and the many landscapes it makes. Sediment is implicated in many other landscape-making practices, of course, and those practices are in turn also entangled with economic structures. Dams, for instance, have been viewed largely through the lenses of flood control and power generation, but are now being reconsidered (and, in cases like the Elwha River in Washington, dismantled) because of the way that they have interrupted sediment transport, blocked aquatic migrations, and unmade cultural landscapes.[34] But looking at dredging closely gives us a representative window into the present trajectory of this change in values and an understanding of how much work lies ahead. The way that sediment is valued is changing dramatically. These changes are being enshrined in federal law, as with the 2020 Water Resource Development Act's charge for the Corps' federal standard calculations to consider the "economic and environmental benefits" that might be obtained through the intelligent placement of dredged material.[35] The new practices and landscapes that are emerging through changes in dredging begin to point toward the broader new structures that are needed. We discuss some of these, from the emergence of "beneficial use" as a category of dredged material placement to islands built with dredged material in the Chesapeake Bay and a Louisiana river. But these examples are isolated glimpses of nascent futures. They pale in scale compared to the vast transformations wrought from coast to coast by the past two hundred years of dredging, channel-straightening, harbor-making, dumping, and filling. As pilot projects, examples like the ones in this chapter are testing new ways of landscape-making that are desperately needed.[36] In light of the climate crisis, though, they are not, on their own, adequate. Change must scale up, soon.

2. The Mud Dump Site

Dredging and sediment management have had massive impacts on the ecological structure and function of coasts and waterways. In ports around the world, the jagged geometries of docks and piers, of berths and slips, signal a common history of dredging, filling, and bulkheading. Ports have often been sited in estuaries or other low-lying regions with large tidal expanses, like marshes or mudflats. Both port construction and the larger processes of urban expansion that accompany port growth have displaced these ecologically fecund landscapes.[37] (On the other hand, the human acceleration of upland erosion—through agriculture and deforestation, in particular—has often led to rapid sedimentation in estuaries, producing new wetlands.[38])

These processes and impacts can be traced far into the past. The Roman Empire altered coastlines and carved ports out of (or near) river mouths at sites like the Portus Claudius and the later hexagonal Portus Trajanus, where grain and olive oil imports for the city of Rome were unloaded, or Leptis Magna in North Africa, where ships bound for Rome were loaded with those agricultural

32 Rockström, et al., "A Safe Operating Space for Humanity."
Raworth, *Doughnut Economics*.

33 See for instance T7-T10 in Massumi's 99 *Theses on the Revaluation of Value*.

34 East et al., "Large-scale Dam Removal on the Elwha River."
Milligan, "Hybrid Migrations and Design of Deluge."

35 Water Resources Development Act of 2020, Section 125.

36 We discuss experimentation, technology, and piloting in more detail in "Technology, Pace, and Adaptation."

37 Masoud's *Terra Sorta Firma* is a global atlas of this process of urban land-making and ecological displacement.

38 Kirwan et al., "Rapid Wetland Expansion."
Syvitski and Kettner, "Sediment Flux and the Anthropocene."

39 Urbanus, "Rome's Imperial Port."

40 Mundy, *The Death of Aztec Tenochtitlan, the Life of Mexico City*, 34–39.

41 Headrick, *Humans Versus Nature*, 131–32, 202–3.

42 Di Palma, *Wasteland*.

43 Marks, *China*, 172–77.

44 Di Palma, *Wasteland*.

45 On the transformation of perceptions and attitudes toward wetlands, see further discussion in "Aesthetics and Publics."

46 Hooke, "On the History of Humans as Geomorphic Agents."

47 Davidson, "How Much Wetland has been Lost?"

48 Levinson, *The Box*.

products.[39] Land reclamation—the process of converting wet land into dry, or of raising shallow underwater areas above the waters—has been practiced in many different circumstances, using a wide diversity of methods. The Nahuatl *altepetls* (city-states) in the pre-Columbian Valley of Mexico, like Tenochtitlan, built chinampas and dikes to shape agricultural land out of salty lakes.[40] The Dutch and Frisians converted peat bogs into polders, and powered complex drainage systems with windmills.[41] Dutch drainage experts aided the seventeenth-century English in reclaiming the fens of Lincolnshire, Cambridgeshire, and Norfolk.[42] Yuan and Ming dynasty Chinese settlers accelerated the formation of the Pearl River Delta by aggregating shallow sand bars into *shatian*, or "sandy fields."[43]

Whenever land was reclaimed or ports carved, ecosystems were reshaped—or eliminated. Often, this was seen as beneficial: in many regions of the world, marshes and other wetlands hosted malaria-bearing mosquitoes, and European culture, in particular, shifted decisively in the early modern period toward viewing wetlands as unproductive "wasteland."[44] (Only recently have the positive aspects of wetlands been widely appreciated.[45]) Other habitats, like mudflats and shallow-water areas, were rarely considered at all.

The scale and extent of such transformations has accelerated significantly in the past two centuries.[46] Technological change, including the exploitation of fossil fuels, the development of steam engines, and the invention of machines like bucket and clamshell dredgers, radically increased the capacity of port cities to dredge navigation channels and build new land. Since 1900, well over half—by some estimates, closer to three-quarters—of the world's wetlands have been lost.[47]

In "Choreographing Sediment," we discussed how one of New York City's tidal bays was intensively restructured by decades of dredging, including the efforts of the 1907 Jamaica Bay Improvement Commission to build what would have been the world's largest port. Though the port was never completed, the restructuring produced a bay that today is characterized by a greatly shrunken marsh island complex, surrounded by open water and circled by dredged navigation channels.

While Jamaica Bay was the portion of New York City most radically altered by filling with dredged material, over the course of the nineteenth and twentieth centuries, the broader coastline of New York City saw its own great transformation. As late as the 1950s, New York City, with its Hudson River piers and port terminals in places like Redhook, was central to the American maritime economy. Roughly a third of America's seaborne trade in manufactured goods passed through the city's port facilities, and it has been estimated that one in six New York City workers' livelihoods depended directly on the port. But these facilities were increasingly antiquated, relatively cramped, and built for decades-old shipping technologies. The state of New Jersey sensed opportunity, announcing in 1955 that a 450-acre tract of tidal marsh in the city of Elizabeth would be developed into the nation's largest port facility, and, in 1956, hosting at the port of Newark the first container ship service to the United States. By 1962, when containerization pioneer Malcolm McLean's Sealand Terminal opened, the shift was irreversible: retrofitting Manhattan's piers to handle the new standardized shipping containers that were transforming the global shipping industry was not possible.[48] The bulk of shipping in the New York-New Jersey harbor would move outward, concentrating in Newark Bay and (for petrochemicals) along Arthur Kill.

As shipping moved, the filling of marshes and transformation of ecosystems followed. Low, swampy land was available, in part because it had never been seen as desirable for agricultural or residential development. Newark Bay, which is fed by two major rivers, the Passaic and the Hackensack, was once a vibrant estuary, home to large populations of fish like shad, benthic organisms like oysters, and migratory seabirds like the Double-crested Cormorant. Dredging, filling, urbanization, and industrialization almost entirely eliminated this ecosystem.[49] Some 21 million tons of material—mostly various wastes, including dredged spoils—were used to fill the salt marshes on the western edge of the bay, making solid ground for the city of Newark and, eventually, its port facilities. This constructed land has encroached so much on Newark Bay that the bay is only about half the size of the bay that European explorers found when they first arrived around the turn of the fifteenth century. Estuary-wide, roughly three-quarters of all tidal marshes and wetlands were either filled or dredged.[50] Underwater, the bay's contours have also been thoroughly resculpted. This dredging radically accelerated with containerization. A bottom that once averaged 17 feet in depth is now sliced by 45-foot-deep navigation channels.

A hundred years earlier, bays like Newark and rivers like the Passaic were becoming increasingly contaminated from the burgeoning industrialization of the estuary. One "newspaper article from 1878 mentioned New York's waters as 'having become impregnated by...kerosene refining factories...' and stated that '...the striped bass...have become so permeated...as to be unfit for the table'."[51] Congress passed the Rivers and Harbors Act of 1899, which introduced the first regulations on dumping, discharging, and filling in American waters.[52] The permits and penalties introduced by the act gave states and municipalities the capacity to determine where dumping, including the placement of dredged material, would take place. In New York's harbors, waste material dump sites began to move offshore, away from the immediate vicinities of river mouths and marshes, toward the deeper waters of the New York Bight.

From far above, the Bight is a wide triangle aiming inland, with the two short sides formed by the coasts of New Jersey and Long Island. The Hudson-Raritan Estuary lies at the apex of the triangle, along a line drawn in from the Atlantic Ocean by the Hudson Canyon, which was incised more than 10,000 years ago in an era of much lower sea levels. The canyon forms a natural deep-water shipping channel pointing into the estuary; it has since been extended into the estuary by the deeply dredged Ambrose Channel, the main channel leading toward the inner harbors.

In 1914, the Supervisor of New York Harbor, working with the Army Corps of Engineers, designated separate dump sites for the various materials that were being deposited into the Bight: garbage and refuse, "cellar dirt" (material produced by the excavation of basements), sewage sludge, incinerated wood, acid wastes, and "mud" (or, in modern terms, dredged material).[53] This designation set the location of the Mud Dump Site, a roughly three square mile area of ocean floor, in waters between fifty and a hundred feet in depth. More than 300 million cubic yards of "mud" were dumped there between 1914 and 1997, building enormous slumping underwater mounds more than thirty feet thick in places.[54]

Through the 1980s and early '90s, pressure increased on the Corps to stop placing contaminated sediments at the dump site, due to growing concerns

49 Iannuzzi and Ludwig, "Historical and Current Ecology of the Lower Passaic River."

50 Crawford et al., "Historical Changes in the Ecological Health of the Newark Bay."

51 USACE DOTS, "Evaluation," 8.

52 USACE DOTS, "Evaluation."

53 Williams, "Geologic Effects of Ocean Dumping."

54 See map in USACE DOTS, "Evaluation," 15.

55 Revkin, "Harbor to Be Dredged."

56 Carney, "Ocean Dumping Still Debated."

57 Revkin, "Harbor to Be Dredged."

58 Hevesi, "New Site Proposed."

59 USACE and EPA, "Site Management and Monitoring Plan (2009)," 23.

60 Revkin, "Harbor to Be Dredged."

61 USACE and EPA, "Site Management and Monitoring Plan."
"Historic Area Remediation Site," USACE New York District.

62 Sharkey, "Ocean Dumping Hearing Set."
Revkin, "Harbor to Be Dredged."

63 USACE and EPA, "Site Management and Monitoring Plan (2020)."

64 USACE New York District, "Dredged Material Management Plan."

about public health and marine life impacts.[55] Fishers, for instance, charged that lobsters were being sickened by pollutants.[56] These contaminated sediments, sometimes referred to as "black mayonnaise" on account of their color and texture,[57] were dredged during the deepening of navigation channels or port berths in the region's harbors.[58] EPA studies of the dump site in the mid-nineties found that organic contaminants like PAH, PCBs, and DDT were bioaccumulating in ocean floor species like bristle worms, which gave the contaminants opportunity to travel through the Bight's food webs.[59]

In response, most contaminated material began to be diverted to alternative placement sites, such as upland pit mines, construction activities (like building the foundations and parking lots of a shopping mall in Elizabeth, New Jersey), and aquatic confined disposal facilities (essentially, submarine burial).[60] The Mud Dump Site itself was notionally decommissioned—but it was immediately replaced by a new, overlapping designation: the Historic Area Remediation Site, or HARS.

From above, HARS is a large rectangle, superimposed over the smaller Mud Dump Site. The core of HARS is the "priority remediation area," a zone of nine square nautical miles where the Corps is in the process of placing a thick layer of new, clean sediment to seal contaminated sediment in place and isolate it from the ecosystem. Since HARS was designated in September 1997, over 75 million cubic yards of sediment, including mud, sand, glacial till, red clay, and occasionally larger dredged rock, has been placed within its bounds.[61] This is almost twice the amount that was originally estimated to be needed to remediate the entire HARS, and yet the Corps reports that only about half of the remediation cells have been adequately capped. One imagines that HARS will remain open for quite some time to come.

This perhaps reflects the entanglement of ecological and economic motivations that is so common in dredging. Certainly the establishment of HARS does. The year before HARS was established, the EPA, working with the Army Corps and the Port Authority, had proposed to expand the Mud Dump Site. Public sentiment was strongly opposed to further placement at the dump site, and the outcry pressured the agencies into the designation of HARS.[62] Capping dirty material with clean sediment is the most common remediation technique utilized for dealing with contaminated sites, whether underwater or on land. It is effective, generally, and testing shows that it is effective at HARS.[63] The ecosystem benefits of capping the Mud Dump Site are real.

Yet it is hard to miss that capping has also provided the port at least some of what it desired in the first place: the opportunity to keep using the cheapest method of disposal available, open ocean dumping. The Mud Dump Site was closed, but the area designated for placement was expanded. Far more dredged sediment finds its way to HARS than to any other placement site in the region.[64] The problems created by dumping are being resolved by more dumping. The flow of split-hull hoppers to the Bight continues uninterrupted.

* * *

Much of the trajectory here is typical. An early industrial economy pollutes and contaminates; ecosystems are harmed; later industrial or post-industrial society attempts to contain and remediate the contamination, while restoring ecosystems.[65]

What is unusual about dredging as an industrial system, though, is that a material (dredged sediment) and process (dredging) that was part of the problem is now a critical ingredient in the repair.[66] At HARS, we see one exceptionally literal example of this, as dumping is a crude and impromptu yet ecological response to the problems created by dumping. We imagine that HARS will be far from the final approach to managing sediment in New York Harbor—indeed, the port has already been increasingly deploying sediment to build marshes and beaches. There are many other examples of the ecological use of dredging and dredged sediment, often more sophisticated than HARS. Marshes in estuaries that have suffered centuries of degradation due to human impacts are being augmented by inch-thin lifts of dredged sediment, sprayed into the air above those marshes and then falling back amid marsh grasses in a process known as thin layer placement.[67] Beaches that have been cut off from their littoral supply of sand by jetties and navigation channels are growing again, fed by dredged sand piled in offshore berms.[68]

This part of the trajectory is much less common. The consequences of metal mining are not remediated by mining more metals. Microplastic pollution will not be resolved by generating and distributing more microplastics. Burning more carbon will not bring down atmospheric carbon concentrations. So what is special about sediment? Why is continuing to shape the movement of sediment an important part of responding to the very crises that moving sediment is so entangled with?

3. Sediment as Ecological Infrastructure

The answers to these questions have to do with both the affordances of sediment as a material and the nature of these crises.

One key dimension of the crises is obviously the acceleration of sea-level rise as a result of anthropogenic climate change. While sea level has always risen and fallen on geologic time scales, in response to global climate shifts, the pace of sea-level rise radically accelerated over the course of the twentieth and early twenty-first centuries. The annual increase in global mean sea level was more than three times greater in 2015 than it was a century before, in 1915.[69] And sea-level rise projections for the coming century show even greater rates of acceleration and, with that acceleration, local impacts that range from serious to catastrophic.[70] These impacts will be (already are being) felt most rapidly in the littoral landscapes where sediment's influence is most prominent.

But sedimentary landscapes are not only threatened by accelerated environmental change. They can also play a key role in adapting to its consequences, including the more frequent and more intense storms that are accompanying climate change. Shorelands like marshes are capable of gaining elevation and adapting topographically to sea-level rise, so long as the rate of accretion—the combined capture of inorganic sediment by a marsh and the generation of organic sediment through the growth, death, and decay of marsh plants—can match or exceed the combined rates of subsidence, erosion, and sea-level rise.[71] The availability of sediment is one of the two most crucial factors determining the capacity of a given shoreland to keep pace with sea-level

65 Landscape architects have frequently discussed the consequences of such shifts for our work in recent decades.
Czerniak, *Formerly Urban.*
Desimini, "From Planned Shrinkage to Formerly Urban."
Meyer, "Uncertain Parks."
Waldheim and Berger, "Logistics Landscape."

66 This should be distinguished from cases where an output from one industrial system is (smartly) re-routed to become an input for another industrial system, as, for instance, in the industrial ecology of Kalundborg in Denmark.
Bélanger, *Landscape as Infrastructure*, 216-19.

67 USACE, "Thin Layer Placement."

68 USACE, "Literature Review of Nearshore Berms."

69 Dangendorf et al., "Persistent Acceleration in Global Sea-Level Rise."

70 NOAA, for instance, has anticipated at least a foot of global sea-level rise by 2100 in even the most optimistic emissions scenarios, and the possibility of six feet or more if the world remains on a high emissions pathway. (These rates vary substantially from region to region.)
NOAA, "Global and Regional Sea Level Rise Scenarios."

71 Reed, "Sea-Level Rise and Coastal Marsh Sustainability."

rise.[72] (The other is the rate of the sea-level rise itself.) Thus, with each foot of additional local sea-level rise, more and more sediment is required for marshes to adapt. But where sediment supply is adequate and sea-level rise slow enough, marshes can grow upward and sustain themselves.

Because of capacities like this, marshes, mudflats, dunes, swamps, coastal forests, and other sedimentary landscapes are increasingly recognized as crucial components of regional climate adaptation strategies. After the "superstorm" Hurricane Sandy, for example, government agencies, private philanthropists, and designers in storm-battered states like New York and New Jersey partnered on a series of projects that illustrate the potential of sedimentary landscapes as climate-adaptation infrastructure. Many of these projects were developed as part of Rebuild by Design, a large-scale design competition associated with the federal Department of Housing and Urban Development and the private Rockefeller Foundation. Between October 2012, when Sandy struck, and October 2014, when Housing and Urban Development awarded $930 million to six Rebuild by Design projects, designers and their allies worked with a panoply of government entities, technical advisors, and communities to envision feasible, buildable infrastructure that might better protect northeastern coasts from the next storm.[73]

The majority of these proposals centered landscape as that infrastructure.[74] The Interboro Team's "Living With the Bay" layered new marsh islands, fluvial floodplains, and oceanfront dunes as elements of a protective strategy for the southern shore of Long Island's Nassau County, while calling out "sediment flow" as a crucial component of their multi-scalar approach.[75] The "New Meadowlands" proposal (team led by MIT Center for Advanced Urbanism and ZUS + Urbanisten) enrolled existing wetlands as infrastructure, trumpeted the adaptive potential of wetland soil accretion, and recommended creating additional sedimentary landscapes, including freshwater marshes and Atlantic White Cedar ponds, as a means of dealing with stormwater flows.[76] SCAPE Landscape Architecture's "Living Breakwaters" redeployed and remade rock breakwaters, a common coastal protection structure, reworking them as habitat (punctuated by "reef streets" and "reef ridges"), studying their potential to host oysters, and placing them where they are most likely to encourage sediment flows to build beaches on the shore of Staten Island.[77] Within each of these proposals, sedimentary landscapes were enrolled to provide protection from storms, to offer habitat, and to become publicly accessible.[78]

While Rebuild by Design's competition format was a relatively unusual means for spurring adaptation design innovation (and the political ecology of the competition has come under scrutiny[79]) the physical and processual shift toward landscape as infrastructure that it embodies is spreading much more broadly across policy, planning, engineering, and design practices.[80] Coastal scientists and engineers have been working to precisely define the protective capacities of features like offshore reefs, barrier islands, and coastal wetlands. While that work is on-going, and a landscape like a marsh performs in ways that can be frustratingly indeterminate relative to expectations of stability and permanence, it is clear that sedimentary landscapes offer real and substantial protection.[81] Similarly, inland along rivers and streams, in floodplains and freshwater wetlands, infrastructural tools like levee setbacks, constructed wetlands, channel deconstruction, and dam removal are increasingly preferred to raising

72 Kirwan et al., "Limits on the Adaptability."
Swanson et al., "Wetland Accretion Rate Model of Ecosystem Resilience."

73 Ovink and Boeijenga, *Too Big*.

74 Rebuild by Design, *Rebuild by Design*.

Other work in the region, like Structures of Coastal Resilience, a 2013–14 university-led design research initiative also funded by the Rockefeller Foundation and executed in close cooperation with the Corps' North Atlantic Division, has similarly focused on landscape-based strategies.
Nordenson, Nordenson, and Chapman, *Structures of Coastal Resilience*.

75 "Living with the Bay—Rebuild by Design," Rebuild by Design.

76 "New Meadowlands—Rebuild by Design," Rebuild by Design.

77 "Living Breakwaters—Rebuild by Design," Rebuild by Design.

78 However, these proposals generally also incorporated elements of "traditional" hard infrastructure: a "flood protection berm" in "New Meadowlands" and new dikes on the bayside of barrier islands in "Living With the Bay," for instance.

79 Some reflections on the political ecology of Rebuild by Design include:
Cohen, "Interviews with Rebuild by Design's Working Group."
Fleming, "Design and the Green New Deal."
Shapiro-Kline, "The Impact of the Public Process."

80 Mossop, *Sustainable Coastal Design and Planning*.
Rossano, *Floodscapes*.

Regarding the institutional networks through which such knowledge and expertise is flowing, see Goh, "Flows in Formation."

81 Morris et al., "From Grey to Green."
Narayan et al., "The Value of Coastal Wetlands."
Saleh and Weinstein, "The Role of Nature-Based Infrastructure (NBI)."
Sutton-Grier et al., "Investing in Natural and Nature-based Infrastructure."

levees, straightening waterways, or burying water flows underground.[82] Policy documents like 2021's *International Guidelines on Natural and Nature-Based Features for Flood Risk Management*, organized by the Corps' Engineering With Nature program, both evidence and continue to drive this transition.[83]

As this work recognizes, sedimentary landscapes have substantial protective value. They can reduce wave energy in a coastal storm or accommodate floodwaters after a heavy rain. That is one core reason for the shift toward landscape as infrastructure. Another is the capacity of these landscapes to rework themselves over time in accordance with the dynamic processes they engage. And, sedimentary landscapes are also increasingly valued for their ability to perform across a multitude of criteria simultaneously: where a levee is single-mindedly effective at holding back storm surge, a wide marsh not only dissipates wave energy, but also stores carbon, supports ecological niches for a diverse array of flora and fauna, sustains human livelihoods through occupations like fishing, and offers people compelling experiences ranging from crouching in a duck blind on a cold morning to a leisurely summer paddle.[84]

Contemplating the scale of transformation that the planet Earth will undergo in the remaining decades of the twenty-first century can induce vertigo. As the climate continues to warm at an unprecedented pace, sea-level rises, ecosystems migrate (or, more frighteningly, collapse), and economies transition, a vast new infrastructural apparatus will emerge, an adaptive life-support system for a new global climate. Landscapes will be destroyed, repaired, and built on a tremendous scale. Within this landscape-making work, many thousands of design decisions will delimit future worlds. What kinds of shores and floodplains do we want? When our children's children walk on the edge of a lake, or the bank of a river, what will they see? Will they be able to touch the water? Will it teem with life? Ebb and flow with time? Or will it, like the ruins of an earlier era, be encased in concrete, hardened, and protected against change and transformation?

This—even more than protective value, even more than multifunctionality, even more than adaptive capacity—is why the landscapes that sediment fosters are so valuable. They are valuable because of how they answer these questions. They are valuable because they are landscapes worth living in. They are valuable because they affirm life itself.

We should invest now in the sedimentary landscapes that will facilitate this transition. Focusing energy and funding on making marshes, augmenting dunes, and returning floodplains to rivers means shifting away from fracture-critical,[85] too-big-to-fail infrastructure, and toward gradients of risk, habitation, and investment that are more appropriate to the dynamism of a coast or a river. The investments we make now will either move us toward a better-adapted, more just, aesthetically rich future—or continue to drag us down an unsustainable path.[86] Ultimately, we should not invest in living infrastructure like sedimentary landscapes because they protect exactly the way we live now—we should invest in them because better worlds are possible.

* * *

Climate change is not the only accelerated parameter of environmental change, though. The twentieth century was—and, so far, the twenty-first

82 Lane, "Natural Flood Management."

83 Bridges et al., *International Guidelines*.

84 Bridges et al., *Use of NNBF*.

85 Fisher, *Designing to Avoid Disaster*.

86 Goh's *Form and Flow* offers a critical account of what she calls "a political ecology of design," focusing on how "contesting visions of urban futures" are produced and how those visions relate to the power to implement or bring them into being. These processes are crucial to whether the climate transition will be just. Similarly, in "For Whom Do We Account in Climate Adaptation?" M'Closkey and VanDerSys discuss in detail one example of the local politics of adaptation of design, drawing from their participation in the Resilient by Design Bay Area Challenge as members of the Bionic team.

century continues to be—a period of stark reversal in the vector of human influence over sediment. For the majority of human history, agriculture and deforestation were the most significant ways that people affected sediment flows. Both of these increase erosion, speed up the movement of sediment, and correspondingly increase sediment loads in downstream waterbodies, including coastal estuaries. In the Chesapeake Bay, for instance, European settlers increased sedimentation rates in estuaries to roughly twenty-five times the pre-settlement rate by aggressively expanding, cutting forests, and employing few soil conservation techniques.[87] The record of this period can be seen today in former colonial ports, like Dumfries in Virginia and Joppatowne in Maryland, that were abandoned as ports when they silted in. (Globally, the story is similar. For example, a number of major river deltas, including the post-1855 Yellow River Delta, the Po, the Ebro, and the Colorado in Texas, were formed by the redirection of rivers by human societies.[88])

But by the early twentieth century industrialization took firm hold of the United States and dam construction spread across the continent.[89] The same new industrial materials like steel and mass-produced concrete used to build those dams also facilitated increasing control of river systems through the construction of river diversions, bank armoring, and the building of massive levee systems, like those on the Mississippi and Sacramento Rivers. (These processes have since spread globally, particularly to the major river basins of Asia, Africa, and South America, where activities like dam construction continue to the present. Worldwide, there were more than 48,000 large dams cataloged in 2011, with another 2,000 under construction at that point. The great majority of these were constructed after the 1940s.[90]) The cumulative effect of dam-building and river-control has been to radically constrain sediment supply, both in the United States and globally. While human influence had once so increased sediment supply as to fill in ports and build new deltas, our collective influence has now decisively shifted toward restriction.[91] At the same time that sediment is in increasing demand as the material substrate for bolstering shorelands, supply is plummeting.

More broadly, the pace of accelerated environmental change is a key dimension of ecological crisis. Sea-level rise is not a new phenomenon; the rates driven by anthropogenic climate change are. Similarly, variability in sediment supply is not new, but the speed and scale of the intense glut and subsequent starvation that characterizes many human-altered sediment systems is novel. Many other important aspects of ecological systems have faced similar accelerations in the past century: methane levels, coastal nitrogen loads, tropical forest loss, ocean acidification, air and waterbody temperatures, species extinctions. This has been directly conjured by a parallel acceleration in anthropogenic actions, from the spread of technological objects like motor vehicles and computers to fishing and deforestation. Global urbanization has grown incredibly dense in so-called "megaregions" and spread tendrils of transport, extraction, and waste disposal across the vast majority of the earth's surface—including, through dredging, waterbodies and waterways.[92] While ecological systems possess great capacity to adapt (or migrate) in response to environmental change, this contemporary pace, labeled by some scientists as the "Great Acceleration," has pushed (and continues to push) many ecological systems beyond their breaking points.[93] The changes are too many and too fast.

87 Pasternack, Brush, and Hilgartner, "Impact of Historic Land-Use Change."

88 Syvitski and Saito, "Morphodynamics of Deltas."

89 Syvitski and Kettner, "Sediment Flux."

90 Syvitski and Kettner, "Sediment Flux," 961.

91 This trend is exacerbating the consequences of sea-level rise: "since dam construction, regional reforestation, and agricultural sediment-control practices continue to lower sediment yields to the coast, we suggest extant coastal wetlands, many of which formed during periods of high sediment deposition in the past, became increasingly prone to ecological collapse as sediment supplies declined."
Kirwan et al., "Limits on the Adaptability."

92 The infrastructural, landscape-making shockwave generated by the expansion of the Panama Canal offers a case-study in how these effects are networked and tied to economic systems.
Davis, Holmes, and Milligan, "Isthmus."

93 McNeill and Engelke, *The Great Acceleration*.
Steffen et al., "The Trajectory of the Anthropocene."

With ecological crisis as an increasingly evident physical reality has come ecological crisis as a corresponding social reality. Regions like the four coasts studied in this volume are widely *perceived* and *described* as being in a state of crisis. Advocacy organizations and publics speak out on behalf of landscapes, protesting the degradation of California's Salton Sea, algal blooms in Lake Erie, and salt marsh loss on the Gulf Coast from Texas to Florida. Governments, agencies, publics, and designers are responding with programs for large-scale ecological restoration, like the Comprehensive Everglades Restoration Plan and the Chesapeake Bay Program, which include substantial infrastructural components.[94] Across the United States, the enrollment and construction of sedimentary landscapes as part of efforts to adapt to the consequences of both sea-level rise and other dimensions of ecological crisis is driving demand for sediment upward and prompting the revaluation of sediment as a resource. Mud has become vital.

4. Silt, Sand, Slurry, Landscape

To focus our study, let's return now to dredging, to see some of the ways that sediment is being deployed to address these interlocking crises.

Historically, dredged sediment was viewed primarily as a waste material. Once dredged, it was described as "spoils," much like the overburden wasted in mining for precious minerals. This characterization placed dredge-formed landscapes in a maligned category together with other wastelands, like toxic dumps and (supposedly) fetid swamps.[95] Waterways around the United States are marked by the remnants of this era of dredging. Florida's Indian River Lagoon, for instance, which stretches from Cape Canaveral south through four coastal counties to the St. Lucie River, is dotted with 137 "spoil islands," the product of the dredging of the Atlantic Intracoastal Waterway from 1951 to 1961.[96] While the Lagoon's spoil islands have become cultural and ecological resources, studied for conservation purposes, managed for habitat, and visited by campers, those valuations have been retroactively developed.[97] The initial placement of the spoils was guided by the logistics of dumping and the expedience of piling spoils quickly along the waterway as it was dredged, not any kind of choreography for future use.

Today, agencies and organizations that manage sediment perceive dredged sediment differently, as a resource that can be deployed with purpose and to a variety of benefits, encapsulated in the term of art "beneficial use." The Army Corps' official guidance prioritizes beneficial use, port operators seek to maximize it, environmental advocates find opportunities within it, and dredging companies see it as a growing market that can reverse the perception of dredging as environmentally harmful.[98] The growth of beneficial use demonstrates the revaluation of sediment

In the 1970s, the Army Corps and its partners began a range of experiments in using dredged material to build habitat. A pair of programs at the US Army Engineer Waterways Experiment Station were established to study the efficacy of these efforts: the Dredged Material Research Program (1973–1978) and Environmental Effects of Dredging Programs (1979–1987).[99] At sites like the Southwest Pass in Louisiana, Apalachicola Bay in Florida, and Miller Sands on the Columbia River in Oregon, the Corps experimented with reshaping

94 Doyle and Drew, *Large-Scale Ecosystem Restoration*.

95 Di Palma, *Wasteland*.
Berger, *Drosscape*.

96 "Spoil Islands," Florida Medical Entomology Laboratory.

97 Hailey, *Spoil Islands*.

98 EPA, "Dredged Material Management Action Agenda."
EPA and USACE, "The Role of the Federal Standard."

Beneficial use has also been enshrined in federal law, as for instance both Section 204 of the 1992 Water Resources Development Act (WRDA) and Section 207 of WRDA 1996 "encourage USACE to incorporate beneficial uses of dredged material into constructing, operating, and maintaining its Civil Works navigation projects."
EPA and USACE, "Identifying, Planning, and Financing."

99 USACE, "Long-Term Monitoring."

100 USACE, "Long-Term Monitoring," 155–58.

101 In these examples, beneficial use involves directly deploying sediment to build or sustain ecologically functional landscapes. But that isn't always the case. The recycling of sand from Erie Pier by private contractors as a construction material constitutes a beneficial use, too. In fact, there are a wide range of economically viable practices like this, involving the excavation of dewatered and sorted sediments. Some, like manufacturing topsoil for agriculture, capping contaminated brownfield soils, or reclaiming abandoned surface mines, can even be said to be ecologically beneficial. Others, like the use of dredged sand and gravel in roadway construction, contribute to the making of landscapes, but without clear ecological benefits. These practices abstract and commodify dredged sediment. They emphasize economic viability, rather than ecological viability. Even so, they also are a part of the shift in understanding of sediment from waste to resource.
EPA and USACE, "Identifying, Planning, and Financing."

102 "Beneficial Uses of Dredged Sediment," DOTS, USACE.

emplaced dredged material to form landscapes with ecological value. At Miller Sands, this meant reshaping an island in the tidal reaches of the lower Columbia River into a series of new landforms—wetland, upland meadow, and dune. This horseshoe-shaped island had been built in 1932, and gradually expanded over time through the placement of dredged material roughly every four years. From 1974 to 1975, this sandy surplus was regraded, prepared as seedbed with farm machinery, and planted with appropriate vegetation, ranging from European beachgrass for the dune to native wetland species like soft rush and slough sedge for the wetland.[100] In Apalachicola Bay, an island confined disposal facility was reconfigured to accept tidal flow, filled to an intertidal elevation with dredged material, and planted with both smooth and saltmeadow cordgrass. The resulting wetland, instigated in 1976, persists to the present. And in 1970, at the Southwest Pass of the Mississippi River, the New Orleans District began its practice of "pumping unconfined dredged material into shallow water areas" to form land—the same practice described at the very beginning of this book.

Such actions, where material dredged from navigation channels is placed with ecological intent, remain the core of beneficial use practices today.[101] From a handful of isolated experiments in the 1970s, beneficial use has expanded to encompass roughly a third of the non-contaminated sediment dredged from federal navigation channels in the United States, or, around 60 million cubic yards a year.[102] The landscapes built (or altered) with this sediment are diverse and broadly distributed around the United States. Port expansions produce enormous one-time surpluses of sediment; from New York's Jamaica Bay to Houston's Galveston Ship Channel to Los Angeles' Upper Newport Bay, these surpluses are increasingly directed toward the construction of new islands and marshes, intentionally formatted for habitat value. On-going maintenance dredging typically provides smaller but steadier streams of sediment, not only to major coastal ports, but also to smaller harbors, like those on the Great Lakes, and inland waterways, like the upper reaches of the Mississippi River or the Tenntom Waterway, which connects the Tennessee and Tombigbee Rivers across Alabama, Mississippi, and Tennessee. Some of the landscapes produced through beneficial use are novel, often new islands like Poplar Island in the Chesapeake, Gaillard Island in Mobile Bay, or Blue Water Atoll in Galveston Bay. In other cases, dredged sediment is used to augment an existing landscape, like the "strategic placement" of sediment as feeder berms close to the shoreline of eroded beaches, or to rebuild a vanished (or vanishing) landscape, as with many of the beneficial use projects that are a part of Louisiana's Coastal Master Plan.

To understand the current state of practice in detail, let's look more closely at three recent examples of living infrastructure built using dredged sediments.

* * *

Move south along the Atlantic coastline from the Hudson-Raritan estuary of New York and New Jersey, and the next major estuary you come to is Delaware Bay. Continue south, and you come to the Chesapeake, which is the largest estuary in the United States. A pair of major ports lie in the Chesapeake: the Port of Virginia, in Hampton Roads, at the southern end of the estuary; and the Port of Baltimore, on the western shore of Maryland, at the northern end of the estuary.

Both Baltimore and the Port of Virginia have a number of major confined disposal facilities associated with them. Virginia's Craney Island, for instance, has three enormous cells, each of which is roughly the same size as New York City's Central Park. For Baltimore, though, dredging and accompanying disposal sites have historically been not only concentrated around the port's actual harbor, but also strung out along the length of the Chesapeake, between the mouth of the Chesapeake in the south and the port in the north.[103]

About fifteen miles south of the Chesapeake Bay Bridge, a towering landmark that connects Annapolis with the Eastern Shore, the newest of these disposal facilities has recently been marked out, diked, and filled with dredged sediment: Poplar Island.

Poplar Island is a new island, in that the material that forms its dikes, uplands, and wetlands has been placed there over roughly the past decade, after being dredged out of Chesapeake navigation channels. But Poplar Island is also not a new island, in that this newly emplaced material has been laid overtop the rough footprint of an existing island.[104] This earlier Poplar Island was, in the early nineteenth century, a long island of over a thousand acres, covered in pine forest. But erosion and sea-level rise ate away at it through the nineteenth and twentieth centuries, so that it first split into three smaller islands and then, by the 1990s, had almost entirely disappeared.[105] Poplar Island's circumstances were not exceptional: the Chesapeake Bay has experienced substantial and accelerating island loss in recent decades. Some 10,000 acres of "remote island habitat"—islands like Poplar Island—have been lost in the Chesapeake over the past century and a half.[106]

This disappearance coincided with the need for the Port of Baltimore and the Corps of Engineers to find new destinations for the sediment they were dredging out of Chesapeake channels. One motivation for this was, as it has been in Duluth, opposition to unconfined open-water placement, which culminated in the 2001 prohibition of open-water placement by the state of Maryland.[107] Another was, again similar to Duluth, that the only major upland placement site near the port, Hart-Miller Island, was nearing full capacity. (It closed at the end of 2009; two smaller facilities, Masonville and Cox Creek, have continued to be operated by the Maryland Port Authority, but they are not capable of handling all of even the local sediment dredged from Baltimore harbor itself, much less the Chesapeake channels.[108]) Dredging was not slowing with the loss of these disposal options, though; around 4.5 million cubic yards of sediment is dredged from the upper Chesapeake every year.[109]

Placement at Poplar Island linked this on-going need to dispose of material being dredged for economic reasons with an emerging mandate for ecological restoration. The Chesapeake Bay Program, begun in 1983, is one of the United States' largest programs of ecological restoration.[110] Slowing or reversing the presumed vector of wetland loss is one of the major aims of the program.[111]

Poplar Island provided a place to do that, and dredged sediment provided the material to do it with. Construction began with a perimeter dike, initially a line of stone over seven miles long, which enclosed the remaining nubs of the earlier Poplar Island and set the limits of the future Poplar Island. With an interior closed off from the bay, infill began, a gradual and incremental process that continues to the present. A key part of this process of infill is the construction of interior dikes, which divide the island into seventeen cells. Five upland cells,

103 This route up the Chesapeake is not the only major shipping route to Baltimore. Just north of Baltimore, the Chesapeake and Delaware Canal cuts across the narrow neck of the Delmarva Peninsula to the Delaware River and, via the Delaware, the Atlantic Ocean. The C&D is more shallow though, and so larger ships headed for Baltimore must traverse the 50-foot-deep channels of the Chesapeake.

104 As with the marsh islands of Jamaica Bay, a particular historic footprint has been selected. In the case of Poplar Island, it is the 1847 footprint.
Dalal, Baker, and Mason, "Environmental Assessment of Poplar Island."

105 Erwin, Miller, and Reese, "Poplar Island Environmental Restoration Project."

106 Derrick et al., "Poplar Island."

107 Poltrack, "The Maritime Industry and Our Environment," 69.
Derrick et al., "Poplar Island," 488.

108 Francingues, Thompson, and Hamons, "Exploring Innovative Reuse."

109 Staver et al., "Tidal Marsh Restoration."

110 Wolflin, "Chesapeake Bay Program."

111 There is some real debate as to whether the Chesapeake Bay is losing wetlands. Certainly its wetlands are migrating, but some researchers argue that the gain from inland migration is actually exceeding the loss at water's edge. (See for instance the work of Matt Kirwan at the Virginia Institute of Marine Sciences and the forthcoming essay "Bay Migrations" by Brian Davis and Erin Putalik for *Places Journal*.) Regardless, the perception of wetland loss has contributed substantial impetus to the Bay Program.

on the west side of the island, will receive enough dredged material to raise them 20 feet above low tide, while wetland cells on the protected eastern side will be filled to tidal and subtidal elevations. As the upland cells are filled, the dikes around and between them must be gradually raised. With the perimeter established in stone, this can be accomplished cheaply by excavating sand from the bay floor (initially from a borrow area beneath the future upland cells).

The filling itself, of course, is done with dredged sediment. Around a thousand acres are enclosed by the initial perimeter dike, and almost 40 million cubic yards of sediment has been placed. It arrives at the island on barges, where it is slurried with bay water and pumped into destination cells. A series of measures, including gravitational flow, weir-regulated drainage, ditching, and evaporation are used to dewater and facilitate the consolidation of the slurried sediment. This process can take a long time, typically between three and eight years for each cell.[112]

For the wetland cells, further sculpting follows consolidation and precedes planting: the surface is carefully graded to the intended range of tidal and subtidal elevations, and designed marsh channels are incised into this surface in imitation of the drainage patterns that form in ordinary marshes.[113] Culverts are opened in the perimeter dike, connecting to these designed channels; the tide flows in and out through them, while the dike road bridges above. Finally, marsh grasses—*Sporobolus patens* in the high marsh, *Sporobolus alterniflora* in the low marsh—are planted, between half a meter and a meter on center.[114]

The resultant island is a mosaic, clearly artificial by virtue of the simple, rounded lines of the perimeter dike, but in continual transformation and seemingly perpetual construction. Dredged sediment cracks in the sun as it dewaters in some cells, while others teem with life as tides rise and fall within them. Recently, the island was extended substantially to the north, another five cells enclosed, along with a large, breakwater-protected embayment that contains a constructed rock reef. The Corps anticipates that this will give the island the capacity to continue to receive an eventual grand total of nearly 70 million cubic yards of dredged material, staying open through a projected completion date of 2044.[115] The island should eventually include 737 acres of wetlands, 840 acres of uplands, and 138 acres in the open-water embayment.[116]

A broad range of animal species have been observed in these constructed habitats. Key species of concern, including the Least Tern, Common Tern, Snowy Egret, Osprey, and diamondback terrapins, have established successful nesting populations, often taking advantage of landscape features designed specifically for them, like the circular, moat-encircled sand-shell nesting islets that have been embedded in many of the wetland cells.[117] Both large predator finfish and forage fish species have established substantial populations in the island's marsh cells and adjacent waters.[118]

The state has also opened the island to the public. Initially, this primarily meant through tours. In one early year about 3,500 members of the public toured the island, guided by staff from the Army Corps of Engineers and the Maryland Port Administration.[119] Since then, public engagement has expanded to include planting the wetland cells with marsh grasses in events organized by the National Aquarium.[120] The Aquarium also runs a program in Maryland schools, "Terrapins in the Classroom," for which hatchling terrapins are collected on Poplar Island in the fall, monitored and cared for at schools around the state,

112 Cornwell et al., "Tidal Marsh Restoration."

113 "Two natural marshes (Hog Island and Chesapeake Bay Environmental Center) were used as a reference for the design of channel morphology in the Poplar Island Marshes." Taddia et al., "High-Resolution Monitoring," 1364.

114 Cornwell, "Tidal Marsh Restoration," 1,674.

115 USACE, "Paul S. Sarbanes Ecosystem Restoration Project."

116 NOAA, "Water Level Variations at Poplar Island, MD."

117 Derrick et al., "Poplar Island."
Erwin, Miller, and Reese, "Poplar Island Environmental Restoration Project."

118 Derrick et al., "Poplar Island."

119 Ibid.

120 "Restoring Valuable Habitats," National Aquarium WATERblog.
"Site Update: Poplar Island," National Aquarium WATERblog.

and then returned to Poplar Island the following spring.[121] Eventually, it seems likely that the island will be opened more fully to the public, as construction activities recede and become limited to the newer northern cells. Perhaps kayakers and recreational boaters will be able to land on the island; perhaps the access roads on the dikes will become walking paths, and new trails will be marked out through the interiors of the upland cells. In the meantime, shorebirds, crabs, and terrapins share the island with excavators and pumping slurry pipes.

* * *

San Francisco Bay suffered a catastrophic loss of wetlands in the late nineteenth and early twentieth centuries. Nearly 90 percent of the wetlands of its larger estuarine complex—the San Francisco Bay-Delta, which includes San Francisco Bay proper, San Pablo Bay, Suisun Bay, and the Sacramento-San Joaquin River Delta—have been lost, being either filled in ("reclaimed") or diked and cut off from tidal action.

One such landscape is the former Hamilton Army Airfield. It sits on the west side of San Pablo Bay, which in turn lies between the main body of San Francisco Bay, to the south, and Suisun Bay, to the east. The Napa and Petaluma Rivers drain their watersheds, which include the wine-producing slopes of the Napa Valley, directly into San Pablo Bay, and the historically sediment-rich waters of the Sacramento and San Joaquin Rivers flow westward into San Pablo from Suisun and the delta. The Airfield was one of a number of early military airbases in the Bay Area, established in the years between World War I and World War II, as military aviation grew in importance. By the 1970s, though, the old runway was largely obsolete, and in 1988 plans were made to close the base.

This closure proved to align well with a mandate for wetland restoration in San Francisco Bay. In 1993, the San Francisco Estuary Project—a collaborative effort established by the Environmental Protection Agency in 1987—published its final report, the Comprehensive Conservation and Management Plan, which marshaled federal, state, and local support behind a "comprehensive, Estuary-wide plan to protect, enhance, restore, and create wetlands in the [San Francisco] Estuary."[122] This led directly to the establishment of the San Francisco Estuary Institute, an independent scientific organization that has been central to the wetland restoration mandate in the Bay Area, and the launch of the San Francisco Bay Area Wetlands Ecosystem Goals Project, a broad, interdisciplinary effort to establish specific goals for regional wetland restoration. In 1999, these initial efforts culminated in the release of the first Baylands Ecosystem Habitat Goals report, which set a target of a net increase of at least 55,000 acres of tidal marsh, which would more than double the amount of tidal marsh in the Baylands.[123] This goal continues to guide bay wetland restoration today.

Before the report was released, plans were already in motion to return the Airfield's land to the waters of the bay. By 2003, when the bulk of the Airfield's land area was given to the California State Coastal Conservancy, the landholder that now stewards the emerging wetland complex, a delicate institutional interchange was already largely complete.[124] The most crucial component of this interchange linked two of the Army Corps' federal missions: navigation and ecosystem restoration.

121 "Terrapins in the Classroom," National Aquarium.

122 Goals Project, "Baylands Ecosystem Habitat Goals," 5.

123 Goals Project, "Baylands Ecosystem Habitat Goals," 93.
Goals Project, "The Baylands and Climate Change," xxviii.

124 Keller, "Where Once there was a Runway."

125 USACE, "Hydrodynamics of a Recently Restored Coastal Wetland."

126 Kress et al., "Alignment of US ACE Civil Works," 204.

127 Keller, "Where Once there was a Runway," 54.
"A Community Project Supporting Stewardship," School Fuel.

128 USACE, "Hydrodynamics of a Recently Restored Coastal Wetland," 64.

At the same time that the Bayland goals were being developed, another major institution, the Port of Oakland, was planning a very different form of growth: deepened navigation channels, to be dredged to the fifty-foot depth that would accommodate the massive post-Panamax vessels that have increasingly come to dominate international shipping. This expansion dredging was projected to generate a large sedimentary surplus.

For Hamilton Airfield to become Hamilton Wetlands, that surplus would be needed. When the wetlands that had existed on the site in the nineteenth century were diked and converted into agricultural land, they were cut off from bay sediment supply. Marshes dried up, their soils began to oxidize, and no new soil formed. Consequently, they subsided substantially over the course of the twentieth century. It was projected that over 9 million cubic yards of material would be needed to reverse the effects of subsidence and bring the airfield landscape back up to an intertidal elevation.[125]

Sediment dredged for Oakland's channel deepening has provided the bulk of that. Slurried sediment was pumped into a set of placement cells that had been constructed behind the existing bayside levee at the airfield. These cells were divided by new, partially temporary berms that wiggled across the straight line of the airport runway, outlining a new watery order that began to overwrite the landscape's recent past. Satellite photos taken between 2007 and 2013 track this transformation. By 2013, the airfield was fully submerged; the Hamilton Wetlands Restoration Project has taken its place. In 2014, with the new sedimentary landforms raised to a mixture of elevations conducive to the formation of tidal marsh, ponds, and higher seasonal wetlands, the next step was to breach the outer levee, which had divided this landscape from San Pablo Bay since reclamation.[126] When the excavators cut the levee, baywater flooded in; the new wetlands were now tidal.

The moment of breach was not the conclusion of the restoration project, though. Rather, it marked a shift from a period when the work of making a wetland was done by dredges, pumping boosters, contractors, and bulldozers to a period when the work is being done by local volunteers and school groups, by sediment carried on the tides, and by pickleweed, bulrush, and cordgrass. With the breach, conditions were suitable to begin transplanting efforts, which have included constructing a plant nursery on-site and engaging volunteers, including both school groups and members of the general public, in planting.[127] The work of tide-borne sediment and plants themselves is more ambiguous, but, at least after the first seven years, appears to be progressing positively. The expectation of the designers of the restoration project was that opening the landscape to the tides would permit the new industrially shaped terrain to reshape itself in the kind of fractal complexity of channels, sloughs, and flats that characterizes historic Baylands; capture sediment drawn in on the tides; and accrete to gain elevation. Some channel formation is clearly taking place, and initial studies indicate that wind, waves, tides, and sediment seem to be interacting to generate a "new inflow of sediment" into the wetlands. A series of smaller berms constructed within the largest placement cells—wiggles within wiggles—were designed to facilitate just this, by subdividing the cells into "disjointed pools" and thereby limiting fetch, or the distance available for wind to generate waves by moving across open water, which in turn limits resuspension and erosion within the nascent wetlands.[128]

Today, Hamilton Wetlands are very much in a social, tidal, and ecological process of becoming. A landscape has been restructured, with the aim of launching it into a new state, where it is capable of self-propagation; but an experiment like this cannot be fully predicted, and so its managers wait and monitor, watching and learning.[129]

* * *

While the Mississippi River Delta as a whole is losing land rapidly—the oft-quoted rate is "about one football field of land every hour"—there are two smaller deltas on the Louisiana coast that are both growing.[130] Both of these deltas are linked to the Mississippi's largest distributary, the Atchafalaya River. One is at the river mouth; the other, the Wax Lake Delta, is located just to the west of the Atchafalaya River Delta, at the mouth of a diversion channel that helps protect low-lying Morgan City when the Atchafalaya floods. Though they are not growing nearly as fast as the larger delta is shrinking, they are growing substantially: the Atchafalaya adds a football field of land about every 11 hours, Wax Lake about every 8 hours.

The formation of this new land is indicative of the volumes of sediment carried by the Atchafalaya River. While it does not carry as much as its parent river, it still carries a massive volume relative to most other North American rivers. And, because Morgan City is a significant port for coastal Louisiana, the Corps has been maintaining—and occasionally deepening—a navigation channel through the Atchafalaya to Morgan City's docks since it was authorized by the Rivers and Harbors Act of June 25, 1910.[131] This channel, and the Atchafalaya River, run roughly perpendicular to the Gulf Intracoastal Waterway, that major commercial vein that stretches from the far southern reaches of Texas' Gulf Coast east through Louisiana, Mississippi, and Alabama to Florida. A further dredged channel runs north from Morgan City, roughly parallel to the Atchafalaya River, along the levee at the eastern edge of the Lower Atchafalaya Basin Floodway System. This channel exits the floodway in a lock and makes its way northeast to Port Allen, on the west bank of the Mississippi across from Baton Rouge, linking Morgan City and the Gulf Intracoastal to the great navigation system of the Mississippi River & Tributaries Project.

To sustain these links, the New Orleans District manages a great deal of dredging in the vicinity of Morgan City. This averages, in recent years, around 13 million cubic yards of dredging per year. This is more sediment than is dredged from any other port or river on the Gulf Coast save New Orleans and the main stem of the Mississippi River. Even much larger regional ports like Houston, Savannah, or Charleston dredge substantially less. Only two other Army Corps districts (Galveston District and Mobile District, both on the Gulf) dredge more in a typical year across their entire districts than is dredged from the channels of the Atchafalaya alone.[132]

The final bend in the Atchafalaya River before the river splays out into its delta is called Horseshoe Bend. Until recent decades, Horseshoe Bend itself was not dredged, because the river channel was deeper than the authorized depth of the navigation channel. But around 1990, the Bend began to fill in and the Corps began to dredge. And once it began to dredge, the quantities of sediment

129 Note that the airfield—the subject of the description here—is only a part of the larger Hamilton Wetlands project, which was expanded from the original area of the airfield to its larger current size of almost 2,600 acres by the acquisition of Bel Marin Keys Unit V, "a 1,612-acre parcel of neighboring farmland to the north" of the airfield. The larger project involves a much larger volume of sediment, over 24 million cubic yards.
Kress, et al., "Where Once there was a Runway," 203.

130 "NASA Studies a Rarity," NASA.
"World of Change," NASA Earth Observatory.

131 EPA and USACE, "Atchafalaya River Bar Channel."

132 These numbers are based on dredging events recorded in data compiled by the US Army Corps of Engineers Navigation Data Center (http://www.navigationdatacenter.us/dredge/drgcorps.htm).

133 USACE, "A Multifactor Ecosystem Assessment," 4.

134 USACE, "A Multifactor Ecosystem Assessment," 5.

135 Ibid.; also Suedel et al., "Creating Horseshoe Bend Island."

136 Foran et al., "Quantifying Wildlife and Navigation Benefits."

137 USACE, "A Multifactor Ecosystem Assessment."

removed, even from this single stretch of river, were vast: between half a million and 2.5 million cubic yards in each dredging event, with events almost every year.[133]

For the first decade, this sediment was placed alongside the river, on eight "wetland development sites"—essentially, stretches of adjacent wetland that were low enough that placing sediment on them could benefit them ecologically, rather than converting them into upland.[134] But by the end of the decade, room was running out. As with Erie Pier in the Great Lakes, as with the revaluation of sediment more broadly, exhaustion of one option prompted the exploration of new possibilities.

The New Orleans District considered three options.[135] First, some or all of the adjacent wetland placement sites could be converted, perhaps by diking, into upland placement sites. Doing this, however, would not only be considerably more expensive than the existing practice of wetland disposal, but could also be seen as ecologically destructive, as any new upland created would come at the expense of existing wetland. Second, a long-distance pipeline could convey dredged slurry from Horseshoe Bend out to the Atchafalaya Bay channel, where a pair of existing ODMDS, or Ocean Dredged Material Disposal Sites, could readily receive vast volumes of sediment. Long-distance pumping, though, rapidly escalates dredged material disposal costs, and is rarely employed for that reason, unless the benefit of transporting sediment outweighs the cost—and there would be no additional benefit to ODMDS dumping.

This left the third option, which was put into practice beginning in 2002. From then, each time Horseshoe Bend was dredged, the dredged sediment was placed in a "midriver open-water placement area," to achieve what the Corps calls "strategic placement."[136] This began with building new but intentionally temporary islands by piling sediment right in the river, to the side of the navigation channel. These piles were placed roughly a mile upriver of a tiny island, now named Horseshoe Bend Island, that had begun to form already as a result of the river's sediment surplus. Each time these new islands were constructed, the river swiftly eroded them, moving much of the eroded sediment south, where it rapidly accelerated the growth of Horseshoe Bend Island, as the Corps engineers had hoped. Rather than placing dredged sediment in a fixed location, then, this third option was indeterminate: the sediment would be placed in each dredging event, but it would be placed with the full knowledge that the river would carry it to a new destination. The general intent of that movement was anticipated, but the particulars—precisely what sediment ended up where, and what landform it would produce when it arrived—were open.

Today, Horseshoe Bend Island has grown to around a hundred acres in size, nourished by this process. A diverse range of ecological communities have formed atop the substrate.[137] Nascent forested broadleaf wetlands, with canopies already twenty to thirty feet high at the time of a series of field surveys in 2013–14, occupy the highest elevations. In these forests, black willows, Chinese tallowtree, and bald cypress have taken root in the island's highly saturated soils, and are occasionally flooded, particularly when the river peaks. The forests are surrounded by scrub-shrub wetlands, where younger willows intermingle with wax myrtle and buttonbush. Emergent wetlands and aquatic beds of vegetation follow at lower elevations, either frequently submerged (emergent wetlands) or permanently inundated (aquatic beds); both are lushly vegetated, with emergent species like cattail and aquatic species like common water hyacinth, alligator weed, and duckweed. This range of habitats supports a broad array of fauna, from colonies of nesting birds like great egret and glossy ibis to a dense collection of infauna, the aquatic invertebrates that live in the soft, submerged substrates along the edges of the island and in the aquatic beds.

All of this seems to have been facilitated by the choice to permit the river and dredged sediment to, in the words of one article published in *Terra et Aqua*, "self-design the island over time."[138] This choice was, in part, logistical and economic. When dredging, it is simplest and cheapest to place sediment in one or two locations, immediately adjacent to the channel being dredged. The temporary piles accomplished this. But this exigency proved aligned with new interest in ecological performance. Permitting the currents to form the island has resulted in landform that is shaped by and thus fitted to actually existing hydrologic conditions, achieving a complexity and variability that would be tremendously difficult to replicate with the blunt instruments of mechanized earthmoving. Comparative ecological surveys seem to bear this out, as they have found that Horseshoe Bend Island is more biodiverse—higher plant species richness, more wading bird nesting colonies, higher infaunal abundance—than Island E, a "traditional dredged material supported wetland island" that lies a mile upriver.[139]

* * *

These three prototypes—Poplar Island, Hamilton Wetlands, and Horseshoe Bend—have been presented according to a particular gradient, from the project with the greatest degree of formal determination to the project that is most open to the influence of landscape processes. At Poplar Island, dikes were constructed first, in closed loops; then those dikes were filled with sediment, and plants were established to shape the ecological character of the island. At Hamilton Wetlands, a once-diked landscape was elevated with dredged sediment, opened to tidal processes, and the landscape was intentionally shaped to facilitate the future accumulation of sediment and, with that accumulation, further growth and transformation. Finally, with the greatest openness, the sediment placed at Horseshoe Bend has been allowed to "self-design," moving between placement site and growing island according to the dictates of flowing water.

This set of choices, of how much will be determined with designed lines and the input of mechanized energy, runs consistently through not just these three projects, but sediment design generally. Sediment choreography is gradually shifting in favor of facilitation and openness. Bulldozers and booster pumps are giving way, in many places, to winds and tides. But it will need to shift much faster. Adaptation to accelerated landscape change will require so many more landscapes like these. We'll need almost unimaginable amounts of sediment to make and sustain them, and we won't be able to rely on industrial machines and fossil fuel energy to move all—or even most—of that sediment.[140]

On these terms, Horseshoe Bend Island is the most transformative. It most clearly figures a necessary post-carbon future, where landscapes are transformed by different means than we typically use today. It is an example of a new, interwoven ecological economy of sediment, where a by-product of

138 Suedel, et al., "Creating Horseshoe Bend Island."

139 Ibid.

140 This is not to say that we should expect to remake all landscapes without machines, nor that loosening control in this fashion is equally desirable or equally possible in all landscapes. Site and social contexts vary substantially, and affordances differ. Recognizing those differences and working intelligently with them is crucial to choreography.

141 "Port of Morgan City," Port of Morgan City.

142 Aucoin, "Louisiana Ports."

143 Haff, "Technology as a Geological Phenomenon." See also discussion in Purdy, *This Land is Our Land*, 76–101.

economic activity becomes an ecologist catalyst. But, as we will soon see, it is also directly tied to the oil economy that is primarily responsible for climate change. (And not only because its sediment is still first mobilized by dredging.) In this paradox, Horseshoe Bend Island points us toward the way that all sediment design is, ultimately, entangled with much larger questions and decisions about our collective planetary future.

5. Crisis and Transformation

Every port specializes. Geography and transportation infrastructure determines much of this: what goods are in demand along the rivers, roads, and railways that stretch inland from a given port; what competing ports might offer easier access to an industry or a consumer base; what materials, resources, and products are likely to be exported through those particular docks.

The sediments that have formed Horseshoe Bend Island were only emplaced because of the desire to maintain the navigation channel to Morgan City, and Morgan City is no exception to the rule of specialization. The names of the companies that own the slips, docks, supply yards, and warehouses along Railroad Avenue make this clear: Oilfield Warehouse & Specialties, General Oil Field Trucking Incorporated, New Offshore Incorporated, Oceaneering International Incorporated, Shell Oil. Morgan City is, as the Port of Morgan City's website proclaims, "the birthplace of the offshore oil exploration industry."[141] Today, it serves primarily as a supply hub, logistics base, and fabrication center for the oil and gas industries.[142]

Insofar as climate change, accelerated by the extraction and combustion of fossil fuels, is the most pressing component of the interlocking planetary environmental crises that we currently face, then, Horseshoe Bend Island has to be understood not only as a design demonstrating choreographic ingenuity, but also as a part of the larger assemblage of infrastructures—the "material technosphere"—that is unwittingly imperiling life on our planet.[143] This is not to discard or discount that ingenuity, nor to set aside the genuine worth of the vibrant lives that willow, wax myrtle, and egret are making on the island. But it does make clear that innovation in sediment design alone cannot resolve the tension between our contemporary economy and life on Earth.

* * *

Dredged navigation channels largely follow the paths of older watercourses, which have long been central to human networks of trade and transportation. The same routes that modern barges and container ships now ply were once used by sailing ships, steamboats, log rafts, and canoes. Further into the past, the rivers, estuaries, and inlets those lines are traced across were originally formed by the erosive work of water and became migration routes for species like salmon or sturgeon, home waters for others. Today's navigation routes are supported by many other infrastructure networks. GPS navigation systems rely on orbiting satellites to precisely calculate ships' positions. Pipelines convey and tanks store petroleum that fuels ships' engines. Rail lines and roads fan outward from the ports that ships unload at, interlocking at intermodal switching yards. Mines provide the metal that shipyards craft vessels out of, and shipbreaking sites—in Brownsville, Texas, on the shores of Arabian Sea in Alang, India—tear those vessels apart again for their metal when they reach the end of their service lives. Such networks of physical connection can be traced for any infrastructure. No infrastructure exists in isolation.

Infrastructure networks are not only physical though. They are also thoroughly social. Corporate employees sit at desks with spreadsheets and balance books, shaping the logistical calculus that directs the movement of goods and, in turn, the production of infrastructure to facilitate that movement. Government officials review Environmental Impact Statements, determining whether a new dredging project will violate existing laws and regulations. Politicians seek corporate donations and popular support, weighing their statements and votes accordingly. Residents on a barrier island complain that the adjacent navigation channel is trapping the sand that would otherwise nourish the beaches in front of their homes. Across both physical and social dimensions, infrastructures participate in continual processes of *world-making*, simultaneously shaping the worlds they are entangled with and being shaped by those worlds.[144]

For Morgan City, the navigation channels of the Atchafalaya, and Horseshoe Bend, the most intimate of these ties are to the world of offshore oil drilling. Offshore oil in the Gulf of Mexico provides only a small fraction of the planet's fossil fuel supply—its annual share of carbon emissions is measured in millions of tons, while collective global carbon emissions are measured in gigatonnes, or billions of tons.[145] Yet that fraction is nonetheless a massive amount, and it is extracted, transported, refined, and stored by an enormous and enormously complex network of infrastructures: oil platforms, offshore pipelines, onshore pipelines, storage facilities and tank farms, refineries.

The carbon that is emitted as a result of the work of this Gulf network of infrastructures—and infrastructures like them from Alaska's North Slope to the seas around Norway, from the Persian Gulf to the Niger Delta's basin—has helped drive an unprecedentedly rapid transformation in the planetary climate. That transformation has accelerated in recent decades.[146] Surface temperatures are rising and heatwaves are becoming more frequent and intense. Oceans are warming, acidifying, and losing oxygen. Precipitation is increasing. Glaciers and Arctic sea ice are in retreat. Sea level is rising. Climate zones are shifting toward the poles, and plants are migrating in response. The primary cause of this transformation is the increasing concentration of greenhouse gases in the atmosphere, which carbon emissions are by far the single largest contributor to.

The impacts of offshore oil drilling are thus planetary. But they are also local, and unevenly distributed. In Louisiana, these impacts are concentrated along the length of the Mississippi River between Baton Rouge and New Orleans, a stretch often referred to as "Cancer Alley."[147] There, along the river, extracted oil is transformed into gasoline, plastics, fertilizers, asphalt, and other petrochemical products. This is done at massive chemical facilities owned by corporations like Shell, ExxonMobil, and DuPont, facilities which have replaced agricultural plantations once worked by enslaved Africans. Today, many of the descendants of those enslaved people live along this chemical corridor, often in towns that originated as "freedmen's towns," small settlements built on land granted to the emancipated at the end of the Civil War. As white-owned plantations were converted into chemical facilities in the mid-twentieth century, these towns endured in place, where they suffer from the toxins the facilities

144 This reciprocity is one of the most basic and crucial insights of infrastructure studies. A pair of case studies, both closely related to the topic of navigation channels, can be found in Cohen's *The Deadly Life of Logistics* and Carse's *Beyond the Big Ditch*.

145 OilChange International and Greenpeace, "The Climate Change Costs of Offshore Oil Drilling."

146 IPCC, "2021: Summary for Policymakers."

147 Misrach and Orff, *Petrochemical America*.

148 See "Choreographing Sediment."

149 Bullard, *Unequal Protection.*
Levine, *Love Canal.*
Mohai, Pellow, and Roberts, "Environmental Justice."
Nixon, *Slow Violence and the Environmentalism of the Poor.*

150 It is important to be clear here that this is not only a description of how strata have been built, but also of relationships that exist today. Those relationships can be changed. They must change if a different world is going to be made. Karp, "History as End."

151 We discuss Keller Easterling's notion of infrastructural disposition in "Technology, Pace, and Adaptation."

152 The concept of a world-system is used in social science to describe extremely large social and material realities, "comprised of interconnected nations, firms, households, classes, and identity groups of all kinds," with a particular focus on understanding the historical phenomena that produced and continue to shape the modern, liberal, and capitalist world. Wallerstein, *World-Systems Analysis.*

153 Harvey, *A Brief History of Neoliberalism.*
Klein, *On Fire.*
Aronoff et al., *A Planet to Win.*

Writing in *The Guardian*, historian Adam Tooze notes that "this is the basic thrust behind neoliberalism, or the market revolution - to depoliticise distributional issues, including the very unequal consequences of societal risks, whether those be due to structural change in the global division of labour, environmental damage, or disease."
Tooze, "Has Covid Ended the Neoliberal Era?"

154 See for instance Ellul's *The Technological Society* for a critique of technique, and Scott's *Seeing Like a State* for a critique of modernism's rationalizing tendencies. Bélanger extends Scott's critique to the domain of infrastructure in *Landscape as Infrastructure.* See also Moore, "The Capitalocene Part II." We discuss the role of technology in greater depth in "Technology, Pace, and Adaptation."

155 Demuth's *The Floating Coast* offers a vivid account of (in part) how diametrically opposed political and economic systems—American capitalism and Soviet communism—held common foundational assumptions about the role of extraction and fossil fuels in modern society.

release into the air, water, and land, which have been linked to the high cancer rates experienced by the towns' largely Black populations. Both poverty and racial inequity are directly linked to the distribution of the local impacts of oil, from these toxins to the displacement via sea-level rise of indigenous and impoverished communities like the Isle de Jean Charles Band.[148]

Why are these particular groups of people in harm's way? Why, as the extensive discourse of environmental justice describes, do such impacts fall largely on impoverished communities and communities of color, not only in Louisiana, but across the United States and beyond?[149] Contemporary inequities rest on thick strata of historical processes, of past decisions, of prior exercises of power. These strata are materialized in infrastructure, including the infrastructure of ports, dredging, and sediment management.[150] None of us, individually, have the freedom to opt out of the dispositions of these systems.[151] We do not, individually, have the capacity to decide we are tired of these harms and end them by restructuring our lives. You might make a personal choice to walk to work, to avoid bringing groceries home in plastic bags, to eat less carbon-intensive meat, or none at all. But even if you make those virtuous choices—which existing infrastructures often make difficult—the technosphere will burrow forward, its vast inertia almost entirely undisturbed. Oil will still be extracted, fossil fuels still burned for power and transportation (and dredging), petrochemicals still transformed into products like plastics for packaging. The worst excesses of this system will continue to be inflicted on those with the least power to alter it.

It can be difficult to describe precisely why the *world-system* that our infrastructure materializes is so locked into extractive and exploitative relationships.[152] The causes are multiple, they are thickly interwoven, and their descriptions are hotly contested.

Capitalism forms the most important economic dimension of these causes, though perhaps less as the ideology that "-ism" implies than as accumulated history and actions, lending inertia to institutional and collective dispositions toward patterns of behaviors and power relations. Neoliberal political theory and action has re-entrenched the rawest forms of capitalism over roughly the last fifty years, providing cover for both the multinational corporations who drive fossil fuel extraction and their stakeholders who profit from it.[153] Colonialism is another important facet of these interwoven causes; colonialist actions, like land theft and enslavement, set in motion many of the historical trajectories that placed communities like the Isle de Jean Charles Band and the towns along the Mississippi River in harm's way. Perhaps somewhat less obviously, the rationalizing tendencies of Western modernism and the ideology that philosophers of technology call technique both share a bent toward instrumentalization, toward understanding things only in terms of their usefulness.[154] This crude utilitarianism—and what it has, historically, left out of its calculus, like the worth of living beings and ecological systems—is also deeply embedded in the infrastructures we have inherited. And—no less important—the material technosphere has, despite the nascent growth of renewable energy technologies, a deeply embedded dependence on fossil fuel energy and petrochemical products, a dependence which no nation, economic system, or political movement has yet been able to effectively undo.[155]

Collectively, these forces are oriented toward extraction and exploitation. They have proven this orientation over time.[156] The material and the living—people, animal lives and plant communities, the movements of sediment and water that sustain life—are overlooked in favor of abstractions. These abstractions, like money and Gross Domestic Product, are elevated into ends-in-themselves, rather than being understood as tools that should serve life, constrained by real material limits.[157] And the voraciousness of these forces is responsible not only for the unevenness of local impacts, like those noted in Louisiana, but drives the larger processes presently terraforming the planet toward calamity.[158]

Sediment design, like any form of design, necessarily takes place within a world that is shaped by networks of commodified infrastructure that have these qualities embedded in them. When navigation channels are dredged, they are dredged to support the role of ports in the global economy as it exists now, as it has been structured by forces like neoliberalism, modernism, and fossil-fuel dependence. Any new landscape made with such dredged sediment is woven into this world. There is no outside or without, only the possibility to make new things now and, through them, to build a different after.[159]

The fact that there is no escaping this entanglement can be, counterintuitively, hopeful, in at least one sense: no individual project, community, or collaborative of designers should bear the impossible burden of achieving disentanglement on their own. We can and should celebrate steps forward, like Horseshoe Bend Island.

But we must also recognize that what we are entangled with is a linked ecological and economic crisis, and maintain the urgency that crisis requires. Species are going extinct, mosquito-borne diseases are spreading, heat and drought are breaking down ecosystems and causing crop failures, water scarcity is increasing on land even as sea-level rises, storms grow more intense, and floods grow more destructive. Forced climate migration is already reshaping international politics. Only rapid decarbonization of the global economy—which will require not only new technology, but a thorough reinvention of the fundamental orientation and power structures of the world's economic systems—can slow the pace.[160] Yet, cruelly, these disasters are concentrated in the parts of the world that are least responsible for the carbon emissions creating them. Because those low-income nations have less resources for adaptation and transformation, wealthy countries like the United States have an urgent responsibility to eliminate their emissions as fast as possible.[161] If there is a danger of despairing fatalism that might be born of recognizing entanglement, there is also a danger of complacent gradualism; both must be avoided.

The present crisis demands a transformation in both values and the infrastructures—including sedimentary landscapes—that embody those values.[162] Regeneration, reconciliation, and reciprocity must replace extraction and exploitation. Reparations must be made, across a broad spectrum of injuries, including but not limited to the wounds being cut now as a result of climate change, in order to begin processes of repair. Work of care and cultivation must be valued without being linked to an imperative for endless growth. Above all, life and the land that sustains it must be valued. No individual can bear the burden of materializing these values alone, but we, collectively, must.[163]

156 Fossil fuels embody this orientation literally, as they are definitionally extracted and not renewable. A finite store of energy is being drawn down, and its further expenditure is directly at odds with the planet's health. Modernism's generalized orientation toward efficiency has also played a substantial role in this extraction and exploitation. Today, while the European settler-colonialism of the modern era may be in retreat, the transfers of wealth from Global South to settler-colonial societies that it facilitated, the formations of nation-states it built, and the patterns of power and dispossession it structured persist. Capitalism itself is characterized by voraciousness, abstraction, and detachment, which further generates, as a by-product of its operations, extraction and exploitation of both peoples and the environment. This tendency was noted in the earliest sustained critiques of capitalism, as in Marx's (recently much-debated) notion of metabolic rift.
Foster, "Marx's Theory of Metabolic Rift."
Foster and Burkett, "Value Isn't Everything."
Hickel, *The Divide.*
Moore, *Capitalism in the Web of Life.*

157 John Bellamy Foster notes: "an even more momentous epochal crisis than the one that produced the transition from feudalism to capitalism, I will argue, is occurring today, arising from the unlimited expansion of a capitalist system geared to a process of abstract wealth creation." This detachment of capitalist use-value from real use-value lies at the core of the voraciousness of capitalism. Real use-value has material limits, while capitalist use-value expands without limit.
Foster, "The Epochal Crisis."

158 Consider the peculiar shape of fossil-fuel capitalism. The wealth that exists in companies like Exxon-Mobil, BP, and Shell is based on the anticipated future burning of fossil fuels—oil reserves and coal deposits only have value if we expect that they will be burned. Transition away from fossil fuels would cause that wealth to disappear, and thus is intensely resisted by those who possess it, much as the confluence of wealth and power (literally) embodied in American chattel slavery fueled resistance to abolition.
Aronoff et al., *A Planet to Win*, 36–41.

159 Purdy, *This Land is Our Land*, 144–50.

160 Rockstrom et al., "A Roadmap for Rapid Decarbonization."

161 Aronoff et al., *A Planet to Win*, 4.

162 This partially hinges on how we understand the concept of value itself. Is it understood in narrowly economic terms, and, in particular, the terms of capitalist economy ("commodity value," in Marxist analysis)? Or is it understood more broadly, in terms of intrinsic worth? If the former, then it will seem that the task is to make sure that the things we care about—like

ecological activity—are properly priced, are accounted for in terms of commodity value. But if the latter understanding is better, then we have a bigger task before us, a task that is more cultural than scientific. From this perspective, as Purdy notes in *This Land is Our Land*, "the people [working to price ecological activity] are good, sincere, and making forms of incremental progress that can help to improve things, or at least slow their degradation"—but "the real questions—what work is good to do, what life is good to live?—are open only at the margins, when we have paid our debts and earned our keep in this planetary workhouse...our world... functions as a kind of totalitarian system of value, imposed through the mechanics of price and through the material infrastructure of our ecological interdependence and harm...We do not really choose the uses we make of the world or of ourselves and one another." To make a world of reciprocity and care, these questions must move from the margins to the center.

163 As noted in "Dredge Matters," we are not all equally responsible for climate change or crisis, and some of us owe much more than many others. Yet planetary crisis does present an opportunity to make a democratic "we" at a new scale, a "we" defined not by equal responsibility for the conditions we face, but by solidarity and reciprocal care for our collective future. Olúfẹ́mi O. Táíwò discusses one approach to future-oriented repair, linking climate justice and reparations, in *Reconsidering Reparations* (Oxford University Press, 2022).

In the chapters that follow, we work toward this. In "Aesthetics and Publics," we explore how this collective agency might emerge; in "Technology, Pace, and Adaptation," we consider how our relationship to the tools that we use to shape landscapes may need to shift. Finally, in "Why Design?" we offer Public Sediment for Alameda Creek as an example of one step toward an ecological economy of sediment. There, our aim is to show that the way forward is not through critique alone, but through the slow, messy work of building other worlds, together, landscape by landscape.

Bibliography:

Ted R. Angradi, Mark S. Pearson, David W. Bolgrien, Brent J. Bellinger, Matthew A. Starry, and Carol Reschke, "Predicting Submerged Aquatic Vegetation Cover and Occurrence in a Lake Superior Estuary," *Journal of Great Lakes Research* 39, no. 4 (2013): 536–46.

Kate Aronoff, Alyssa Battistoni, Daniel Aldana Cohen, and Thea Riofrancos, *A Planet to Win: Why We Need a Green New Deal* (Verso Books, 2019).

Paul Aucoin, "Louisiana Ports: The Industry That Drives All Others," (presentation, Louisiana Transportation Conference, Baton Rouge, LA, February 25–28, 2018).

Pierre Bélanger, *Landscape as Infrastructure: A Base Primer* (Routledge, 2017).

Alan Berger, *Drosscape: Wasting Land in Urban America* (Princeton Architectural Press, 2006).

Rodger Brannan, James Skurla, and Stephen Castleberry, "Erie Pier: Finding Sustainable Markets for Dredged Materials," *Journal of Business Case Studies* 10, no. 2 (2014): 121–30.

Dan Breneman, "Habitat Restoration Effectiveness," presentation to Areas of Concern Conference, Cleveland, OH, September 12, 2019.

Todd S. Bridges, et al., *Use of Natural and Nature-Based Features (NNBF) for Coastal Resilience*, ERDC SR-15-1 (Vicksburg, MS: US Army Engineer and Research Development Center, 2015).

Todd S. Bridges, Jeffrey K. King, Jonathan D. Simm, Michael W. Beck, Georganna Collins, Quirijn Lodder, and Ram K. Mohan, eds., *International Guidelines on Natural and Nature-Based Features for Flood Risk Management* (Vicksburg, MS: US Army Engineer and Research Development Center, 2021).

Steven Brossart, US Army Corps of Engineers, Detroit District, "Erie Pier Briefing" (presentation given November 19, 2014).

Robert D. Bullard, *Unequal Protection: Environmental Justice and Communities of Color* (Sierra Club Books, 1997).

Sean Burkholder, "Designing Dredge: Engaging the Sediment Landscapes of the Great Lakes Basin," *Journal of Landscape Architecture* 11, no. 1 (2016): 6–17.

Sean Burkholder, Brian Davis, Rob Holmes, Justine Holzman, Tim Maly, Brett Milligan, and Gena Wirth, "Dredge Dump Dike," *Limn 10: Chokepoints*, eds. Ashley Carse, Jason Cons, and Townsend Middleton (2018), https://limn.it/issues/chokepoints/.

Leo Carney, "Ocean Dumping Still Debated," *New York Times*, Nov. 7, 1982.

Ashley Carse, *Beyond the Big Ditch: Politics, Ecology, and Infrastructure at the Panama Canal* (MIT Press, 2014).

Gene Clark, "Duluth-Superior Harbor Erie Pier: CDF to PRF Conversion" (presentation given to Dredging 2012 Conference, San Diego, CA, October 2012).

Daniel Aldana Cohen, "Interviews with Rebuild by Design's Working Group of Experts," *Public Culture* 28, no. 2 (2016): 317–50.

Deborah Cohen, *The Deadly Life of Logistics* (University of Minnesota Press, 2014).

Jeffrey C. Cornwell, Michael S. Owens, Lorie W. Staver, and J. Stevenson, "Tidal Marsh Restoration at Poplar Island I: Transformation of Estuarine Sediments into Marsh Soils," *Wetlands* 40, no. 6 (2020): 1,674.

D. W. Crawford, N. L. Bonnevie, C. A. Gillis, and R. J. Wenning, "Historical Changes in the Ecological Health of the Newark Bay Estuary, New Jersey," *Ecotoxicology and Environmental Safety* 29, no. 3 (1994): 276–303.

Julia Czerniak, ed., *Formerly Urban: Projecting Rust Belt Futures* (New York, NY: Princeton Architectural Press, 2013).

V. P. Dalal, J. E. Baker, and R. P. Mason, "Environmental Assessment of Poplar Island Dredged Material Placement Site, Talbot County, Maryland," *Estuaries* 22, no. 3 (1999): 770–84.

Sönke Dangendorf et al., "Persistent Acceleration in Global Sea-Level Rise since the 1960s," *Nature Climate Change* 9, no. 9 (2019): 705–10.

Nick Davidson, "How Much Wetland has been Lost? Long-term and Recent Trends in Global Wetland Area," *Marine & Freshwater Research* 65, no. 10 (2014).

Brian Davis, Rob Holmes, and Brett Milligan, "Isthmus," *Places Journal* (2015).

Bathsheba Demuth, *The Floating Coast: An Environmental History of the Bering Strait* (W.W. Norton & Company, 2019)

Peggy Derrick, Jeffrey McKee, Scott Johnson, and Mark Mendelsohn, "Poplar Island Environmental Restoration Project: Project Successes, Lessons Learned, and Future Plans," in *Proceedings of the World Dredging Congress* 1 (2007): 487–500.

Jill Desimini, "From Planned Shrinkage to Formerly Urban: Staking Landscape Architecture's Claim in the Shrinking City Debate," *Landscape Journal* 33, no. 1 (2014): 17–35.

"Beneficial Uses of Dredged Sediment," Dredging Operations Technical Support, US Army Corps of Engineers, https://budm.el.erdc.dren.mil/, retrieved June 18, 2021.

Mary Doyle and Cynthia A. Drew, *Large-Scale Ecosystem Restoration: Five Case Studies from the United States* (Island Press, 2008).

"Port History," Duluth Seaway Port Authority, https://duluthport.com/about-us/port-history/.

Amy E. East, et al., "Large-scale Dam Removal on the Elwha River, Washington, USA: River Channel and Floodplain Geomorphic Change," *Geomorphology* 228 (2015): 765–86.

Jacques Ellul, *The Technological Society*, trans. John Wilkinson (New York: Vintage, 1964).

R. Michael Erwin, Jason Miller, and Jan G. Reese, "Poplar Island Environmental Restoration Project: Challenges in Waterbird Restoration on an Island in Chesapeake Bay," *Ecological Restoration* 25, no. 4 (2007): 256–62.

Arturo Escobar, "Transiciones: A Space for Research and Design for Transitions to the Pluriverse," *Design Philosophy Papers* 13, no. 1 (2015): 13–23.

Thomas Fisher, *Designing to Avoid Disaster: The Nature of Fracture-Critical Design* (Routledge, 2012).

Billy Fleming, "Design and the Green New Deal," *Places Journal* (2019).

"Spoil Islands," Florida Medical Entomology Laboratory, University of Florida, https://fmel.ifas.ufl.edu/general-information/natural-habitats-at-fmel/spoil-islands/, accessed Oct 19, 2020.

Christy M. Foran et al., "Quantifying Wildlife and Navigation Benefits of a Dredging Beneficial-Use Project in the Lower Atchafalaya River: A Demonstration of Engineering with Nature®," *Integrated Environmental Assessment and Management* 14, no. 6 (2018): 759–68.

John Bellamy Foster, "Marx's Theory of Metabolic Rift: Classical Foundations for Environmental Sociology," *American Journal of Sociology* 105, no. 2 (1999): 366–405.

John Bellamy Foster, "The Epochal Crisis," *Monthly Review*, October 1, 2013.

John Bellamy Foster and Paul Burkett, "Value Isn't Everything," *Monthly Review*, November 1, 2018.

N. R. Francingues, D. W. Thompson, and F. L. Hamons, "Exploring Innovative Reuse of Dredged Material from Baltimore Harbor," in *Proceedings of the Western Dredging Association Technical Conference and Texas A&M Dredging Seminar, Nashville, Tennessee* (2011).

Tony Fry and Adam Nocek, eds., *Design in Crisis: New Worlds, Philosophies and Practices* (Routledge, 2020).

Great Lakes Dredging Team, "Beneficial Use of Dredged Material in the Great Lakes," by Gene Clark and David Knight (2013), https://repository.library.noaa.gov/view/noaa/34654/noaa_34654_DS1.pdf.

Goals Project, "Baylands Ecosystem Habitat Goals," A Report of Habitat Recommendations prepared by the San Francisco Bay Area Wetlands Ecosystem Goals Project (San Francisco, CA/Oakland, CA: US Environmental Protection Agency and SF Bay Regional Water Quality Control Board, 1999).

Goals Project, "The Baylands and Climate Change: What We Can Do: Baylands Ecosystem Habitat Goals Science Update 2015," A Report of Habitat Recommendations prepared by the San Francisco Bay Area Wetlands Ecosystem Goals Project (San Francisco, CA/Oakland, CA: US Environmental Protection Agency and SF Bay Regional Water Quality Control Board, 2015): xxviii.

Kian Goh, "Flows in Formation: The Global-Urban Networks of Climate Change Adaptation," *Urban Studies* 57, no. 11 (2020): 2,222–40.

Kian Goh, *Form and Flow: The Spatial Politics of Urban Resilience and Climate Justice* (The MIT Press, 2021).

Ramachandra Guha, *Environmentalism: A Global History* (Longman, 2000).

Peter Haff, "Technology as a Geological Phenomenon: Implications for Human Well-Being," in "A Stratigraphical Basis for the Anthropocene," eds. C. N. Waters, et al., *Geological Society, London, Special Publications* 395, no. 1 (2014): 301–9

Charlie Hailey, *Spoil Islands: Reading the Makeshift Archipelago* (Lexington Books, 2013).

David Harvey, *A Brief History of Neoliberalism* (Oxford University Press, 2005).

Daniel Headrick, *Humans Versus Nature: A Global Environmental History* (New York, NY: Oxford University Press, 2020).

Dennis Hevesi, "New Site Proposed for Dioxin-Tainted Sand," *New York Times*, Jan. 11, 1995.

Jason Hickel, *The Divide: Global Inequality from Conquest to Free Markets* (W.W. Norton & Company, 2018).

Roger Hooke, "On the History of Humans as Geomorphic Agents," *Geology* 28, no. 9 (2000): 843–46.

Timothy J. Iannuzzi and David F. Ludwig, "Historical and Current Ecology of the Lower Passaic River," *Urban Habitats* 2, no. 1 (2004): 147–73.

IPCC, "2021: Summary for Policymakers," in *Climate Change 2021: The Physical Science Basis, Contribution of Working Group I to the Sixth Assessment Report of the Intergovernmental Panel on Climate Change*, eds. V. Masson-Delmotte, et al. (Cambridge University Press, 2021).

Matthew Karp, "History as End: 1619, 1776, and the Politics of the Past," *Harper's Magazine*, July 2021.

Edward Keller, "Where Once there was a Runway: Returning the Former Hamilton Army Airfield to Nature," *The Military Engineer* 107, no. 693 (January–February 2015): 52–54.

Matthew L. Kirwan et al., "Limits on the Adaptability of Coastal Marshes to Rising Sea Level," *Geophysical Research Letters* 37, no. 23 (2010).

Matthew Kirwan, A. Brad Murray, Jeffrey P. Donnelly, and D. Reide Corbett, "Rapid Wetland Expansion during European Settlement and its Implication for Marsh Survival under Modern Sediment Delivery Rates," *Geology* 39, no. 5 (2011): 507–10.

Naomi Klein, *On Fire: The (Burning) Case for a Green New Deal* (Simon & Schuster, 2019).

Marin M. Kress et al., "Alignment of US ACE Civil Works Missions to Restore Habitat and Increase Environmental Resiliency," *Coastal Management* 44, no. 3 (2016).

Stuart N. Lane, "Natural Flood Management," *Wiley Interdisciplinary Reviews: Water* 4, no. 3 (2017): e1211.

Adeline Gordon Levine, *Love Canal: Science, Politics, and People* (Lexington Books, 1982).

Marc Levinson, *The Box: How the Shipping Container Made the World Smaller and the World Economy Bigger* (Princeton University Press, 2006).

Robert Marks, *China: An Environmental History* (Rowman & Littlefield Publishers, 2017).

Fadi Masoud, *Terra Sorta Firma* (Actar, 2021).

Brian Massumi, *99 Theses on the Revaluation of Value* (University of Minnesota Press, 2018).

Karen M'Closkey and Keith VanDerSys, "For Whom Do We Account in Climate Adaptation?" in *A Blueprint for Coastal Adaptation: Uniting Design, Economics, and Policy*, eds. Carolyn Kousky, Billy Fleming, and Alan M. Berger (Island Press, 2021).

P. McCallister and R. Kavalar, "Confined Disposal Program for Polluted Maintenance Dredging in the Great Lakes," *Oceans* 82 (1982): 1,042–45.

J. R. McNeill and Peter Engelke, *The Great Acceleration: An Environmental History of the Anthropocene since 1945* (Belknap Press, 2016).

Elizabeth Meyer, "Uncertain Parks," in *Large Parks*, eds. Julia Czerniak and George Hargreaves (New York, NY: Princeton Architectural Press, 2007).

Brett Milligan, "Hybrid Migrations and Design of Deluge," in *Bracket 3: At Extremes*, eds. Lola Sheppard and Maya Przybylski (Actar, 2015).

Minnesota Pollution Control Agency, *Environmental Assessment Worksheet: 21st Avenue West Aquatic Habitat Restoration Project*, by Dan Breneman, July 2013.

Richard Misrach and Kate Orff, *Petrochemical America* (Aperture, 2014).

Paul Mohai, David Pellow, and J. Timmons Roberts, "Environmental Justice," *Annual Review of Environment and Resources* 34 (2009): 405–30.

Jason W. Moore, *Capitalism in the Web of Life: Ecology and the Accumulation of Capital* (Verso Books, 2015).

Jason W. Moore, "The Capitalocene Part II: Accumulation by Appropriation and the Centrality of Unpaid Work/Energy," *The Journal of Peasant Studies* 45, no. 2 (2018): 237–79.

Rebecca L. Morris, Teresa M. Konlechner, Marco Ghisalberti, and Stephen E. Swearer, "From Grey to Green: Efficacy of Eco-engineering Solutions for Nature-based Coastal Defence," *Global Change Biology* 24, no. 5 (2018): 1,827–42.

Elizabeth Mossop, ed., *Sustainable Coastal Design and Planning* (Boca Raton, FL: CRC Press, 2018).

Barbara E. Mundy, *The Death of Aztec Tenochtitlan, the Life of Mexico City* (University of Texas Press, 2021).

Siddharth Narayan, et al., "The Value of Coastal Wetlands for Flood Damage Reduction in the Northeastern USA," *Scientific Reports* 7, no. 1 (2017): 1–12.

"NASA Studies a Rarity: Growing Louisiana Deltas," NASA, by Carol Rasmussen, accessed May 1, 2022, https://www.nasa.gov/feature/jpl/nasa-studies-a-rarity-growing-louisiana-deltas.

"World of Change: Growing Deltas in Atchafalaya Bay," NASA Earth Observatory, accessed May 1, 2022, https://earthobservatory.nasa.gov/world-of-change/WaxLake.

Terrapins in the Classroom," National Aquarium, accessed September 23, 2021, https://aqua.org/education/programs/terrapins-in-the-classroom.

"Restoring Valuable Habitats," National Aquarium WATERblog, accessed September 23, 2021, https://nationalaquarium.wordpress.com/2009/11/04/restoring-valuable-habitats/.

"Site Update: Poplar Island," National Aquarium WATERblog, accessed September 23, 2021, https://nationalaquarium.wordpress.com/2012/01/27/site-update-poplar-island/.

National Oceanic and Atmospheric Administration, "Global and Regional Sea Level Rise Scenarios for the United States," NOAA Technical Report NOS CO-OPS 083, by William V. Sweet et al. (Silver Spring, MD: NOAA, 2017).

National Oceanic and Atmospheric Administration, "Water Level Variations at Poplar Island, MD," NOAA Technical Report NOS CO-OPS 076, by Jena Kent (April 2015).

Rob Nixon, *Slow Violence and the Environmentalism of the Poor* (Harvard University Press, 2011).

Catherine Seavitt Nordenson, Guy Nordenson, and Julia Chapman, *Structures of Coastal Resilience* (Island Press, 2018).

OilChange International and Greenpeace, "The Climate Change Costs of Offshore Oil Drilling," by Tim Donaghy (June 2016).

Henk Ovink and Jelte Boeijenga, *Too Big: Rebuild by Design: A Transformative Approach to Climate Change* (Rotterdam: nai010 publishers, 2018).

Gregory B. Pasternack, Grace S. Brush, and William B. Hilgartner, "Impact of Historic Land-Use Change on Sediment Delivery to a Chesapeake Bay Subestuarine Delta," *Earth Surface Processes and Landforms* 26, no. 4 (2001): 409–27.

Sean Poltrack, "The Maritime Industry and Our Environment: The Delicate Balance of Economic and Environmental Concerns, Globally, Nationally, and within the Port of Baltimore," *University of Baltimore Journal of Environmental Law* 8, no. 1 (Fall 2000).

"Port of Morgan City," Port of Morgan City, accessed July 29, 2021, https://www.portofmc.com/.

Jedediah Purdy, *This Land is Our Land* (Princeton University Press, 2019).

Kate Raworth, *Doughnut Economics: Seven Ways to Think Like a 21st-Century Economist* (Chelsea Green Publishing, 2017).

"Living Breakwaters—Rebuild by Design," Rebuild by Design, accessed April 28, 2022, https://www.rebuildbydesign.org/work/funded-projects/living-breakwaters/.

"Living with the Bay—Rebuild by Design," Rebuild by Design, accessed April 28, 2022, https://www.rebuildbydesign.org/work/funded-projects/living-with-the-bay/.

"New Meadowlands—Rebuild by Design," Rebuild by Design, accessed April 28, 2022, https://www.rebuildbydesign.org/work/funded-projects/new-meadowlands/.

Rebuild by Design, *Rebuild by Design*, eds. John Gendall, Josh Bisker, Amy Chester, Tara Eisenberg, Scott Davis, and Henk Ovink (2015).

Denise J. Reed, "Sea-Level Rise and Coastal Marsh Sustainability: Geological and Ecological Factors in the Mississippi Delta Plain," *Geomorphology* 48, no. 1-3 (2002): 233–43.

Andrew Revkin, "Harbor to Be Dredged, but Much Tainted Mud Lacks Home," *New York Times*, May 12, 1997.

Johan Rockstrom et al., "A Roadmap for Rapid Decarbonization," *Science* 355:6331, March 24, 2017.

Johan Rockström, et al., "A Safe Operating Space for Humanity," *Nature* 461, no. 7263 (2009): 472–75.

Paul Roncken, Sven Stremke, and Maurice Paulissen, "Landscape Machines: Productive Nature and the Future Sublime," *Journal of Landscape Architecture* 6, no. 1 (2011): 68–81.

Frédéric Rossano, *Floodscapes: Contemporary Landscape Strategies in Times of Climate Change* (Rotterdam: nai010 publishers, 2021).

Firas Saleh and Michael P. Weinstein, "The Role of Nature-Based Infrastructure (NBI) in Coastal Resiliency Planning: A Literature Review," *Journal of Environmental Management* 183 (2016): 1088–98.

"A Community Project Supporting Stewardship of the Hamilton Wetlands," School Fuel: the Novato Foundation for Public Education, accessed May 1, 2022, https://novatoschoolfuel.org/a-community-project-supporting-stewardship-of-the-hamilton-wetlands/.

James Scott, *Seeing Like a State* (Yale University Press, 1999).

Justine Shapiro-Kline, "The Impact of the Public Process in Rebuild by Design" (Master's Thesis, Columbia University, 2014).

Joe Sharkey, "Ocean Dumping Hearing Set," *New York Times*, May 4, 1996.

L. W. Staver et al., "Tidal Marsh Restoration at Poplar Island: II. Elevation Trends, Vegetation Development, and Carbon Dynamics," *Wetlands* 40, no. 6 (2020): 1,687.

St. Louis River Estuary, "21st Avenue West Restoration Site," https://stlouisriverestuary.org/restoration_21w.php.

Will Steffen, Wendy Broadgate, Lisa Deutsch, Owen Gaffney, and Cornelia Ludwig, "The Trajectory of the Anthropocene: the Great Acceleration," *The Anthropocene Review* 2, no. 1 (2015): 81–98.

Steven L. Stockton, memorandum, "Federal Standard Clarification Regarding Federal Dredging Mission and Interactions with Non-Federal Agencies," October 21, 2015, https://planning.erdc.dren.mil/toolbox/library/MemosandLetters/15Oct-FederalStandardClarification.pdf.

Burton Suedel et al., "Creating Horseshoe Bend Island, Atchafalaya River, Louisiana," *Terra Et Aqua* 140 (2015): 26–31.

Ariana E. Sutton-Grier et al., "Investing in Natural and Nature-based Infrastructure: Building Better along our Coasts," *Sustainability* 10, no. 2 (2018).

Kathleen M. Swanson et al., "Wetland Accretion Rate Model of Ecosystem Resilience (WARMER) and its Application to Habitat Sustainability for Endangered Species in the San Francisco Estuary," *Estuaries and Coasts* 37 (2014): 476–92.

James Syvitski and Albert Kettner, "Sediment Flux and the Anthropocene," *Philosophical Transactions of the Royal Society A: Mathematical, Physical and Engineering Sciences* 369, no. 1938 (2011): 957–75.

James Syvitski and Yoshiki Saito, "Morphodynamics of Deltas under the Influence of Humans," *Global and Planetary Change* 57, no. 3-4 (2007): 261–82.

Yuri Taddia et al., "High-Resolution Monitoring of Tidal Systems Using UAV: A Case Study on Poplar Island, MD (USA)," *Remote Sensing* 13, no. 7 (2021).

Cameron Tonkinwise, "Design for Transitions: From and to What?," *Design Philosophy Papers* 13, no. 1 (2015): 85–92.

Adam Tooze, "Has Covid Ended the Neoliberal Era?," *The Guardian*, September 2, 2021.

Jason Urbanus, "Rome's Imperial Port," *Archaeology* 68, no. 2 (2015): 26–33.

US Army Corps of Engineers, "A Multifactor Ecosystem Assessment of Wetlands Created Using a Novel Dredged Material Placement Technique in the Atchafalaya River, Louisiana: An Engineering With Nature Demonstration Project," ERDC TR-17-5, by Jacob F. Berkowitz et al. (Vicksburg, MS: USACE Engineer Research and Development Center, 2017).

US Army Corps of Engineers, "Army Corps Awards Contract for Restoring Black Wall and Rulers Bar Marsh Islands in Jamaica Bay, New York," April 2, 2012, https://www.nan.usace.army.mil/Media/News-Releases/Article/487946/army-corps-awards-contract-for-restoring-black-wall-and-rulers-bar-marsh-island/.

US Army Corps of Engineers, "Hydrodynamics of a Recently Restored Coastal Wetland: Hamilton Wetlands, California," ERDC/CHL TR-20-11, by Richard Styles, et al. (Vicksburg, MS: USACE Engineer Research and Development Center, 2020).

US Army Corps of Engineers, "Literature Review of Nearshore Berms," ERDC/CHL SR-19-2, by Katherine E. Brutsche, Brian C. McFall, Duncan B. Bryant, and Ping Wang (Vicksburg, MS: US Army Engineer Research and Development Center, 2019).

US Army Corps of Engineers, "Long-Term Monitoring of Eleven Corps of Engineers Habitat Development Field Sites Built of Dredged Material, 1974–1987," Dredging Operations Technical Support Program Technical Report D-89-1, by Mary C. Landin, James W. Webb, and Paul L. Knutson (Vicksburg, MS: Corps of Engineers Waterways Experiment Station, December 1989).

US Army Corps of Engineers, "Paul S. Sarbanes Ecosystem Restoration Project at Poplar Island, Talbot County, MD," Fact Sheet as of Feb. 1, 2021.

US Army Corps of Engineers, "Thin Layer Placement: Technical Definition for U.S. Army Corps of Engineers Applications," ERDC TN-19-1, by Jacob F. Berkowitz, Candice Piercy, Tim Welp, and Christine VanZomeren (Vicksburg, MS: US Army Engineer Research and Development Center, 2019).

US Army Corps of Engineers, Dredging Operations and Technical Support Program, "Evaluation of the 1980 Capping Operations at the Experimental Mud Dump Site, New York Bight Apex," Technical Report D-83-3, by Joseph M. O'Connor and Susan G. O'Connor (Vicksburg, MS: U.S. Army Engineer Waterways Experiment Station).

US Army Corps of Engineers, Engineer Research and Development Center, "Sustainable Confined Disposal Facilities for Long-term Management of Dredged Material," ERDC TN-DOER-D10, by Susan E. Bailey et al. (Vicksburg, MS: Engineer Research and Development Center, 2010).

US Army Corps of Engineers, Great Lakes & Ohio River Division, "Confined Disposal Facilities on the Great Lakes," by Jan Miller (Chicago, IL: US Army Corps of Engineers, 1998).

"Dredged Material Management Plan," US Army Corps of Engineers New York District, accessed March 12, 2022, https://www.nan.usace.army.mil/Missions/Navigation/Dredged-Material-Management-Plan/.

"Historic Area Remediation Site," US Army Corps of Engineers New York District, accessed March 12, 2022, https://www.nan.usace.army.mil/Missions/Navigation/Historic-Area-Remediation-Site-HARS/.

US Army Corps of Engineers and US Environmental Protection Agency, "Great Lakes Confined Disposal Facilities," ed. Jan Miller (2003).

US Army Corps of Engineers and US Environmental Protection Agency, "Site Management and Monitoring Plan for the Historic Area Remediation Site," (New York, NY: USACE and EPA, 2009).

US Army Corps of Engineers and US Environmental Protection Agency, "Site Management and Monitoring Plan for Historic Area Remediation Site (HARS)," (New York, NY: USACE and EPA, 2020).

US Environmental Protection Agency, "Dredged Material Management Action Agenda for the Next Decade (Based on a Workshop Sponsored by the National Dredging Team)", EPA 842-B-04-002 (Washington, DC: July 2003).

US Environmental Protection Agency and US Army Corps of Engineers, "Atchafalaya River Bar Channel, Louisiana, Site Management Plan for the Maintenance Dredging, Ocean Dredged Material Disposal Sites East and West" (EPA and USACE, 2013).

US Environmental Protection Agency and US Army Corps of Engineers, "Identifying, Planning, and Financing Beneficial Use Projects Using Dredged Material," EPA 842-B-07-001 (Washington, DC: October 2007).

US Environmental Protection Agency and US Army Corps of Engineers, "The Role of the Federal Standard in the Beneficial Use of Dredged Material from U.S. Army Corps of Engineers New and Maintenance Navigation Projects," EPA 842-B-07-002 (Washington, DC: October 2007).

US Fish and Wildlife Service, *An Ecological Design for the 21st Avenue West Remediation-to-Restoration Project*, by George Host, Paul Meysembourg, Carol Reschke, Valerie Brady, Gerald Niemi, Annie Bracey, Lucinda Johnson, Matthew James, Jay Austin, and Elissa Buttermore, June 2013.

Charles Waldheim and Alan Berger, "Logistics Landscape," *Landscape Journal* 27, no. 2 (2008): 219-246.

Immanuel Wallerstein, *World-Systems Analysis: An Introduction* (Duke University Press, 2004).

Water Resources Development Act of 2020, H.R. 7575, 116th Congress, Section 125 (2020).

S. Jeffress Williams, "Geologic Effects of Ocean Dumping on the New York Bight Inner Shelf," in *Ocean Dumping and Marine Pollution*, eds. Harold D. Palmer and M. Grant Gross (Dowden, Hutchinson, and Ross, Inc., 1979).

John P. Wolflin, "Chesapeake Bay Program: A Watershed Approach to Management," in *Sustainable Use and Development of Watersheds*, eds. I. Ethem Gönenc, Angheluta Vadineanu, John P. Wolflin, and Rosemarie C. Russo (Dordrecht: Springer, 2008), 435–51.

Cat Island (Sean Burkholder/Above All Ohio)

Inland Chokepoint

The Great Lakes

North America's Inland Coast faces the unique challenge of having an abundance of sediment and few options for where to put it. As most of the current dredge disposal facilities reach capacity, the Great Lakes region is under immense pressure to innovate to find new sediment practices.

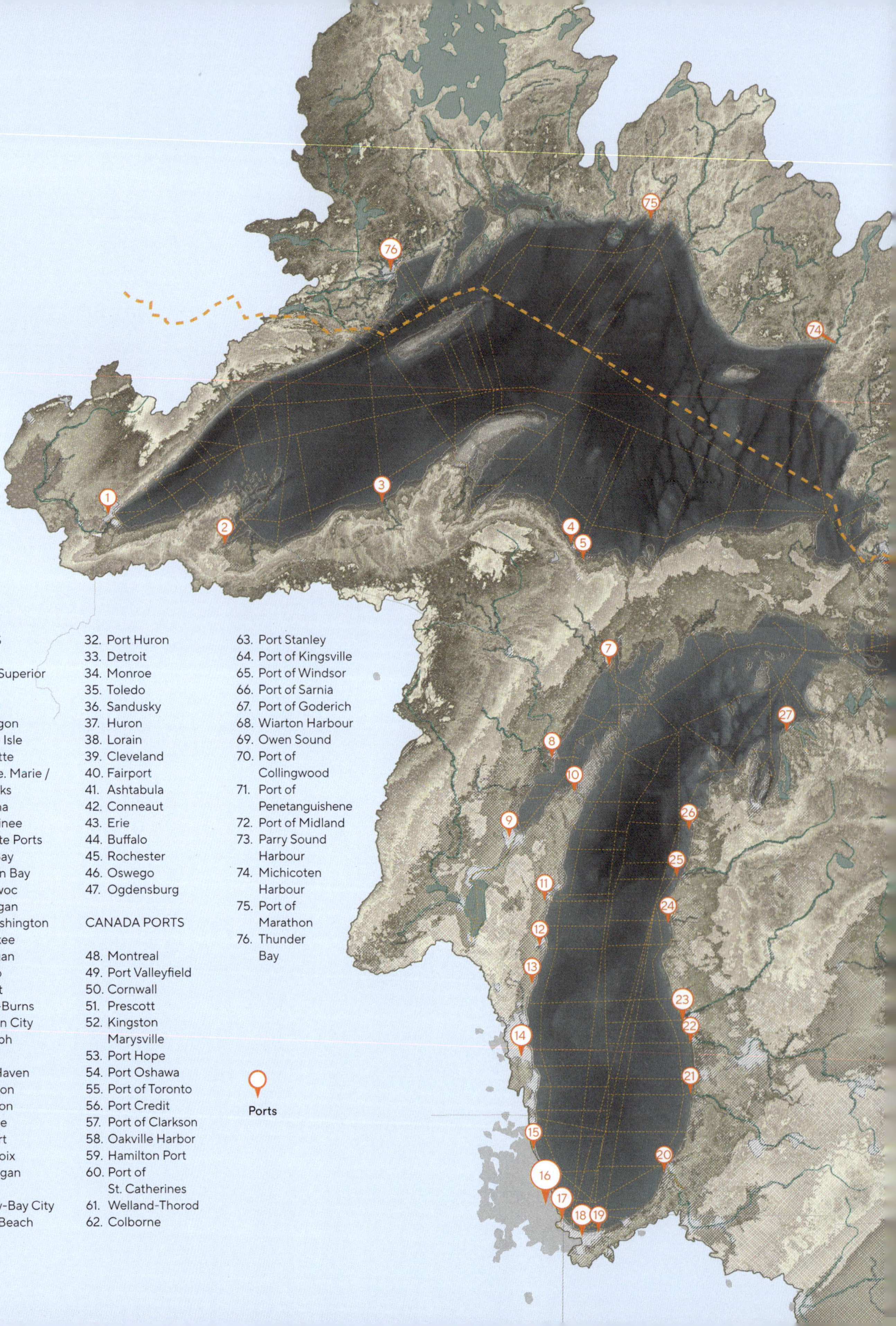

USA PORTS

1. Duluth-Superior Harbor
2. Ashland
3. Ontonagon
4. Presque Isle
5. Marquette
6. Sault Ste. Marie / Soo Locks
7. Escabana
8. Menominee Marinette Ports
9. Green Bay
10. Sturgeon Bay
11. Manitowoc
12. Sheboygan
13. Port Washington
14. Milwaukee
15. Waukegan
16. Chicago
17. Calumet
18. Indiana-Burns
19. Michigan City
20. St. Joseph
21. Holland
22. Grand Haven
23. Muskegon
24. Ludington
25. Manistee
26. Frankfort
27. Charlevoix
28. Cheboygan
29. Alpena
30. Saginaw-Bay City
31. Harbor Beach
32. Port Huron
33. Detroit
34. Monroe
35. Toledo
36. Sandusky
37. Huron
38. Lorain
39. Cleveland
40. Fairport
41. Ashtabula
42. Conneaut
43. Erie
44. Buffalo
45. Rochester
46. Oswego
47. Ogdensburg

CANADA PORTS

48. Montreal
49. Port Valleyfield
50. Cornwall
51. Prescott
52. Kingston Marysville
53. Port Hope
54. Port Oshawa
55. Port of Toronto
56. Port Credit
57. Port of Clarkson
58. Oakville Harbor
59. Hamilton Port
60. Port of St. Catherines
61. Welland-Thorod
62. Colborne
63. Port Stanley
64. Port of Kingsville
65. Port of Windsor
66. Port of Sarnia
67. Port of Goderich
68. Wiarton Harbour
69. Owen Sound
70. Port of Collingwood
71. Port of Penetanguishene
72. Port of Midland
73. Parry Sound Harbour
74. Michicoten Harbour
75. Port of Marathon
76. Thunder Bay

A Coast Rich in Mud

Each year, 2 to 5 million cubic yards of bottom sediments are dredged from the edges of the Great Lakes to maintain navigation channels, remediate contaminated areas, and aid waterfront construction.[1]

This dredged substrate presents a disposal challenge. Moving dredge from the more than 100 commercial ports and harbors that line the Great Lakes is no easy task. Currently, more than half of this material is placed in nearshore landfills, or Confined Disposal Facilities (CDFs).[2]

These facilities are located close to river mouths and thus to the source of sediment, but they are at capacity and unable to accommodate future dredging projections. A surprising amount of dredged material is released in open water, and smaller amounts are beneficially reused for beach nourishment, upland construction, or habitat creation.[3]

Dealing with Dredge

Sediment disposal in the Great Lakes is at an inevitable crossroads, as the CDF technology expires and as the remaining active CDFs near capacity.

Of the 45 CDFs built since 1970, only 20 are fully active, and the system is at 80% capacity.[4] While actions such as CDF harvesting and elevating can incrementally prolong their lifespan, the era of confined disposal is over. Two questions arise: what to do with this excess material, and what to do with the "waste" landscapes that have emerged? New models for beneficial reuse must be developed to transform the dumping of dredge into a cyclical and ecologically beneficial sediment practice.

In 1994, Site 3 was expanded to a total area of 397 acres. It still serves as an active CDF for the Port of Toledo.

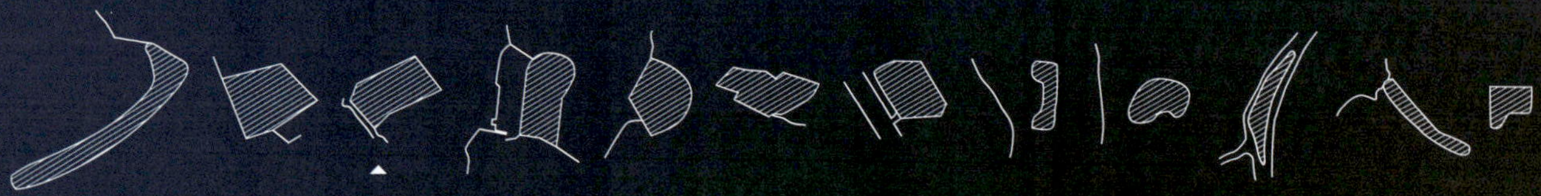

Great Lakes CDF Shapes

Toledo CDF (Sean Burkholder/Above All Ohio)

Unintended Ecosystems

CDFs are created by policy, not design. Yet they are not singular or static infrastructures—they are shaped over time by ecological and social dynamics.

CDFs take multiple forms based upon the different operating procedures of the eight states and two countries that share the Great Lakes region. They grow into novel ecosystems of the lakeshore, hosting unintended habitats and human activities like hunting as they mature.

Some CDFs have transformed into popular recreational areas, others into Audubon sanctioned nature preserves providing habitat to vulnerable species. Given the potential ecological and social implications of CDFs, design should play a substantial role in reimagining them for future generations.

Confined Disposal Facility (CDF) Types

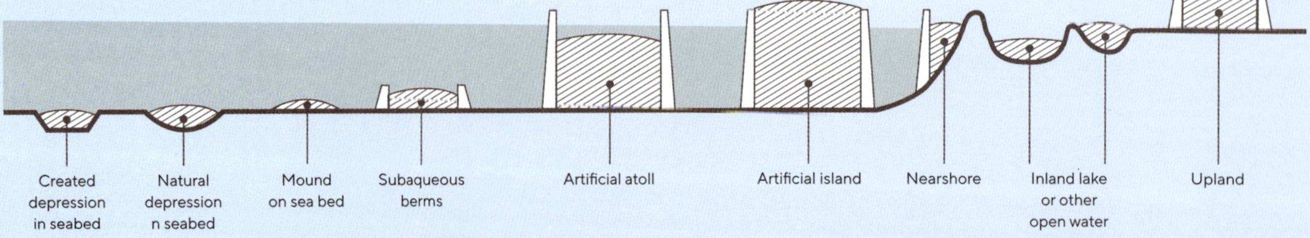

TIMES BEACH
BUFFALO, NY

A CDF facility constructed atop a former recreational beach, Times Beach was used as a placement area for dredge material from Buffalo Harbor. The arrival of migratory birds led to the closure of the site, leaving the CDF partially filled. Today, Times Beach is used as a nature preserve and outdoor research laboratory.

(Sean Burkholder)

ERIE HARBOR CDF
ERIE, PA

The Erie Harbor CDF is an in-lake facility attached to open land. The CDF can be accessed by roadway or by water.

(Sean Burkholder/Above All Ohio)

DIKE 14
CLEVELAND, OH

Dike 14 was created in the 1970s as a CDF recieving dredged sediment from the Cuyahoga River. Dredge placement stopped in 1999, and the facility transitioned to a place for wildlife and recreation, now known as the Cleveland Lakefront Nature Preserve.

(Sean Burkholder)

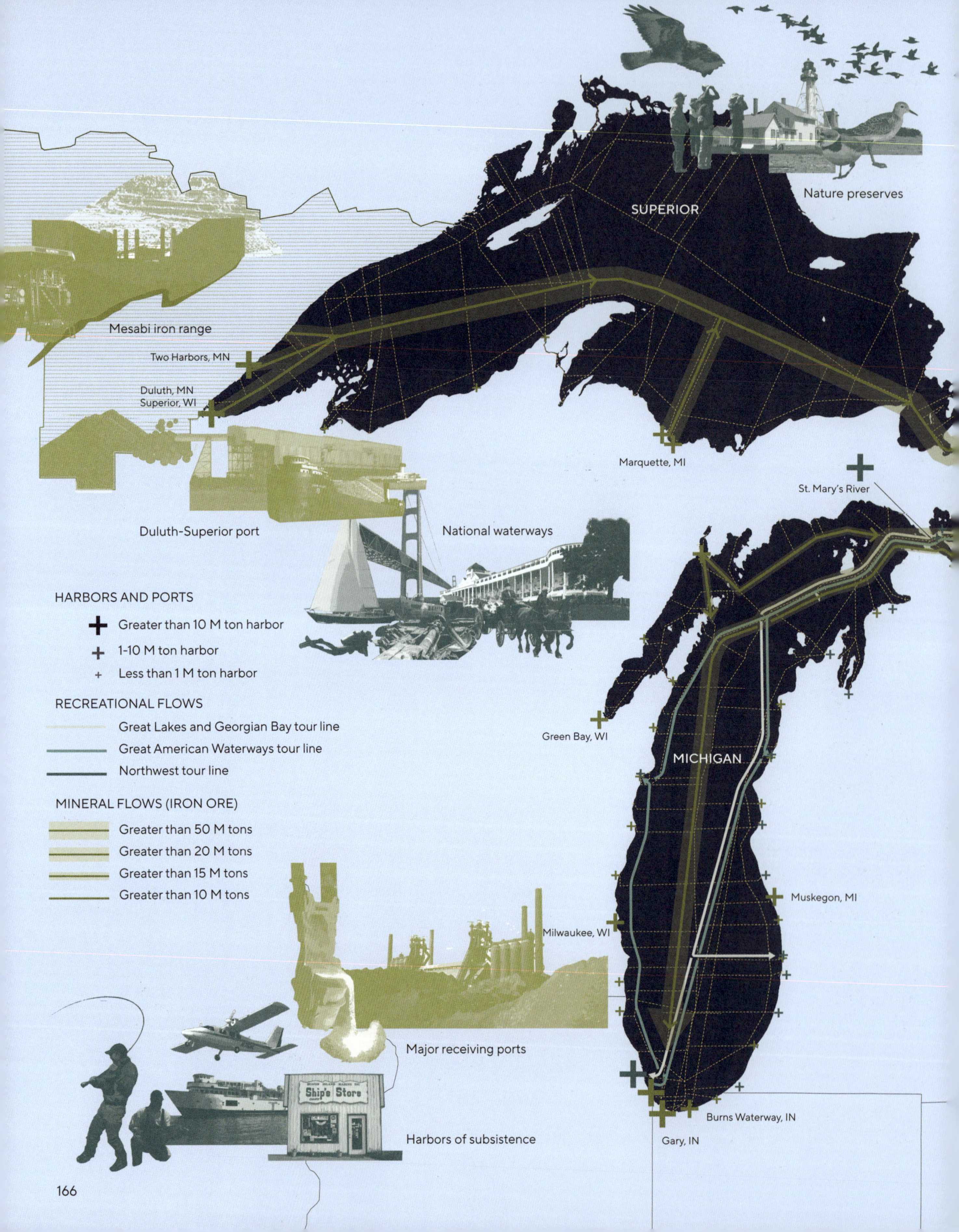
Nature preserves
SUPERIOR
Mesabi iron range
Two Harbors, MN
Duluth, MN
Superior, WI
Marquette, MI
St. Mary's River
Duluth-Superior port
National waterways
HARBORS AND PORTS
Greater than 10 M ton harbor
1-10 M ton harbor
Less than 1 M ton harbor
RECREATIONAL FLOWS
Great Lakes and Georgian Bay tour line
Great American Waterways tour line
Northwest tour line
MINERAL FLOWS (IRON ORE)
Greater than 50 M tons
Greater than 20 M tons
Greater than 15 M tons
Greater than 10 M tons
Green Bay, WI
MICHIGAN
Muskegon, MI
Milwaukee, WI
Major receiving ports
Harbors of subsistence
Ship's Store
Burns Waterway, IN
Gary, IN

Deep and Shallow Draft Economies

While the mechanics of dredging are simple, dredging is not an isolated practice and is deeply tied to local, regional, and global economies.

Deep draft dredging responds to the need to move bulk commodities from and between inland sites, urban centers, and the Atlantic Ocean. From geologic seam to pellet to furnace, iron ore is the dominant bulk commodity transported within the basin system, and it requires its own particular network of navigation infrastructure to accommodate these geographically isolated flows.

Overlaid onto the larger, deeper, industrial navigation network is a shallower, more localized dredging regime that enables recreational use of the Great Lakes system. As people turn to the Great Lakes for leisure, shallow-draft dredging enables experiences like day trips, shipwreck scuba dives, and week-long island paddles. These two economies work in concert to activate and excavate the diverse edges of the lakeshore.

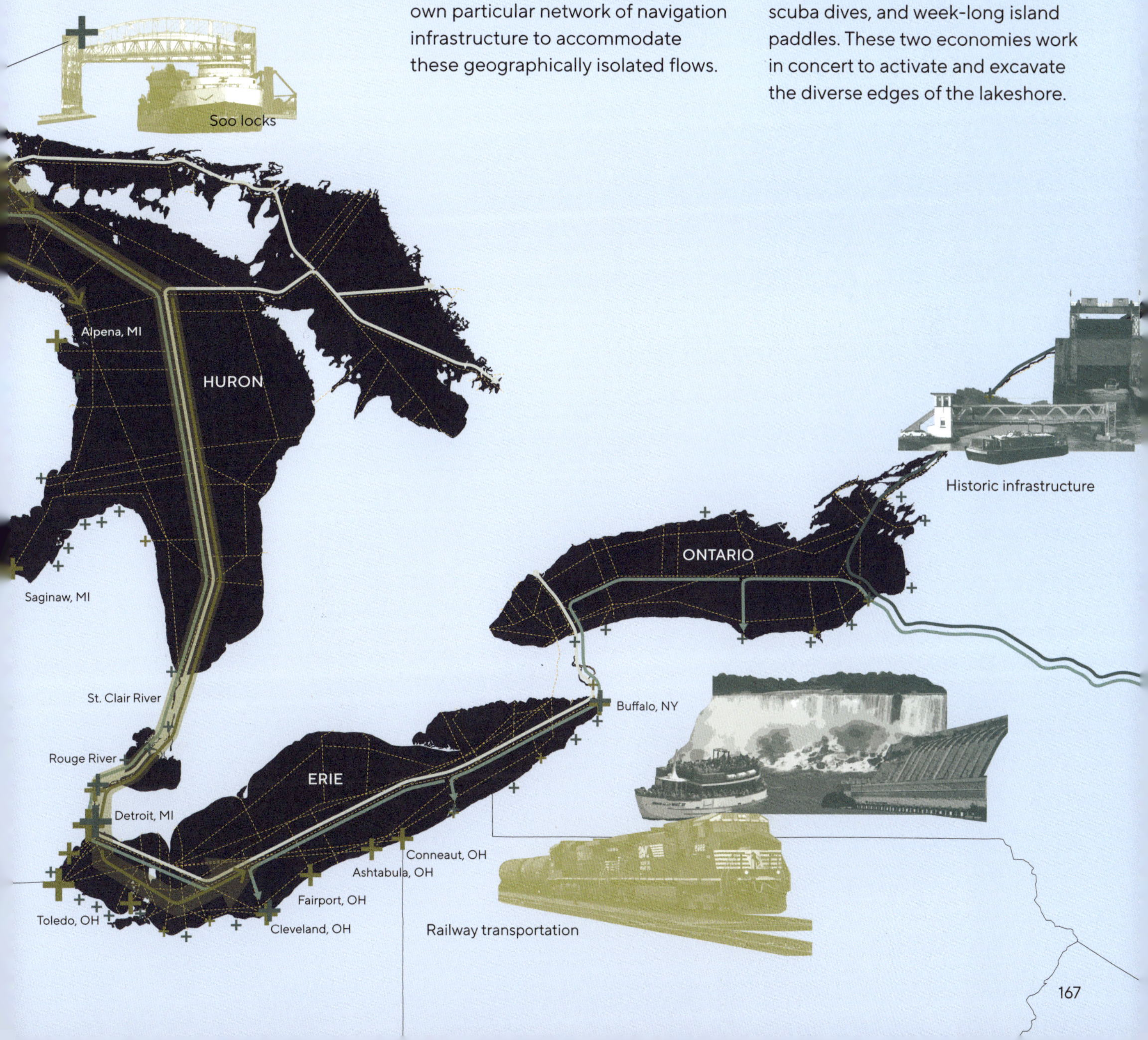

Toxic Legacies

Sediment contamination limits placement and beneficial reuse options for dredged material. Contamination can be introduced to the Great Lakes along industrial tributaries that feed the lakes and connect to upstream sedimentsheds that host industries of the past and present.

ST. LOUIS RIVER WATERSHED
Disturbed Industrial Uplands

ST. LOUIS RIVER WATERSHED
The Agricultural Midlands

Great Lakes Basin
Forest Vegetation

PORT OF DULUTH
Lowlands, Infrastructural Obsolescence, and CDFs

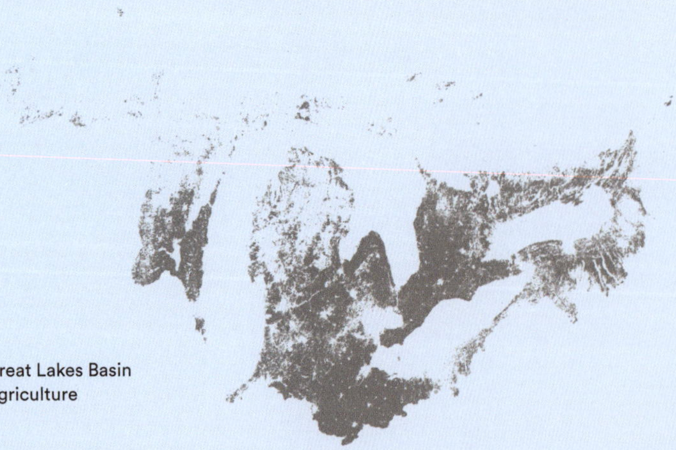

Great Lakes Basin
Agriculture

Mine Tailings
Mine tailings contribute suspended and dissolved sediment loads

Logging
Clear cutting and selective harvest reduces duff layer in forests, accelerating runoff and sedimentation processes

Confluence of Rivers
Tributaries accumulate sediment, a process that has been accelerated due to anthropogenic extractive forces

Sediment Loading
Stream bank erosion and lack of substantial stream buffers around agricultural landscapes increase sediment volumes in rivers

Agricultural Land Use
Less dense vegetative cover and exposure of bare soil accelerate runoff and sedimentation into river tributaries and wetlands

Shrinking Wetlands
Drained and degraded wetlands across the Great Lakes sedimentshed lack the capacity to handle the large sediment volumes coming from agricultural runoff

Chicago Area CDF (Mary Pat McGuire, 2022)

Environmental Justice

Waterfront communities, often low-income communities of color, face increasing pressure to live with dredged and contaminated sediment as storage capacity narrows. In the Southeast Side of Chicago, dredging is at the core of an ongoing environmental justice conflict.[5]

The Chicago Area CDF receives dredge material from the Calumet Harbor and River and was slated to be at capacity and closed in 2022, when it would be transformed into a public park.[6] The Army Corps recently proposed to continue dredge placement operations and implement "life extension measures," including vertical expansion of the facility by over 20 feet in height, for 20 additional years, due to challenges identifying alternative placement sites.[7]

This plan faces strong opposition from community advocates like Friends of the Park and the Alliance of the Southeast, who say their neighborhood is already overburdened by industrial operations and historically has lacked access to the lakeshore.[8] This conflict reveals the social impacts of dredging and the need for equitable and environmentally just sediment management practices and decision-making processes.

LAKE MICHIGAN

Shoreline length:	1,640 miles
Volume:	1,180 cubic miles
Retention Time:	62 years
Population:	12.05 M US

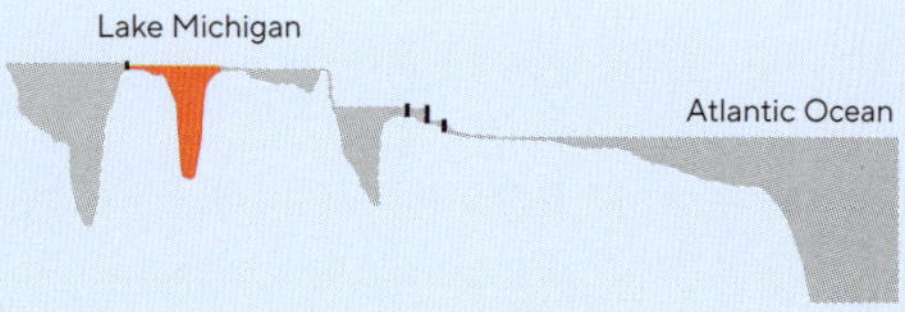

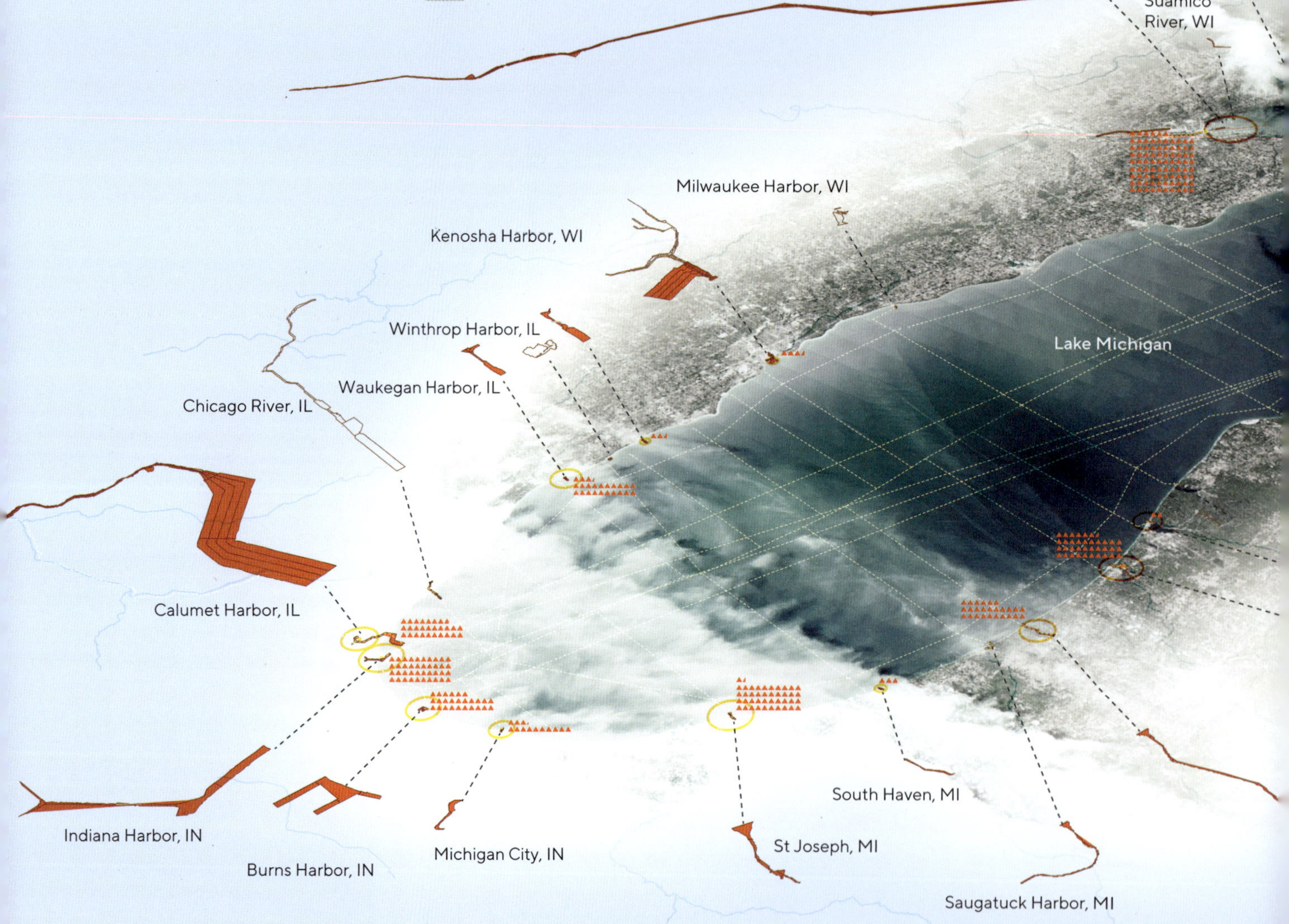

- ▲ 10,000 cy dredged material by USACE, 2009-2013
- USACE maintained channels
- USACE dredging contract zone, 2009-2013
- ······ Waterways / shipping routes

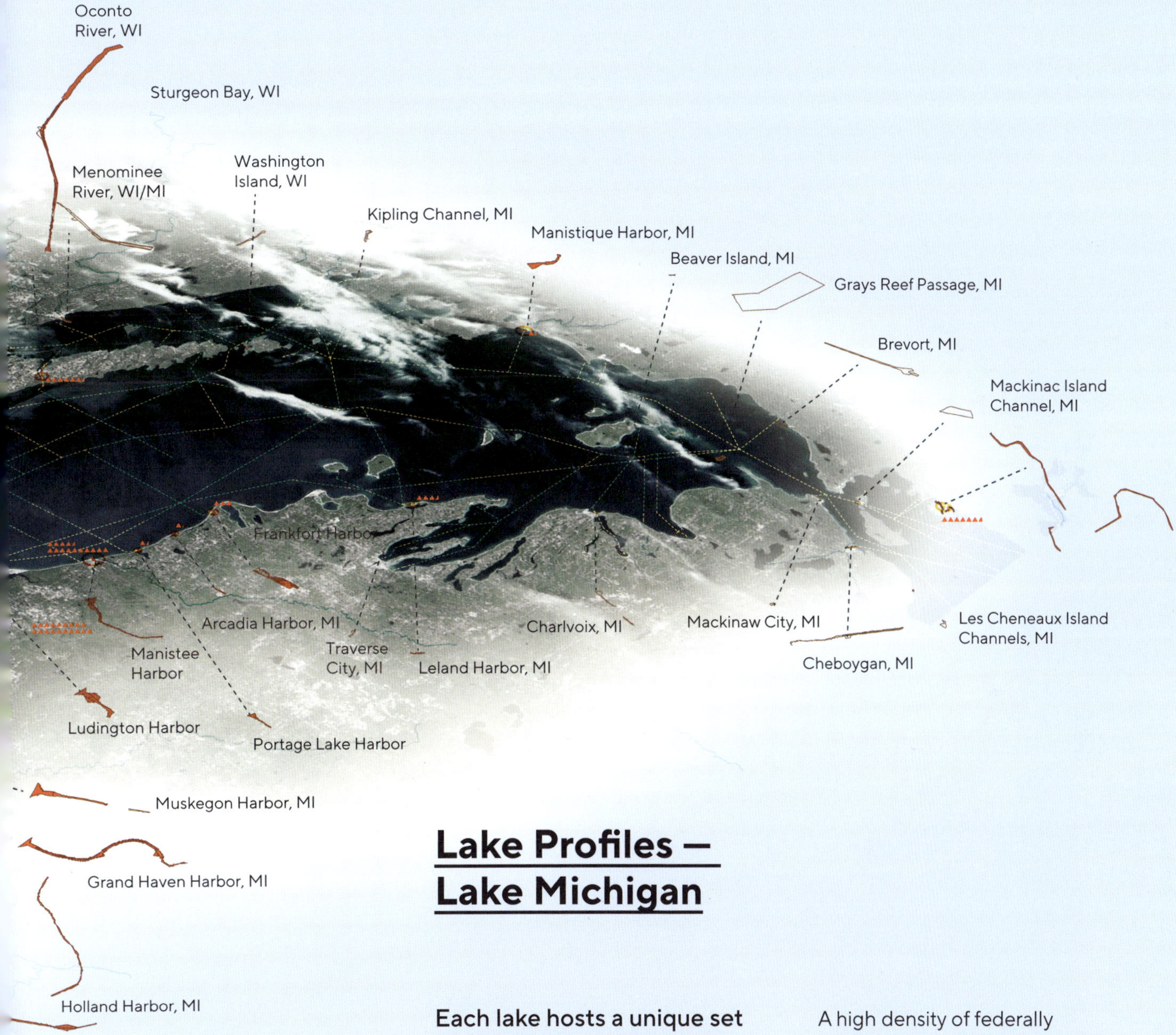

Lake Profiles – Lake Michigan

Each lake hosts a unique set of sediment management conditions. Lake Michigan is the only Great Lake that is entirely within the boundary of the United States, and therefore primarily managed by the US Army Corps of Engineers.

A high density of federally maintained navigation channels perforate its shoreline, serving both recreational and commercial interests. The northern portion is colder and sparsely populated, while the southern part is temperate and highly urbanized. A unique "cul-de-sac" circular pattern of water flow around the lake creates a long retention time within the system, impacting the residence time of contaminants within the waterbody.

Lake Erie

Dredge practices in Lake Erie are highly concentrated around large commercial ports such as Detroit, Cleveland, and Toledo.

Lake Erie is the shallowest Great Lake and smallest by volume, warming rapidly in the spring and summer and freezing quickly in the winter. A majority of the watershed is urban or agricultural, and the lake suffers from runoff pollution. Most of the water arrives via the Detroit River, and "wind setups"—or wind-driven pushing of water—produce short-term water level changes of up to 16 feet from east to west. Shallow, warm conditions in the summer make Erie one of the most biologically productive of the Great Lakes. Dredging is very concentrated here due to the location of large commercial ports.

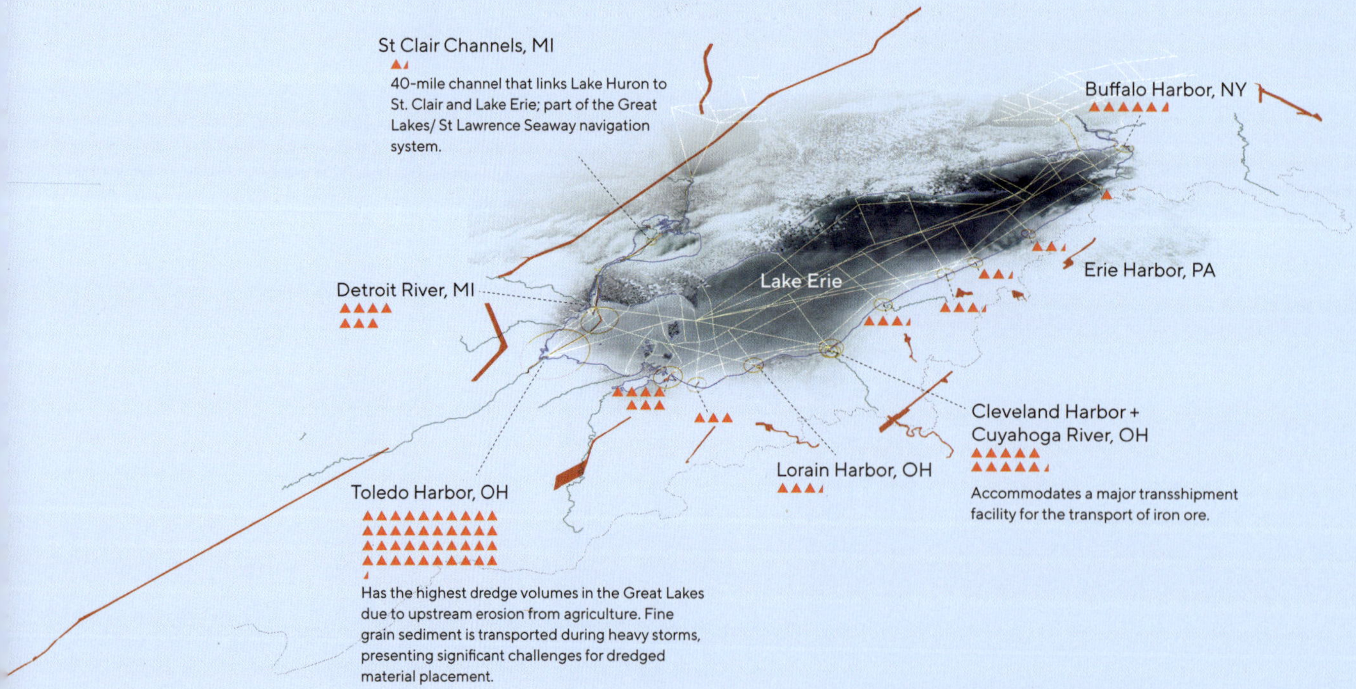

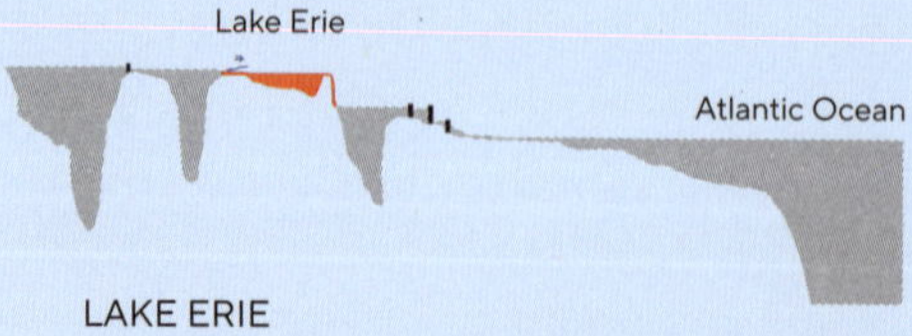

▲ 100,000 cy dredged material by USACE, 2009-2013

USACE maintained channels

USACE dredging contract zone, 2009-2013

Waterways / shipping routes

LAKE ERIE

Shoreline length:	871 miles
Volume:	116 cubic miles
Retention Time:	2.7 years
Population:	10.5 M US / 890k CAN

Lake Superior

Despite its sparse population, Lake Superior hosts the largest port by tonnage in the Great Lakes, the Port of Duluth-Superior.

Shipping to and from this bulk commodity port is enabled and limited by the Soo Lock and St. Mary's River, at the eastern edge of the lake. While Lake Superior is the largest and deepest of the five lakes, St. Mary's River has been identified by shipping companies as the limiting depth segment for the entire Great Lakes navigation system, and thus a pinch point of national commerce.

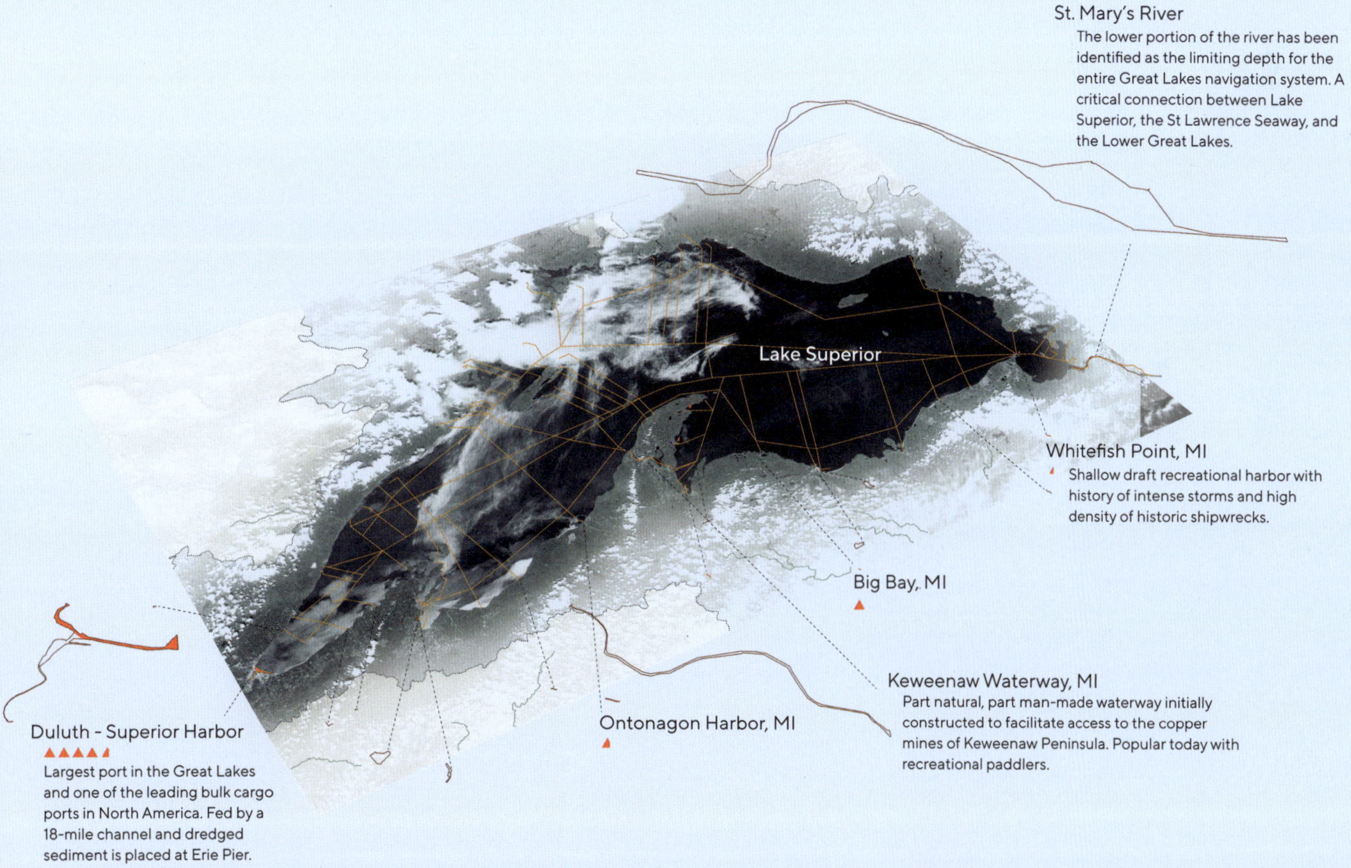

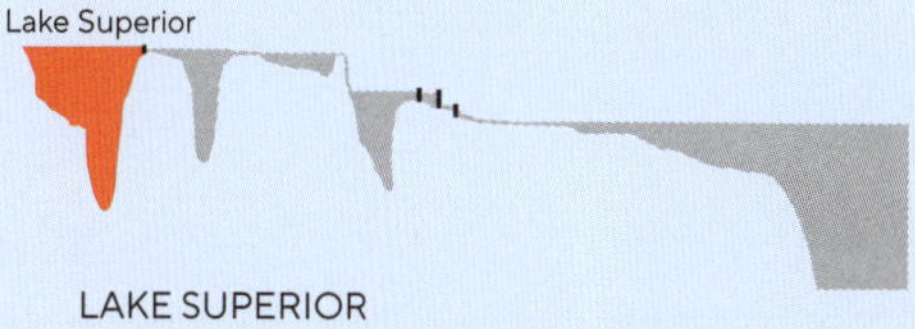

LAKE SUPERIOR

Shoreline length:	2,730 miles
Volume:	2,935 cubic miles
Retention Time:	173 years
Population:	444K US / 216k CAN

Lake Ontario

Lake Ontario provides a critical link between the Great Lakes and the Atlantic Ocean through seaways and channels jointly managed by Canada and the US.

The United States shoreline is less developed than the Canadian side, which borders the most populated province, including the urban industrial centers of Hamilton and Toronto. These cities have historically commanded higher dredge volumes than the US federal shores, though the lake itself provides a critical connection to the St. Lawrence Seaway and the Atlantic Ocean beyond.

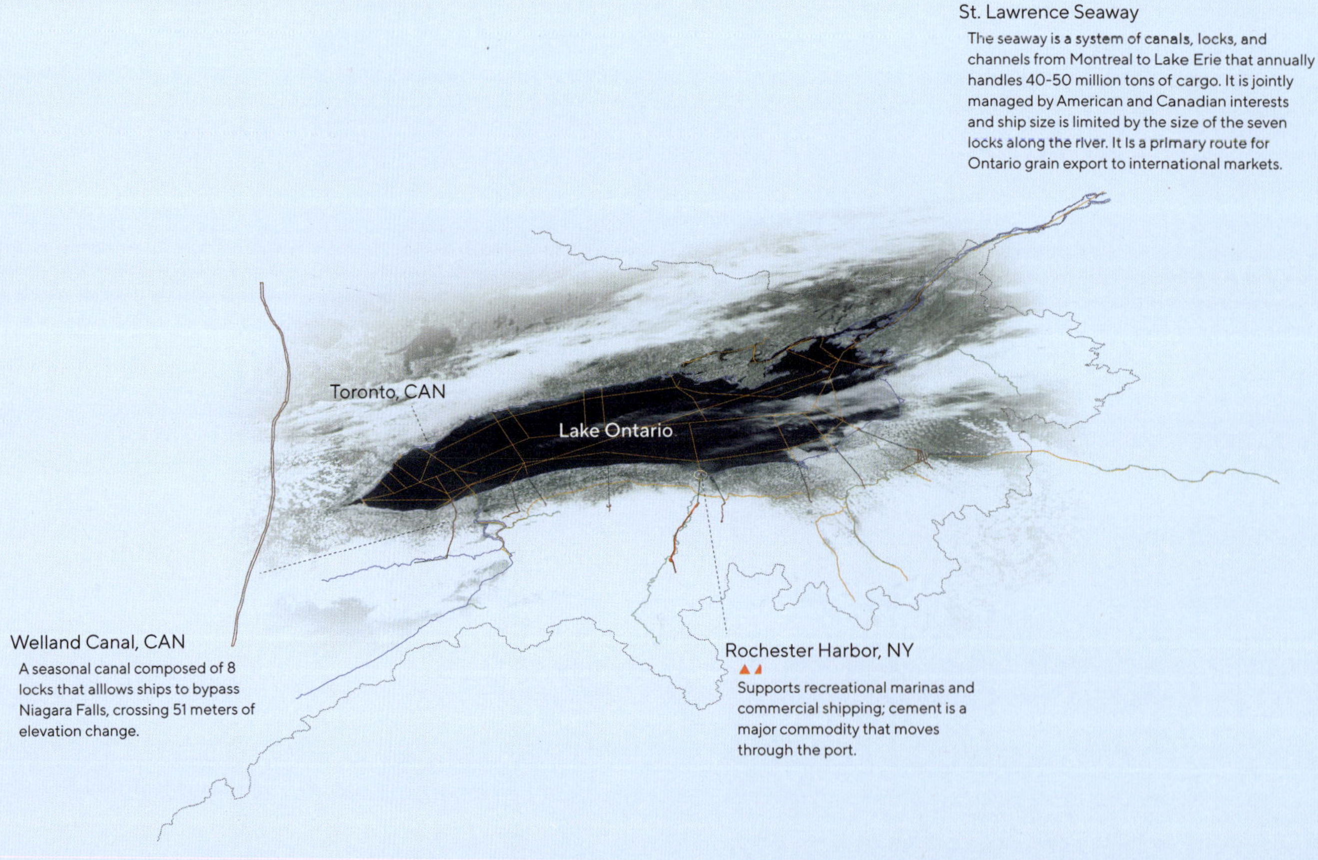

▲ 100,000 cy dredged material by USACE, 2009-2013

USACE maintained channels

USACE dredging contract zone, 2009-2013

Waterways / shipping routes

LAKE ONTARIO

Shoreline length:	712 miles
Volume:	393 cubic miles
Retention Time:	6 years
Population:	7.8 M US / 5 M CAN

Lake Huron

While Huron has the longest shoreline of the Great Lakes, dredging volumes are relatively low compared to the rest of the basin.

Lake Huron is characterized by sandy beaches, rocky shorelines, and over 30,000 islands, including the world's largest freshwater island, Manitoulin. Lake Huron and Lake Michigan are considered to be one lake hydrologically, due to their connection at the Straits of Mackinac. Near this connection, a unique configuration of 7.5 miles of dredge channel services the "subsistence harbor" of the Les Cheneaux archipelago, the sole link to the mainland for this island home community.[9]

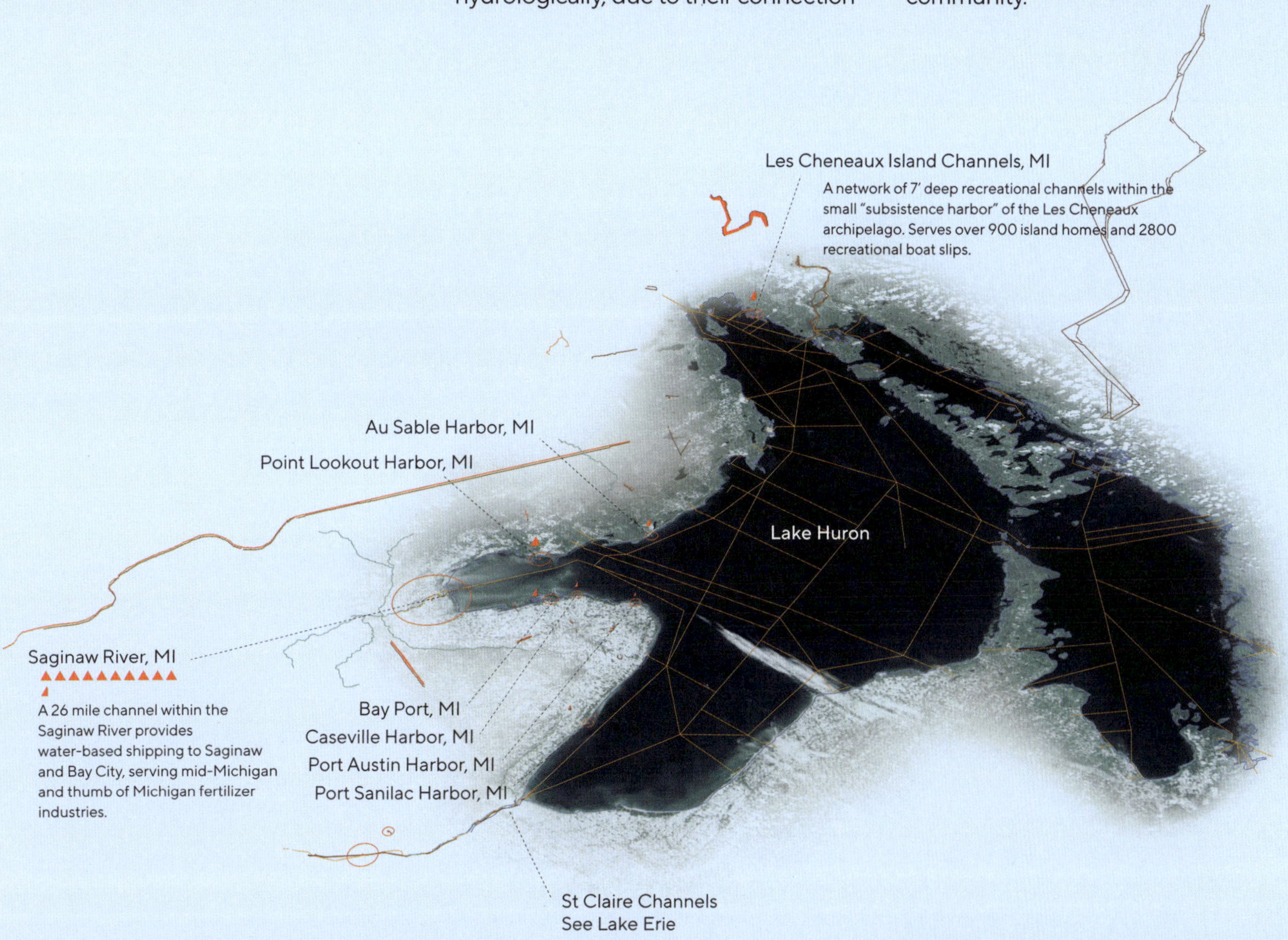

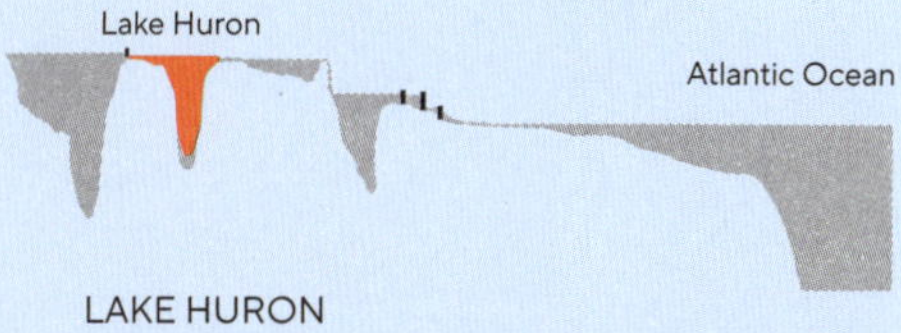

LAKE HURON

Shoreline length:	3,830 miles
Volume:	849 cubic miles
Retention Time:	21 years
Population:	1.5 M US / 400k CAN

Rethinking Placement

While each lake has its unique conditions, CDFs continue to be the common response for dredge placement despite the inherent physical, ecological, and social challenges to this approach. The following pages feature Great Lakes projects that treat clean sediment as a resource, testing new ways to introduce it back into environments and active systems.

In Port Bay, New York, a pilot project tests passive dredge management for a small navigation channel adjacent to an active marina.[10] Clean dredged material was placed in a bell-shaped form along the shore, designed to erode and nourish the adjacent barrier bar shoreline that is susceptible to erosion. Within the cycle of a single storm, the material was transported along the shore and reworked by the waves and wind. The pilot includes ongoing community engagement, habitat monitoring, and remote sensing.

This project was developed by the Healthy Port Futures initiative, a collaboration initiated by members of the Dredge Research Collaborative, Sean Burkholder and Brian Davis, along with Research Associate Tess Ruswick.

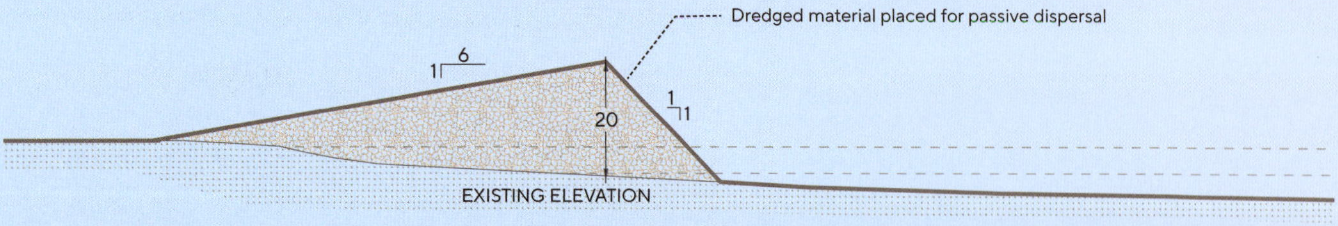

Dredged material placed on 4/29/2020 (Lindsey Gerstenslager)

Passive management area on 5/5/2020 (Lindsey Gerstenslager)

Dredged material placement on 4/29/2020 (Healthy Port Futures / Ramboll Engineering)

Sorting Dredge

The Placement and Reuse Facility (PRF) at Erie Pier in Duluth, MN highlights a potential future for dredged sediment. Rather than a site of long-term disposal, the facility separates dredged sediment into graded material that can be resold and reused locally.

DredgeFest Great Lakes tour of Erie Pier

Dredge as Habitat

The Cat Island Dredged Materials Disposal Facility demonstrates the shifting environmental sensibility of dredge management in the Great Lakes. A 4.2-mile constructed squiggle serves as a wave barrier and protects what will ultimately be 1,225 acres of shallow water and wetland habitat created through the placement of clean dredge.[11]

The first CDF built to store dredge from the Port of Green Bay, Renard Island, was built in 1979 and is at capacity only 17 years later.

Each of the reconstructed "islands" are contained on three sides by dikes. The dikes act simultaneously as a framework and wave barrier, encouraging new wildlife areas as the islands are slowly filled in with dredged sediment over the next 25 years.

The Cat Island Dredged Materials Disposal Facility is located on the footprint of the historic islands, before they eroded away.

Placement of dredged material to build Cat Island (Sean Burkholder/Above All Ohio)

Slowing Shoreline Erosion

Sediment can be dredged where it accumulates in a navigation channel, or it can be thoughtfully managed closer to the source through shoreline stabilization and erosion reduction.

North of Chicago, a project at Illinois Beach State Park tests rubble ridges as a low-impact method of shoreline stabilization that can potentially reduce the need for future dredging. Unlike conventional riprap revetments that harden the shore, the rubble ridges aim to reduce erosion of the park edge by slowing wave action offshore.[12] The rubble ridges work together as a system, creating a wide feature where no single row must do too much. This innovation is intended to make the feature less prone to failure, creates a wider zone of nearshore complexity that may become habitat, and enables significant cost savings because the construction techniques are simpler.

This project was developed by the Healthy Port Futures initiative, a collaboration conceived by members of the Dredge Research Collaborative, Sean Burkholder and Brian Davis, along with Research Associate Tess Ruswick.

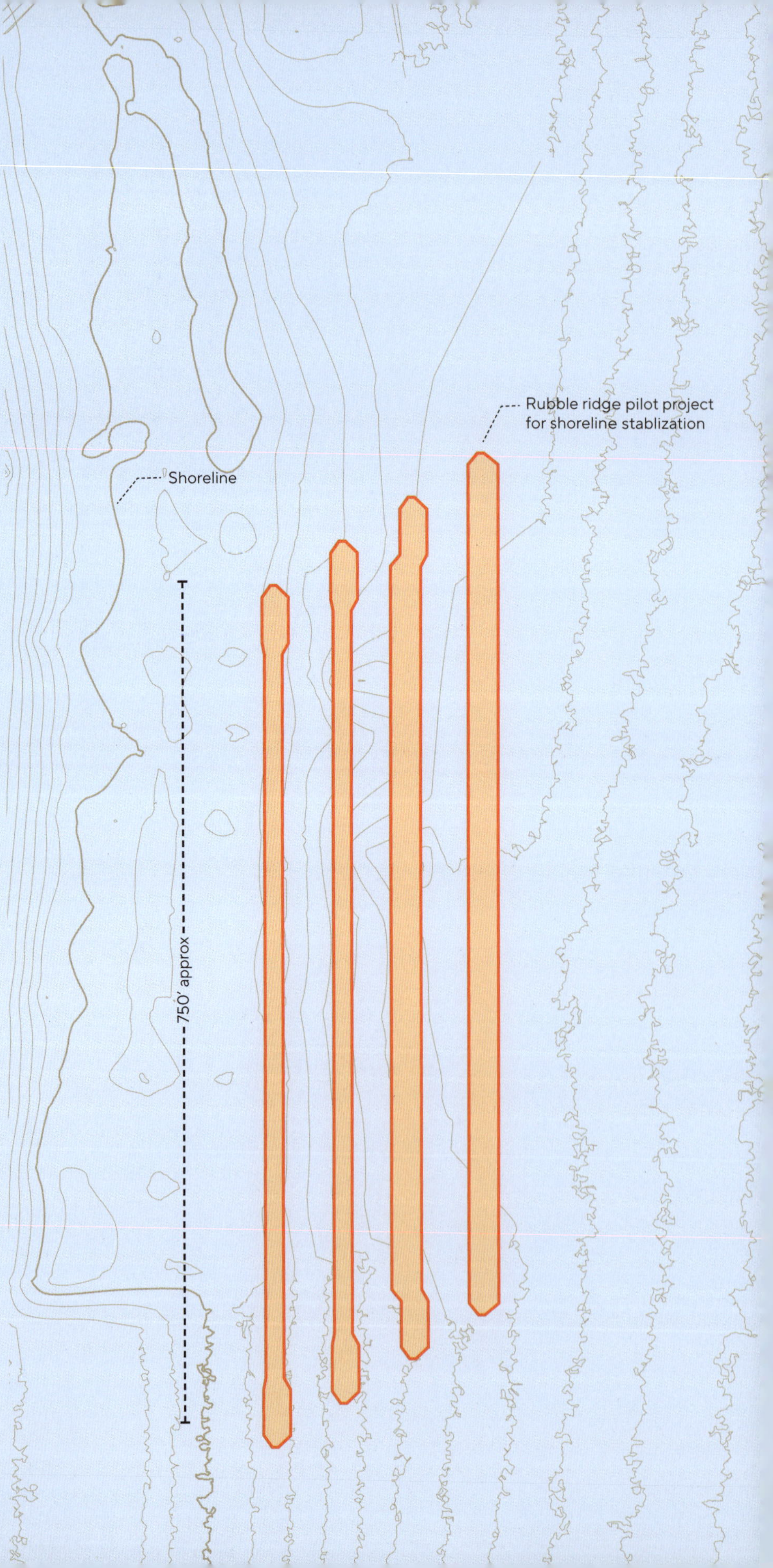

NAD 1983 (2011) StatePlane Illinois East FIPS 1201 (US Feet), IGLD85

Rubble ridges for shoreline stabilization (Sean Burkholder, 2021)

Endnotes:

1. Great Lakes and Ohio River Division, US Army Corps of Engineers. Great Lakes System Dredged Material Management Long Term Strategic Plan. April 2012. https://www.lre.usace.army.mil/Portals/69/docs/Navigation/Great%20Lakes%20Dredged%20Material%20Management%20Long%20Term%20Strategic%20Plan.pdf

2. Ibid.

3. Ibid.

4. Ibid.

5. "Confined Disposal Facility (CDF)." Friends of the Parks. Accessed April 22, 2022. https://www.fotp.org/confined-disposal-facility-cdf.html

6. Ibid.

7. "Calumet Harbor and River/Chicago Area Waterway System Dredged Material Disposal Facility (DMDF)." Chicago District, US Army Corps of Engineers. Accessed April 22, 2022. https://www.lrc.usace.army.mil/Missions/Civil-Works-Projects/Calumet-Harbor-and-River/.

8. Lee, William. "Residents Fight Army Corps Plan to Expand Sludge Tank at the Mouth of the Calumet River Instead of Building a Park." Chicago Tribune, March 26, 2021. https://www.chicagotribune.com/news/environment/ct-environmental-justice-sludge-tank-calumet-river-20210325-z5pfoshoorfuvnnakacxy5bdqm-story.html.

9. US Army Corps of Engineers, Final Environmental Statement: Confined Disposal Facility and Maintenance Dredging of the Lex Cheneaux Island Federal Navigation Channels, Michigan. Detroit, Michigan. January 1979. https://apps.dtic.mil/sti/pdfs/ADA107092.pdf

10. Sean Burkholder, Brian Davis, Theresa Ruswick. Healthy Port Futures: Port Bay Sediment Engine. August 2021. Accessed 22 April 2022. https://healthyportfutures.com/wp-content/uploads/2022/02/PortBayReport.pdf

11. "Restoring the Cat Island Chain in Green Bay, Wisconsin." US Department of the Interior, September 3, 2020. https://www.doi.gov/restoration/restoring-cat-island-chain-green-bay-wisconsin.

12. Barnes, Laura L. "Underwater Innovation at Illinois Beach State Park to Help Mitigate Coastal Erosion." Illinois Sustainable Technology Center blog, February 23, 2022. https://blog.istc.illinois.edu/2022/02/18/underwater-innovation-at-illinois-beach-state-park-to-help-mitigate-coastal-erosion/.

Image Credits:

Spread 1:
Burkholder, Sean and Above All Ohio. "Cat Island." (photograph).

Spread 2:
Great Lakes Aquatic Habitat Framework. Stream Networks, Reach Catchments, Watershed Boundaries: Great Lakes Hydrography Dataset. Shapefile. 2015. https://www.glahf.org/watersheds/
US Bureau of Transportation Statistics. US Ports. Shapefile. 2016. https://koordinates.com/layer/22813-us-ports/
United States Army Corps of Engineers (USACE), US Department of Transportation (USDOT)/Bureau of Transportation Statistics (BTS). Navigable Waterway Lines. Shapefile. 2006. https://data-usdot.opendata.arcgis.com/datasets/c91bc22014744f9389488f8ca203b4d4_0/about

Spread 3:
Burkholder, Sean and Above All Ohio. "Toledo CDF." (photograph).

Spread 4:
CDF diagram adapted from:
"Options for Dredged Material Placement / Decontamination Technologies." Environmental Protection Agency. Accessed April 22, 2022. https://archive.epa.gov/region02/water/dredge/web/html/disposal3.html.
Burkholder, Sean. "Times Beach." (photograph).
Burkholder, Sean and Above All Ohio. "Erie Harbor CDF." (photograph).
Burkholder, Sean. "Dike 14." (photograph).

Spread 5:
Great Lakes Aquatic Habitat Framework. Stream Networks, Reach Catchments, Watershed Boundaries: Great Lakes Hydrography Dataset. Shapefile. 2015. https://www.glahf.org/watersheds/
US Bureau of Transportation Statistics. US Ports. Shapefile. 2016. https://koordinates.com/layer/22813-us-ports/
United States Army Corps of Engineers (USACE), US Department of Transportation (USDOT)/Bureau of Transportation Statistics (BTS). Navigable Waterway Lines. Shapefile. 2006. https://data-usdot.opendata.arcgis.com/datasets/c91bc22014744f9389488f8ca203b4d4_0/about
American Steamship Company. Accessed April 23, 2022. http://www.americansteamship.com/great-lakes-trade-patterns.php.

Spread 6:
(left)
US Geological Survey. National Landcover Dataset. Shapefile. 2016. https://www.mrlc.gov/data?f%5B0%5D=category%3Aland%20cover&f%5B1%5D=year%3A2016

Spread 7:
McGuire, Mary Pat. "Chicago CDF." (Photograph). 2022.

Spread 8-10:
All lake profile maps were developed with the following data sources:
US Army Corps of Engineers, Institute for Water Resources. "Dredging Contracts Awarded." 2009–2013. https://usace.contentdm.oclc.org/digital/collection/p16021coll2/id/3046
US Bureau of Transportation Statistics. US Ports. Shapefile. 2016. https://koordinates.com/layer/22813-us-ports/
United States Army Corps of Engineers (USACE), US Department of Transportation (USDOT)/Bureau of Transportation Statistics (BTS). Navigable Waterway Lines. Shapefile. 2006. https://data-usdot.opendata.arcgis.com/datasets/c91bc22014744f9389488f8ca203b4d4_0/about
US Army Corps of Engineers. National Channel Framework. 2017. Shapefile. https://geospatial-usace.opendata.arcgis.com/maps/9227967a2748410983352b501c0c7b39/about
Great Lakes Aquatic Habitat Framework. Stream Networks, Reach Catchments, Watershed Boundaries: Great Lakes Hydrography Dataset. Shapefile. 2015. https://www.glahf.org/watersheds/
"About Our Great Lakes: Lake by Lake Profiles." About Our Great Lakes. NOAA Great Lakes Environmental Research Laboratory - Ann Arbor, MI, USA. Accessed April 22, 2022. https://www.glerl.noaa.gov/education/ourlakes/lakes.html.

Spread 11:
(left)
Burkholder, Sean. "Dike 14." (photograph).
Gerstenslager, Lindsey. "Port Bay: Dredged Material Placed on 4/29/2020." (Photograph). 2020.
Gerstenslager, Lindsey. "Port Bay: Passive Management Area on 5/5/2020." (Photograph). 2020.
(right)
Healthy Port Futures and Ramboll Engineering. "Dredged Material Placement on 4/29/2020." (Photograph). 2020.

Spread 12:
Milligan, Brett / Dredge Research Collaborative. "DredgeFest Great Lakes Tour of Erie Pier." (Photograph). 2015.

Spread 13:
Burkholder, Sean and Above All Ohio. "Placement of Dredged Material to Build Cat Island." (photograph).
Detroit District, US Army Corps of Engineers. Phase II Report: Dredge Material Management Plan Study, Green Bay Harbor, Wisconsin. Vol 1. July 2011. 9. https://www.lre.usace.army.mil/Portals/69/docs/PPPM/PlanningandStudies/GBDMMP/GreenBayDMMP2.pdf

Spread 14:
(left)
Adapted from technical drawing provided by Healthy Port Futures.
"Illinois Beach State Park." Healthy Port Futures. Accessed April 22, 2022. https://healthyportfutures.com/project/illinois-beach-state-park/.
(right)
Burkholder, Sean. "Rubble ridges for shoreline stabilization." (Photograph). 2021.

Juanita Irizarry: Parks, Sediment and Environmental Justice

Juanita Irizarry is the executive director of Friends of the Parks, a parks advocacy organization in Chicago. She has spent her lifetime in the realms of political and community organizing and development in favor of healthy, mixed-income communities, largely inspired by her lifelong social location in Chicago's Humboldt Park community.

DRC: Can you briefly tell us about the mission of Friends of the Parks and your role there?

JI: Friends of the Parks is about 47 years old. We started in the context of Chicago's park system being under-maintained, with community residents and advocates getting together to advocate for better stewardship of our park spaces for public use. We also recognized that fighting for good parks also meant fighting government corruption. In the early 80s, the Department of Justice came in and sued the Chicago Park District for discrimination against minority communities and won that case, which lead to the Park District functioning under a consent decree to try to remedy some of those things for a while.

During that process, Friends of the Park was contracted to work across the city to build out a system of Park Advisory Councils to have the community more actively engaged in looking at what was going on in our local parks; being aware and engaged and stewarding and advocating. So a big part of how we do our work is connected to those local Park Advisory Councils or other park partners in order to understand what the issues are in their communities and inform what things we advocate about. We are also really trying to listen to the voices of the community around what the important issues are around parks, and certainly environmental justice has been a big call lately—particularly in the southeast side of Chicago.

The work of the [Chicago] Confined Disposal Facility is related to that work because of its location, but it also happens to be smack in the middle of our vision for completing a lakefront park and path system.

DRC: Can you tell us about the Chicago CDF and its impacts on local communities?

JI: The Chicago CDF was created in the early 1980s and was supposed to be closed 10 years later. It's there because the confluence of the Calumet River and the lakefront is an important location as shipping comes into Chicago, into the Illinois International Port District. However, the river needs to be dredged to make it deeper, and create more space for that shipping to come through. It's also in a location that has a deep, deep history of steel industry in Chicago.

And while the steel mills around that area are closed, there's a lot of pollution left behind, and the area has been zoned for a lot of industrial work that continues to contribute to pollution in the area... pollution washes into the river still these days. The dredge is toxic and has to be put somewhere, and the Army Corps has been allowed —despite the General Assembly's original intent that [the CDF] only be for 10 years—to continue and continue and continue to use this site right on our lakefront, which is now also subject to massive erosion challenges.

DRC: Has there been a public, open decision-making process for the expansion?

JI: We have public processes all the time; whether we think they're effective, useful, or open is a whole other question, right? We and others who are moving forward were under the understanding that 2022 was the latest goal date for closure of this CDF site, and it was a surprise to us when the Army Corps announced, after the public process that they did, that they decided to keep the CDF on site and expand it. The public process was probably more "public" in terms of how many people were engaged because of our engagement and attention to this site, because we have been waiting for the day when [the CDF] would finally be capped and turned over to the Park District to become a park. The Corps went through the steps that they were required to go through, but we had to force them to make more information available to the public that was supposed to be public information, that they did not automatically provide. So it's been a process of lots of engagement, and now we're bringing lawyers to the table as well to keep forcing appropriate public engagement and government oversight.

DRC: What do you see is missing in that public engagement?

JI: As with many community public processes, you make an announcement, you don't really care if anyone actually shows up, you have a meeting that may or may not be of the type that's actually open to the public. And then you say you did it, even if you didn't get very much engagement, right?

What was communicated to the community [in these meetings] was that they were going to move [the CDF], and that's what we were all understanding. The question was, "How do we get to the right place in terms of where we're gonna move?" But then, "Oh, we're not moving, we're just staying here and building higher" happened, and that was in a very tight timeframe in terms of when they were going to run out of space. Very tight.

...the actual idea of keeping the CDF where it is and building it 25' higher on our lakefront was not put on the table as one of the options. The community only vetted the [alternate sites]. But the alternate sites did not offer any alternatives outside of moving to another site, whether within the community or another environmental justice community. So then they just came back and said, "Well, then this is what we have to do." Wait a minute, that wasn't even part of the conversation. And if you had told us that that was part of the conversation, maybe that would have influenced other parts of the conversation.

DRC: So where are things now? Are other sites being sought, is it moving, is this having an effect, what you're doing, to actually adjust the process?

JI: We and the Alliance of the Southeast have engaged Environmental Law and Policy Center as our lawyers, and together we have engaged the Illinois EPA in their process of reviewing the permit that the Army Corps put forth to expand. And per what we forced during that process, the Illinois EPA went back and changed their original approach. They have made the Army Corps apply for additional permits because our lawyers helped to lift up the fact that they weren't actually applying for the right permits and the Illinois EPA was about to let them get away with that. We've got six months right now before we will get to another moment where we know if the Illinois EPA is going to let them go forward. And in the meantime, we as advocates are talking to—as we have been for the last two years—every decision-maker that we can to try to get others to step in and stop it. One of the challenges is that the city of Chicago has signed off as a cost-share partner. So, the city could decide not to do that and pretty much make them go back to the drawing board. But our assessment is that the city doesn't want to alienate the Army Corps of Engineers that they depend on deeply for other lakeshore erosion management things that need to be done along our lakefront.

DRC: Are there alternative sites or management practices that Friends of the Park have identified as preferable to this site, or the other environmental justice community sites?

JI: Friends of the Park is trying really hard not to be the one who says what should happen because the community should say that. The community has been very clear that they don't want to advocate for it to be moved from there and just dumped in another environmental justice location.

The solutions talk about a number of things. Number one, limiting the amount of dredge that actually needs to go into the dump, because it's not all toxic; there's beneficial reuse that could be happening with some of it that the Army Corps isn't bothering to figure out. There are also strategies that could happen to minimize the amount of new dredge, new material going into the river that needs to be dealt with in terms of the amount of toxic dredge. We're trying to get them to think about that as well because that would change the solutions.

DRC: How do you try to get people mobilized to take an interest in these things?

JI: Well, we have never said to anyone, "Come get excited about sediment!" We spend a lot of time just being sensitive to what people's entry points are.

Some of our folks are very much environmentalists and very much interested in nature, generally.

Other people are interested in their parks because their kid needs a playground or sports field or program or summer camp. I grew up in a community, Humboldt Park, which is the heart of the Puerto Rican community in Chicago, and the park is an extension of our culture. It's our cultural expression, it's our way of showing that "we shall not be moved" in the midst of gentrification, so people in my community may associate the park with our cultural expression more than the natural features of the park. That and baseball, baseball, baseball, baseball in the Puerto Rican community. And four generations of family out to watch your loved one play in the baseball finals, the playoffs. So we try to be very receptive to the fact that people are going to have all of their different reasons for why they care about the park.

DRC: I'm guessing there's also a strong component there about education?

JI: We are [trying to get] kids to care about the lake and get to the lake. 25% of the kids we work with have never been to the lakefront. Just getting them to feel like the parks are theirs, and maybe they'll find their way into a lifestyle that engages it, or maybe even careers that are related to it. That's our younger kids. Our high school kids actually get involved in what we talk about: "parktivism," park activism. In that case we have done projects specifically about the CDF to get them engaged in what it is.

We [also] have a really strong media voice and quick, easy access to the media. So, we do get opportunities to lift up these issues in the community. This has been further bolstered because our CDF is in an environmental justice community where there's a massive fight over this removal of a recycling plant from a wealthy white neighborhood to a black and brown environmental justice community. The timing of that also meant that it was easier to get public attention to other issues of environmental injustice.
We also have a lecture series that we do, and our first lecture of the year in 2020 was specifically about the CDF and what it means for it to be on our lakeshore, being expanded on our lakeshore, in the midst of the highest lake levels we've ever seen.

One more thing about the CDF that I didn't say. The Army Corps keeps on saying that they're required by law to do the least cost alternative, and that's why they're doing this.

DRC: Right, the Federal Standard.

JI: And so some of our conversation with our legislators are, "Is it the least cost alternative when they're using Park District land and not paying the Park District to do that on there? Why is it not in the calculus that they have to count the use of the land that has already been conveyed to the Park District but continues to be used as a dump?" There are things that are not considered part of the "cost." Maybe that should change.

Liminal Lakeshores

Sean Burkholder

In most instances, at least for me, the idea of a "lake" creates an image of a large still body of water. I believe that I am not the only one with this initial image. In a lovely book titled *Still Waters*, Curt Stager laces together tales of a series of lakes from across the world. Throughout the text, the presentation of these lakes as quiet, and at times, downright romantic subjects proliferates. From a peaceful start at Walden Pond reflecting on the misanthropic father of American transcendentalism to his personal account of Siberia's giant Lake Baikal, Stager and his photographs rouse images of mirrors and tranquility. And while it is a delightful book, I think it misses just how sneaky lakes can be. No one goes to the beach and wonders if there will be waves, but a 10-foot swell crashing into the shore of Lake Michigan still seems to surprise people, even people that have lived their whole lives there. I am not sure why this is. It feels like that game where you can only move when the other person is looking in the other direction. While you look at the lake, it sits still, some glorious reflection of the distant mountains or the clouds staring back at you from its mirror surface. Turn your back, and all hell breaks loose. One storm can wash away hundreds of feet of land, but of course you hardly ever see it happen, you just turn your back, and it is gone. Then you turn back around, and things are peaceful again—blue skies and a bald eagle sitting atop the coastal debris that was once your lawnmower shed as if nothing ever happened.

Of course, I am being somewhat hyperbolic here. Ask a landowner with a shoreline on the Great Lakes if the place is always peaceful, and they will likely have some story about how they could play soccer between the house and shoreline when they were kids. Now they are throwing rocks, concrete, cars, and refrigerators into the lake in a desperate attempt to stop the erosion that has moved that shoreline within feet of their home. It took some time, but keep your back turned long enough, and that sneaky lake will make some changes.[1] I am not a lakefront landowner, but I can attest that it does not take much to kick up waves on the shallow puddle that is Lake Erie that are enough to swallow your boat on what had seemed a peaceful afternoon. Days like that make you question just what pounds of fried perch are worth to you. Lakes are anything but *Still Waters*.

In 1975, the 750-foot ore freighter *Edmund Fitzgerald* was sent to the bottom of Lake Superior by open-lake waves over 25 feet tall. Such a wall of water can also come crashing into the shore as a 20-foot wave did during an 1844 seiche in Buffalo, New York, killing near 80 people. A seiche occurs when wind pushes water across the length of a waterbody, creating a wave run-up, or rise in the water level accompanying a wave, on the leeward side that can exceed 10 feet (or at least it can on Lake Erie). Combine that rise with a 10-foot wave, and you are bound to see some damage. And yet, these events are not that uncommon, as I have watched ice chunks the size of cars get tossed (or very slowly heaved) onto the shores of lakefront properties during seiche events, or rivers running in reverse as the wind-driven waters push back up into their watersheds.

And still we think of lakes as quiet things. This assumed quiescence of lakes often underlies and exacerbates misinformed decision-making. Not unlike larger discussions of climate change, the transformation of lakes happens just slow enough to allow your attention to drift elsewhere. This mindset goes beyond assumed inactivity to a foolhardy belief in stability, prediction, and control. This perception, that they are controllable and predictable, dictates the future of so many lake landscapes.

Young and Out of Control

Another image that comes to mind regarding lakes is the image of something that has been around for a very long time. Lakes look old, big, and still, like a grandfather on the sofa. And we assume they have been around forever and will endure somewhere close to eternity. This is, however, not the case. While it is hard to imagine the continent of North America without the massive Great Lakes in the center, there was a time, even a time when bipedal primates roamed the planet, where the Great Lakes did not exist at all. In geologic terms, the Great Lakes are very young.[2] Carved by ice during a series of glaciations, lakes were created, erased, and recast over and over again. Geologists have given them fresh new names every time this happened. Whittlesey, Maumee, Chicago, Warren, Wayne, Duluth, Grassmere, Lundy, Algonquin, and Iroquois are some of the designations given to the continually shifting lacustrine landscape of the region.

For an example of all of this shifting, imagine standing on the shore of Saginaw Bay, in what is now the crotch of Michigan's "mitten." From there, what looks like a clear and objective shoreline has moved nearly 50 miles both inland (as the coast of Lake Saginaw) and outward into the bay (as Lake Stanley) over only 5,000 years, thanks to the regular glacial rasping of the Pleistocene. Geologically, this is freakishly fast and correlates with a huge volume of sediment being moved around by both water and ice. But in the minds of residents, policymakers, and land managers, even fast geology goes unnoticed and is seldom considered. So we get bored or distracted by faster moving things, turn our back, and well, you know what happens next.

Modern North Americans generally assume that we possess the power to control lakes. The water levels of two of the Great Lakes are said to be "controlled" or "managed" by way of locks and dams at their outflow location. Lake Superior, the first (or highest) lake in the system, has its levels regulated by the Soo Locks leading into Lake Huron, while Lake Ontario's levels are managed by the International Joint Commission through the locks of the St. Lawrence Seaway.[3] This infrastructure leads local residents to believe that stability is possible and that constant lake levels are desirable. But after 16 years of scientific study, the IJC concluded in 2016 that the fluctuating lake levels of Lake Ontario were more valuable (and realistic) than attempting to maintain a constant water elevation as the initial plan from 1958 had intended. Both plans aimed to protect coastal communities and infrastructure from extreme water levels. Still, the strictly maintained water levels associated with "Plan 1958D" have been shown to reduce ecosystem diversity and allow an overpopulation of *typha* in the coastal wetlands of Lake Ontario.[4] And this 2016 finding was not the first time that the IJC exhaustively studied the possibility of controlling lake levels. In the 1970s and 1980s they conducted several studies, including one specifically for Lake Erie, that also concluded that the complete control of lake levels was not technically feasible.[5]

Lake politics have been shaped by a series of competing interests and misinformation, as observed in fights in newspapers, social media, and public meetings. Landowners are most fearful of erosion—losing their land and property to the destructive forces of water and waves. The shipping industry, by contrast, needs deep water to transport goods, and hydropower operators need specific flow rates for effective power generation at peak time; so those folks (assuming they do not own a cottage on the lake somewhere) tend to advocate for higher water levels. Federal and state agencies have to find ways to both protect property and streamline transport, in a climate where water levels have been very difficult to predict or to control.

The beginning of each year in the Great Lakes Region is welcomed with a series of Farmers Almanac-like predictions about whether the water will be higher or lower than the year before, and then when these predictions do not describe what happens, the IJC or the USACE are targeted as the failing "controllers" of water levels. Yet the past century of water level measurements in the Great Lakes, transcribed in chart form, reads like an EKG of a failing heart, with no obvious long-term pattern to be found and the stochasticity increasing in recent years.

With this littoral dynamism comes entire industries associated with erosion control and sediment management, both of which are responses to the unpredictable water levels and the weather that moves them all around. These industries tend to remain in a constant state of reaction, unable to completely predict what will be needed each year, and so bracing for what the lakes throw at them. Coastal infrastructure projects deal with this indeterminacy by being scaled for the most extreme events. When sediment management work does get done, it is usually an emergency response, funded by this or that federal or state program as a one-time capital event separated from or ignoring ideas of long-term engagement. This disaster-based way of working can overlook basic issues such as the negative effects of the operations themselves on both local people and ecosystems.[6] The coasts of lakes across the world are bedazzled with the various products of this emergency management, and in the North American Great Lakes, the Confined Disposal Facility stands as an obvious example.

One of the most unique sedimentary conditions that sets lakes apart is their contained, insular, and self-reinforcing nature. In the Great Lakes, this characteristic has led to the development of a menagerie of landscapes of confined disposal, designed to isolate contaminated sediments near where they were initially dredged. Sediment, pollution, or invasive species cannot simply be shunted off into deep water and forgotten. It takes about 300 years for water to move from Lake Superior out through Lake Ontario. Yes, we can throw the stuff we do not want into the attic, putting it out of our minds for a bit, but it is still there in the house, and, eventually, all of that garbage is going to fall through the ceiling onto the dinner table. The decisions we make on or around lakes are often decisions we have to live with for a very long time.

One of the more influential fall-through-the-ceiling moments (at least in relation to the entanglement of sediment management and coastal design) came about with the realization that the sediment dumped into the open lakes was harming water quality. This seems obvious now (although there are still plenty of folks advocating for open-lake placement), but back in the early 1960s, the connection between the toxins carried in dredged sediment and the quality of the water was either non-existent or ignored. With the passing of the Rivers and Harbors Act of 1970 and the Clean Water Act in 1972, dredged material from many of the ports around the Great Lakes was required to be placed in confined disposal facilities—effectively rock-rimmed bathtubs for the containment of dredged sediment—and there was plenty of money to do so.[7]

While the insularity that lakes face could be framed as a problem, it also could be understood as a tremendously valuable opportunity to reconnect communities of all sorts to the processes taking place at the shore. Those confined disposal facilities are going to be around for a while and exist as landscapes in and of themselves, with plants and animal communities leveraging their coastal location and community organizations working together to protect them. Across the Great Lakes, there are many cases of this: Dike 14 near Cleveland became the Cleveland Lakefront Nature Preserve, Point Mouillee in Michigan near Detroit, is a vast waterfowl hunting area, and the CDF in Manitowoc, Wisconsin is a straight-up urban public space, complete with a dog park. The physical location of many of these facilities—near rivermouths, in large cities—means that they also have the opportunity to provide valuable landscape amenities to urban residents. A recent example comes by way of the facility in southern Chicago used to hold material from the Calumet and Chicago Rivers. According to the post-closure plans for the facility, it was slated to be turned into a public park, something that the community around the facility had been anxiously awaiting. However, when the facility reached capacity, and there was still a need for confined disposal, it was decided that the least problematic solution would be to increase the dikes at the facility and continue its use. The community was understandably upset, while the USACE felt justified because they did not have to permit and construct a new facility somewhere else, likely with another unhappy community. Landscapes of sediment management are political, and they have the expectations of both communities and administrators cast upon them. As lakefront urban landscapes, confined disposal facilities provide a particular opportunity for these varied expectations to play out in real-time.

A Lacustrine Sediment Experience

Imagine yourself walking along a compacted soil path, sunken among mullein and goldenrod, toward a giant expanse of water. This path quickly transforms into a rocky dike of ankle-twisting stones leading into the water. The rocks are dark and pitted, not what you would imagine from a piece of coastal infrastructure, which might be expected to consist of bright gray chunks of strategically placed limestone, forming a perfect trapezoid. Instead, the stones here look like hardened lava or even rocks from the moon. Growing between the stones and lining the water's edge in the seeming soil-less substrate is a collection of native dogwood shrubs, their red stems making the path feel almost planted for effect. Turning back from the water, you can see the buildings of downtown Buffalo, New York, standing above the forest of cottonwood trees that constitute the upland region of the landscape. In the foreground is a mottled marsh of *typha* and *phragmites* and a large patch of open water, crowded with loud migrating shorebirds. Deciding to complete the circuit, you come to the remnants of a concrete weir of some kind, made from actual pieces of concrete and limestone. It no longer functions. This flat area, and what could be assumed as shallower water, has provided an opportunity for a stand of *salix* trees to take hold, making a tall fuzzy spot along the dike that is clearly seen from shore. Finishing the walk through the cottonwoods, you surprise a family of whitetail deer, that after quickly hopping up, slowly walk away into the monoculture of *populus* trees.

As the crackling of the *phragmites* obscures the sounds of the cars on the skyway, this walk along the dike wall at the Times Beach Nature Preserve offers an experiential mixing of the ecological with infrastructural attempts at dominance. Like all CDFs, its history is quite context-dependent, even if it was constructed by a federal agency dealing with federal mandates. The Times Beach disposal area was constructed in 1972 atop a beach that had resulted from the construction of the Buffalo Lighthouse and breakwater in the 1820s. Before it was turned into a CDF, the beach had a long history of public use and proposed projects. Known at the turn of the century as "Buffalo's Bohemia" or "The Island," the beach was the location of a shantytown that at its height had a population of near one thousand people. Efforts to evict the squatters on the beach led the local newspaper, the *Buffalo Times*, to begin a campaign to turn the site into a public bathing beach. By 1931 there were lifeguards on duty at the "Times" beach. However, after only a couple of years, the beach was closed due to bacterial contamination. It embarked on a journey through several park proposals and sales as industrial land before being chosen as a sediment placement site in 1967.

If the tumultuous past of the Times Beach were to be any indication of its future, its quiet existence as a CDF would likely not go unchecked. And so it was, in 1976, the local Audubon group, with its strong political affiliations at the time, made a case for the value of the migratory birds that had begun using the site almost immediately after construction, justifying the cessation of the project and its conversion to a preserve. While there are examples of many CDFs becoming nature preserves after they are filled, the half-filled and inactive facility at Times Beach is unique. What exists today would not have been possible without some response to dredging needs and the water pollution crisis, but also not without those initial residents that used and enjoyed this site or the highly connected squad of bird lovers. With its dikes made of the old slag from the adjacent steelworks, emergency breach repairs, and wide-ranging habitat types that came about through nothing more than a sedimentary gradient from upland to open water, Times Beach is a palimpsest of legible histories and agendas. It provides a sediment-contingent experience that connects people back to the processes that make and remake the coastal environment. Leaving the site, you observe a bird-watcher sitting casually on a short and battered retaining wall next to the street. This unassuming piece of infrastructure poking from the ground is a remnant section of the old seawall that was built first in 1840, destroyed by the seiche of 1844, and then reconstructed in 1866. This wall provided the necessary protection of those initial residents to claim this sediment-scape and call it home. The aesthetic here is unapologetically unnatural, ambiguous, troubling, peaceful, and, to many, quite beautiful.

If landscapes like Times Beach teach us anything, it is that these fluctuating places—fluctuating hydrologically, socially, politically, and ecologically—will continue this way. Expectations that they will "grow up and settle down" do not take the past seriously, nor do they acknowledge our rapidly changing and unpredictable future. It would be in our best interest to plan for that fluctuating as opposed to fighting against it. I personally would like to believe that there is something unique about the aesthetic experience of landscapes of sediment management in general, and those of the lake environment more specifically. I would also like to believe that these experiences leverage what makes living on or near a lake so special. Many sediment-contingent landscapes provide us with this opportunity to foster this experience of sedimentary dynamism and all it brings with it ecologically and culturally. We simply have to recognize their potential and dig in.

1 Charles Carter, William Neal, William Haras, and Orrin Pilkey, Jr., *Living with the Lake Erie Shore* (Durham: Duke University Press, 1987). This book nicely illustrates the changing nature of the lakeshore, with a particular focus on erosion.

2 William Ashworth, *The Late, Great Lakes: An Environmental History* (New York: Alfred A. Knopf, 1986).

3 Superior is not really the first lake, as the headwaters of the system are made up of many smaller lakes, including the massive Lake Nipigon, but if only considering the five primary lakes, Superior is the "first."

4 International Joint Commission, "Plan 2014," https://www.ijc.org/en/plan-2014-factsheet.

5 International Great Lakes Levels Board, "Regulation of Great Lakes Water Levels" (International Joint Commission, 1972) https://www.ijc.org/en/regulation-gl-water-levels-report. International Lake Erie Regulation Study Board, "Lake Erie Water Level Study" (International Joint Commission, 1981) https://ijc.org/sites/default/files/K43.pdf.

6 This issue has been discussed elsewhere, particularly considering larger disasters, and probably best by Naomi Klein in her book *The Shock Doctrine* (Picador, 2008).

7 Sean Burkholder, "Designing Dredge: Engaging the Sediment Landscapes of the Great Lakes Basin," *Journal of Landscape Architecture* 11, no.1 (2016): 6–17.

Aesthetics and Publics

1 USACE, "Fact Sheet—Jamaica Bay Marsh Islands, NY."

2 For other pieces of this story, see "Dredge City," "Choreographing Sediment," and "Mud is Vital."

3 Beneficial uses of dredged material typically cost more than standard disposal or stockpiling techniques due to the additional transport required (getting the material to the desired site) and/or the higher costs of emplacing the material.

1. A Party; All are Invited

In 2012, as part of their operations and maintenance dredging program, the New York District of the US Army Corps of Engineers transported and emplaced a quarter million cubic yards of sand in Jamaica Bay.[1] The *beneficial use* of this dredged channel material sought to reverse systemic wetland loss occurring in the bay due to multiple factors, including channel deepening, land filling, hardening and paving, and high rates of pollutant and nutrient runoff.[2] It created about 30 new acres of tidal salt marsh in Jamaica Bay, poured and molded into the form of two islands: Rulers Bar and Black Wall.

None of this sand would have been brought here without local partners, such as the New York State Department of Environmental Conservation and the New York City Department of Environmental Protection, to shoulder the additional cost.[3] And since there was little federal or state funding available to vegetate the new and barren islands, other organizations stepped in, including the Jamaica Bay Littoral Society, Jamaica Bay Ecowatchers, and the National Park Service, along with youth groups like the Girl Scouts, Boy Scouts, 4-H clubs, church groups, and local high schools. More than 500 public volunteers (including 300 young people) were trained by the environmental organizations to perform the labor for this landscape restoration. Platoons of volunteers harvested, processed, and transported marsh grass seeds. They installed and maintained erosion control fencing. They participated in on-site planting and seeding parties. And they monitored the effectiveness of all these efforts to collectively create healthy salt marsh landscapes.

In a typical planting party at Rulers Bar, a crew of between twenty and forty volunteers arrived by boat. Their journey was timed to low tide, when the exposed intertidal zone could be planted with plugs of cordgrass. Some volunteers stood on handmade, wooden "plug templates" to make small holes in the sand. Others filled these holes with the plugs. Over the course of a few hours, these efforts generated a gridded band of plantings along the tidal edge of the island. Each successive planting party extended the work of the one before it, until a continuous planted perimeter wrapped the nurtured, nascent marsh.

In these efforts, diverse and active public engagement supported the making of dredge landscapes with clearly defined intentions. Sediment design here combined the "how-to" knowledge of restoration experts with the physical labor of volunteers. Through social and playful activities that facilitated education

4 For more on civic ecology and co-creative relationships between publics and landscapes, see Krasny and Tidball, *Civic Ecology*, and Maddox and Tidball, *Grassroots to Global*. For an example of how these relationships play out across multiple decades, specific to the work of landscape architects, see Anne Whiston Spirn's work with the West Philadelphia Landscape Project, as documented (for instance) in Raver, "Where the Water Was."

5 For dredging-related examples of this in the southeastern US, see Carse and Lewis, "New Horizons for Dredging Research."

and enjoyment, designers and their allies formed new civic connections that disrupted the status quo of land-making in the bay. These are dredge landscapes that people can physically access, care for, and experience, even if that access requires a boat.

After participating in the planting parties, members of the public not only *knew* about these vulnerable islands, but also had a hand in physically creating them. Through collective actions, they momentarily became part of the land and its formative development. Such experiences carry forward in myriad, unpredictable ways: it might affect what college degree or profession a person pursues; their future engagement in advocacy or land stewardship; or their sense of geographic belonging or ownership or pride. Young people growing up in cities might develop better physical health over the long term, simply by being exposed to a bay ecosystem and developing an affinity for kayaking. These kinds of community-based stewardship practices have come to be called *civic ecology*, and through them new co-evolutionary relationships between people and landscapes are set in motion.[4]

Co-evolution is the mutual or reciprocal evolution that occurs among interacting species and things. As a process, it is fundamentally about the transformative nature of relations and experiences. We humans act on and engage with landscapes; landscapes, in turn, act on those doings and us. At Rulers Bar, people emplaced a massive pile of mud in the middle of a bay and meticulously covered it in plugs of vegetation. And when the parties were over, landscape forces further molded that crude, chia-pet base into a rich tidal marsh ecosystem that now hosts a multitude of inhabitants.

Designing landscapes with sediment is not only about fostering more robust ecological relationships between sand, plants, invertebrates, and birds. It's also about fostering valued and equitable relationships with people, who would often otherwise have limited access to these amphibious landscapes.

* * *

In addition to their material qualities, sediments act socially and politically. Which is to say that they are affected by sociopolitical processes and produce sociopolitical effects, many of which cannot be known or predicted in advance.[5] Dredging practices and dredge landscapes are engineered in specific, deliberate ways, such as through abstract cost-benefit analyses and techno-institutional habits. In turn, this work reformats conditions for living beings and landscape processes—oxygen levels and water temperature for fish, nesting habitats for birds, changes in tidal velocities and inundation patterns—all of which are dependent on relationships that envelop them. These effects extend far beyond the small number of people directly involved in their making to much larger publics. Yet sediments clearly do not affect everyone equally or in the same way. Rather, their effects are unevenly distributed and specific to local geographic, political, and cultural conditions, such as whether one dwells in a floodplain, an upland hillside, a polluted canal, or on an eroding coastline. Collectively, these are issues of environmental justice.

In "Choreographing Sediment" and "Mud is Vital," we have focused on understanding the predominant technocratic systems that govern sediment management as it exists today. Both now and historically, these management systems have overshadowed the role of publics as constituencies and actors in how dredge landscapes are created and experienced. We contend that crucial, effective means for improving the design of these landscapes lie in diversifying and expanding experiential access to them. Designers can and should intervene in the politics of access: whether or not it is provided, and what its qualities, limits, and potentials are. Who does and does not have access to participation and decision-making in the planning and design of these landscapes? Who can and cannot physically access and experience them? Who is subject to their effects, benefits, and vulnerabilities?

Sediment is a form of infrastructure that is granular, aggregative, malleable, and adaptive. It is the literal *grounding* on which so much depends, be it property values, wetlands, deep water shipping channels, or dredge-nourished beaches. Its soft moldability invites human alteration, such as the deepening of a shipping channel. At the same time, it resists and responds to human desires according to its own material proclivities, like its aggregate disposition to flow into and fill deepened channels at higher rates than before dredging occurred, which in turn can trigger any number of human actions to exert more control over the situation.

All infrastructures (and landscapes) are inherently political, in that particular values, motives, conditions, and functions are selected and designed for.[6] To alter or act on sediment and its flows is no exception. The actual making of infrastructure depends upon the timely enrollment of a diverse, relational assembly of materials and actors for it to manifest and perform.[7] As such, infrastructure is always temporal and always plural, made of interactions between many disparate things, like hydraulic dredges, sands and gravels, construction schedules, trained personnel, booster pumps, levees, management protocols, nursery plantings, environmental regulations and the attendant values driving them, and so forth. Scholars have framed infrastructures as socio-technical systems in order to highlight the obscured yet pivotal components of social organization, situated values, and political action involved in their creation and maintenance.[8]

Such scholarship moves beyond understanding infrastructure as material "hardware"—such as steel and concrete—to understand infrastructures from a variety of ontological positions: as ecologies, as social structures and contracts, as aspirational fictions, as sites of political struggle, as works of maintenance and repair, as landscapes. Uniting all these positions is an understanding of infrastructure as a co-creative process between makers and what is made. People make infrastructures, and those infrastructures change people. Over the past three decades, across various disciplines, critical infrastructure studies have emerged to investigate these co-evolutions and the politically contingent experiences they inscribe.[9] In our work, we seek to ground this understanding of sedimentary infrastructure within the malleable landscapes that they co-create.

Dredging has been a widespread human endeavor for millennia, but only since the industrial revolution, with the exploitation of fossil fuels and the rise of techno-capitalism, have sediments been subject to brute force at massive, tectonic scales,[10] and stripped of value beyond narrowly defined engineering goals. *Efficiency*, as variously interpreted, has been a determining value in this industrialized earthmoving.[11] As discussed in "Mud is Vital," sediments have been regarded as waste, as inefficiencies, as material to be moved out of the way,

6 The infrastructural scholarship on this point is extensive and beyond the bounds of a sidenote. For two of the early, pivotal manuscripts on the topic that gave rise to many more, see:
Star, "The Ethnography of Infrastructure."
Graham and Marvin, *Splintering Urbanism*.

For a useful compendium of how that initial scholarship has evolved across a range of disciplines, see Harvey, Jensen, and Morita, *Infrastructures and Social Complexity*.
More recently, the Green New Deal has brought new vigor into discussions about the socio-political complexities, imaginaries, and possibilities in designing landscapes and infrastructures, including advocacy for designers to be more aware and actively engaged in building political constituencies for equitable public outcomes.
Fleming, "Design and the Green New Deal."

7 Much contemporary scholarship has sought ways to study and understand how these dynamic and complex formations come together, persist, upgrade, decay, breakdown, or never come to be.
Latour, *Aramis*.
Gandy, *The Fabric of Space*.
Hetherington, *Infrastructure, Environment, and Life in the Anthropocene*.
DeLanda, *A New Philosophy of Society*.

8 Harvey, Jensen, and Morita, *Infrastructures and Social Complexity*.
Anand, Gupta, and Appel, *The Promise of Infrastructure*.

9 The Critical Infrastructure Studies Collective, "Critical Infrastructure Studies.org.

10 Vanderostyne and Cohen, "From Hand-Drag to Jumbo."

11 Vanderostyne and Cohen, "From Hand-Drag to Jumbo."

12 There are many pre-industrial precedents for this kind of multiple valuation—for instance, in the cultivation of the Pearl River Delta via shatian practices in the Yuan dynasty period, or the landschop of Frisia. The modern, industrial valorization of efficiency is not a universal human practice, and here we are speaking specifically of transitions within that context. For an example of how agencies such as USACE are working toward evaluating dredging projects within multi-value and multi-stakeholder perspectives, see Collier et al., "Stakeholder Engagement in Dredged Material Management Decisions."

13 For a contemporary global example of this—the expansion of the Panama Canal—see Davis, Holmes, and Milligan, "Isthmus."

14 "Bayou St. John Urban Marsh," Pontchartrain Conservancy.

15 "Bayou St. John Urban Marsh Monitoring Report 2014-2019," Pontchartrain Conservancy.

and discarded as "spoils." Dredged sediments generally had negative values tied to them, since they were seemingly abundant, costly to move, and rarely considered for any other purpose. Only in the last few decades has sediment assumed additional ecological, social, and economic values.[12]

Socially, sediment's negative value was sustained and reinforced by the limited number of siloed, institutional actors involved in decision-making processes—typically the federal and state public agencies, ports and other quasi-public agencies, and contractors and capital-intensive corporations doing the actual work. Within these institutionalized structures, sediment itself had no effective advocates. Sediments were "externalities" that fell outside of logistical calculations.[13]

The dredging industry's economic calculus has been shifting. Institutional actors are embracing more diverse socio-ecological performance criteria, sometimes out of sheer necessity and sometimes for the added gain. But to fully actualize this transition will require a more thorough expansion of the social domain to democratize both experiential access and design decision-making. Can we have more public planting parties? Design more public events timed to moments of mass sediment flows, like boisterous openings of spillways and dams during floods? Recruit more of the people and vulnerable communities affected by dredging and sediment management to participate in their planning and design, to help diversify and improve the objectives that drive infrastructural decisions? Make landscapes that offer richer, more legible, and more diverse experiences of sediment? Can sediment have a more significant and tangible presence in public life, as befits the wisdom and adaptive capacity it helps create?

2. Valuing Experience

Muddy, marshy, land-making parties are not unique to the coastal outskirts of New York's Jamaica Bay. In New Orleans, along Bayou St. John, over 100 public volunteers worked with local organizations to create a half acre of marsh near the mouth of the bayou. This is a highly engineered and highly degraded river waterway, more akin to an urbanized, leveed flood control canal than the wild bayou it once was. But there are ongoing efforts to revive it through a scenic river designation, which mandates "protection and preservation of its aesthetic, scenic, recreation, fish, wildlife, ecological, archeological, botanical, and other natural and physical features."[14]

Near the bayou's protective floodgates, a shoal of sand built up in the center of the channel that impeded water flows during floods. Rather than dredge and dispose of this sediment, the Pontchartrain Conservancy, along with the Lake Pontchartrain Basin Foundation, Restore the Earth Foundation, and the Orleans Levee District, proposed emplacing that sand in the adjacent open sides of the bayou to create vegetated marshes. From 2013 to 2014, volunteers helped build custom sandbag containment walls to hold the dredged sediment in place, and then plant it with over 5,000 plugs of native grasses and sedges.

In the years following those work parties, extensive scientific monitoring of the marshes has deemed them to be quite successful. The evolving plant palette is diverse and continues to expand in coverage,[15] and local anglers and birders have reported more wildlife. The restored pockets of marsh provide migratory refuge and nurseries for a variety of species. The marshes also reduce erosion of the channel's infrastructural bulkheads and levees during flood events, and serve as a living, changing classroom within the urban surrounds.

But let's go back to when the marshes were being made. Imagine walking the trail along the crest of the bayou's levees and seeing a sweaty, muddy gang of volunteers wading around in the bayou's water, wrestling sandbags into a weird formation. Let's say you happen to come by months later and see them at it again, but now you see muddy land behind the sandbags, rather than water. All the people are chatting with each other, shovels in hand, digging holes for little plants. Now switch perspectives and pretend you are one of the volunteers, occasionally looking up to see the curious faces of passersby. Maybe you smile at the looks you are getting, the fun you are having, or the sheer novelty of this experience. Maybe you strike up a conversation, explaining to these hikers, as best you can, the odd work you are doing and why. Shifting perspective again, imagine the local fishers and birders who might have missed these work parties, but who see and inhabit the landscape changes brought about by displacing and emplacing 2,000 cubic yards of sediment, and are thankful for it. Some may be curious enough to investigate why it was done and where else it might be happening. Some might go so far as to get involved with other projects, for their own enjoyment or for their children's prospects.

Clearly we could carry this on. In another time or place, the dredged sand would be simply disposed of, or stockpiled for some more mundane application, like roadway fill. But in this instance, through proactive advocacy and making-do, things went differently. The open-access authoring of this marsh's *becoming* created a sediment public.[16] It created multiple publics, actually, including those involved in the initial advocacy for the design, and those enrolled in the actualized making. Hundreds of people had access to the designing, the making, and the experiencing of the remade landscape. Had civic engagement been more minimal—if it were just industry employees doing this work—the distribution of this sensing would be more limited. We could compare the aesthetic differences between those who have physically and directly experienced Bayou St. John's new marshes, and those who have only done so abstractly, by reading about them on a printed page. The two experiences are radically different, each with their own affordances and limits that create particular realities and worlds. You are in one of those worlds right now.

Why does aesthetics matter here, when we are talking about politics and publics? Let's take a moment to refashion the term itself. One problem with *aesthetics*, in both popular culture and academic realms, is that it is understood as synonymous with visual taste and preferences and associated notions of cultural refinement. It also often assumes a purely visual mode of human sensory perception, limited to certain domains, like art. This is the cultural baggage that comes with the term, but it is *not* what we mean by aesthetics. For us, aesthetics is more akin to experience. It is fundamental to life itself; comprising all that can be sensed and made sense of. Aesthetics encompasses the entire sensorial realm of human and more-than-human experience, and how this collective sensing comes together to form our embodied and situated realities.[17]

Within this aesthetic realm, visual discernment is woven into broader bodily mechanisms through which we think and feel, and influenced by our changing values and knowledge. At any given moment, attention may be more attuned to a particular sense, including the visual, but that attunement happens within

16 Davis, "Public Sediment."

17 For us and many others, aesthetics is not just a privileged domain of humans, and assuming so is itself a limiting and relatively recent cultural habit.
Kohn, *How Forests Think*.
Fuller and Weizman, *Investigative Aesthetics*.

In this book, we provisionally alternate the use of "more-than-human" and "other-than-human" to refer to agents, lifeforms, and materials that are not human. Across many disciplines, both of these phrasings are commonly used in published literature that attempt to bring human sociality into closer and more equitable relationships. Both phrasings have drawbacks. *Non-* and *other-than* can imply othering (as verb) and distancing; *more-than* may imply a reversal of human dominion to that of non-humans taking that position (which some authors may indeed be arguing for). In our use of these terms, we intend neither of these things. We mean only to implicate a diversity of agents in addition to those that are human. These terms are imperfect and likely reflect recent changes in aesthetic and societal values in and of themselves (bringing other-than-human things into the formerly exclusive and superior realm of human agency). The language hasn't yet caught up to where we are headed. Where appropriate and feasible, we try to give names to some of these more-than-human entities, rather than using these catch-all abstractive phrases.

18 Sheppard, "Bridging the Sustainability Gap."

19 Gobster et al., "The Shared Landscape."

20 Massumi, *Politics of Affect*.

21 Erin Manning and Brian Massumi have explored how challenging it is to navigate the world without this reductive coding, such as for many autistic people, who tend to aesthetically experience the world in a more open and less discriminatory way, attending to spatial fields of textured complexity with equal attention. Manning and Massumi, *Thought in the Act*.

22 Massumi, *Politics of Affect*.

a body, which is receiving stimuli to all of its senses, and cannot be detached from other cognitive processes as a "pure" or "abstract" experience. Thus, if you are walking on a levee in New Orleans and see a cohort of sweaty people play-working in the mud, you are not going to just have a "visual" experience. You will immediately start to work out in your mind what is going on and how you might feel about it. You will register other sensory stimuli, like the sounds of voices and machinery and the smell of the mud, which also must be attended to and synthesized in the moment. This more integrative understanding of aesthetic experience is consistent with landscape research that concludes that human responses to environments are simultaneously cognitive, affective, behavioral, and physiological.[18] Context, including the reasons why you might be in a given landscape and the ideas that you bring with you, strongly affects your aesthetic experience of landscapes.[19]

Affect theorist Brian Massumi has focused extensively on the brief transformative moments between pure sensual experience—such as the moment your body first registers a smell—and the rapid, ensuing coding of that experience via thought, language, memory, and emotions. These functions register and categorize that sensation in some way: "Phew! That smell is so foul! It must be all that mud recently dredged from the middle of the Bayou and placed near the base of this levee I'm standing on." The transitional moment is a relational and affective one, wherein an encounter like this is made sensible through an entwinement of thought and feeling. As Massumi states, "bodies think as they feel, on a level with their movements,"[20] and they do so inseparably. As humans, most of us have no choice but to reductively code our sensual experiences—to make sense of sense—since without doing so, the world seems an infinitely complex, unordered, overwhelming field of phenomena. It actually is this, but most of us do not experience it that way because we are evolutionarily wired to reductively encode our sensing of it into manageable data.[21]

How we code experiential information about the landscapes we encounter affects how we relate to them. But how we code experience is also malleable, shaped by our cultural habits, learned patterns, and past experiences, including experiences of landscapes themselves. We might come to really enjoy the rotten-egg smell of anaerobic mud exposed to air, if through experience, we have come to associate that smell with the re-making of lost landscapes that we personally value, like tidal marshes. The same thought-feeling affinity could develop in response to the appearance and sounds of a messy, chaotic construction site where marsh is being made with dredged, earthen material: a scene of large steel pipes strewn across a landscape, spewing a slurry of water and sediment, while loud bulldozers spread mud and the hum of diesel pump engines fills your ears. You might develop an affection for the funky aromas of mudflats at low tide, or you might recoil, depending on your relations with that landscape, both conceptual and physical.

As far back as the philosophy of Baruch Spinoza in the seventeenth century, affect has been defined as the power to affect and be affected. Your ability to selectively change your coding of an aesthetic experience (e.g., how you delimit, define, and act on the smell of mud) means that affect is inherently political or, in Massumi's language, "proto-political." Affect selectively *in-forms* and conditions the nature of relationships with one's surroundings.[22] Affect is the *taking-form* of experience in an encounter. And the openness of encounters—differences in how they come together, how we think about them, and how that depends on our mood at the time—shows the role of chance, choice, and intention in how we individually perceive things, and how things are staged for perception. As we come to experience new phenomena collectively and culturally—like dredged earth being used more often to recreate lost wetlands—a more public and shared sensibility develops that may prove to be more stable.

Thus, *aesthetics* as we use the term assumes that aesthetic preferences are sensitive to place and context; that they encompass the full range of human sensory-cognitive-affective mechanisms in their construction; that they are relational, processual, and mutually affective, meaning that landscapes influence aesthetic sensibilities and those sensibilities influence approaches to landscapes;[23] and, most importantly, that aesthetic perceptions, such as those of landscapes, are dynamic and adaptable, not static.[24]

Approaching aesthetics as the entirety of sensory experience and the way we understand those experiences is crucial to understanding and appreciating dredged and other anthro-geomorphic landscapes. It is also crucial for making planning and design efforts for these lands more effective and equitable.[25] Increasing experiential accessibility can foster greater equity. Environmental justice has fundamental aesthetic dimensions, which involve taking stock of differences in experience and fostering publics and agency around felt issues of concern.[26] Likewise, aesthetics is a key factor in participatory co-design, which aims to democratize, value, and include a wide range of experiences and backgrounds in design processes. As in the public working parties for Jamaica Bay and Bayou St. John, aesthetics are a lever for engaging, forming, and moving publics, as experience shapes politics, and politics shapes experience.[27] What we experience, where we experience, and what is or is not made available for us to experience: these *aesthetics* constitute our lives, inform our collective values, and, in significant part, make us the very diverse humans we are.[28]

The problem for sediment design is that many sedimentary or dredged landscapes present aesthetic challenges. First, there is the unseen or (seemingly) invisible quality of sediments. Sediments act as a stealthy infrastructure, often taken for granted until their altered distribution becomes felt. This is in part because sediments confound clear scale distinctions, fluidly ranging from microscopic particles of silt to vast expanses of mudflat. Sediment is not a concrete, bounded object like a bridge, a runway, or a wastewater treatment plant. Dredged terrain is largely hidden from view. And even when sediments are in plain sight, they are not always legible as infrastructure.

For example, sand on public beaches, like the barrier islands outside Jamaica Bay, is often perceived as "natural." But that sand is routinely imported and spread along the coast by costly industrial means. Vast lengths of coastline in the United States are humanly modified, either through direct, intentional actions, or by unintended offsite effects. The National Oceanic and Atmospheric Administration's National Beach Nourishment Database alone records nearly 400 major projects that have placed around 1.5 billion cubic yards of sand along American coasts.[29] These doings are hard to perceive in the landscape, unless you happen to be at the beach when the dredgers and bulldozers are at work. But conditions such as climate change and sea-level rise are altering this invisibility, bringing mud, sand, and other sediments into broader awareness, forcing us to think with them. An ever-expanding barrage of ever-more

23 Gobster et al., "The Shared Landscape."

24 This discussion and working definitions of aesthetic perception are adapted from an applied study, wherein these definitions were tested through survey assessments of different types of adaptive forest management techniques in California's Sierra Nevada foothills and mountains.
Jenkins, Milligan, and Huang, "Seeing the Forest for more than the Trees."

25 Within landscape architecture, this has been developed extensively, particularly by Elizabeth Meyer and Maria Hellström Reimer.
Meyer, "Sustaining Beauty."
Meyer, "Uncertain Parks."
Reimer, "Unsettling Eco-scapes."

26 Sze, *Environmental Justice in a Moment of Danger*.
Fuller and Weizman, *Investigative Aesthetics*.

27 Rancière, *The Politics of Aesthetics*.
Fuller and Weizman, *Investigative Aesthetics*.

28 Thus, while aesthetics shape publics, are key to democratizing design processes, and are a fundamental component of equity, their value should not be reduced to those dimensions alone, as valuable as each of those dimensions are. Aesthetic expression and judgment are central to being, and should be valued for that.

29 "National Beach Nourishment Database," NOAA.

30 Pilkey and Pilkey-Jarvis, *Useless Arithmetic*.

31 Rancière, *The Politics of Aesthetics*, 7.

32 Ibid, 8.

33 Navigable waters of the United States are "those waters that are subject to the ebb and flow of the tide and/or are presently used, or have been used in the past, or may be susceptible for use to transport interstate or foreign commerce. A determination of navigability, once made, applies laterally over the entire surface of the waterbody, and is not extinguished by later actions or events which impede or destroy navigable capacity." "Code of Federal Regulations," National Archives.

34 The Common Law doctrine of the Public Trust, or the Public Trust Doctrine, is a US legal principle establishing that the government owns and manages certain natural and cultural resources for public use. These typically include waters (such as navigable waters), wildlife, or land. Exactly what the public trust covers has expanded over time and is a point of ongoing debate.
Frank, "The Public Trust Doctrine."

sophisticated coastal inundation mappings implores us to understand that coastlines are not static, and that, as their movement is increasingly accelerated, there's simply not enough sand or money to physically maintain all the beaches as we have been doing.[30]

In contrast, if you live in New Orleans or out among the bayous of the Mississippi River Delta, where this kind of acceleration is readily apparent, where the opening of a water-and-sediment control infrastructure like the Bonnet Carré Spillway is an annual public spectacle (complete with crowds, coolers, and lawn chairs), then you are much more likely to understand and value sediment and its distribution as vital infrastructure for the place in which you live, work, and play—even if you would not necessarily use those terms. Lived awareness shifts how people understand landscapes.

A second aesthetic challenge concerns physical, embodied access to the terrain of sediment. A nourished beach may be publicly accessible, but many floodable, fluvial, and dredge-related landscapes are not, because they are underwater or are very remote, or because access is restricted due to safety and liability concerns. This is often the case for engineered waterways and flood control channels, even in situations where the channel only floods for a few days out of the year. For example, in the San Francisco Bay Area, many of the former creeks that flow through urbanized areas and housing developments and out into the bay were turned into flood control channels by the USACE between 1930 and 1970. The Alameda Creek Flood Control Channel runs twelve miles from the Diablo Hills to its bay terminus, and for that entire stretch the public is not allowed to enter the channel, touch the water, or walk on the sands, silts, and gravels deposited within it. What could be a dynamic, linear public landscape, complete with shifting bars of sediment, is treated as nothing more than an engineered flume for occasional flood waters.

This delimiting of what can be experienced in landscapes is described by French philosopher Jacques Rancière as *the distribution of the sensible*, meaning the "spaces, times, and forms of activity" that determine what is, or is not experientially accessible to citizens, publics, and communities.[31] Occupational status and social position selectively bestow more power, access, and design control of sedimentary landscapes. The "delimitation" of this distribution—which is designed but typically not taken as such—determines "who can have a share in what is common to the community based on what they do and on the time and space in which this activity is performed."[32] To alter the distribution of aesthetic experience is therefore a political act.

As we will explore in the final chapter, public interpretive "creek crawl" events along the Alameda Creek Flood Channel organized by the Public Sediment team were such a political intervention, as they brought participants into greater experiential contact with the creek, expanding the limits of what was aesthetically common, in order to build a constituency for improving the channel and connecting it to local communities.

Many amphibious landscapes described in this book are legally public, as part of our shared commons of "navigable waters,"[33] including coastlines, baylands, mudflats, wetlands, and other littoral zones. The public has a right to these landscapes not only because changes to them can affect nearby communities, but also because they are part of the public trust.[34] When these landscapes are closed off or made inaccessible—another instance of the *enclosure of the commons*[35]—we lose opportunities for engagement, recreation, and civic stewardship. Worse, this foreclosure can become an assumed norm that is then propagated in a multitude of other landscapes.[36] For Rancière, *aesthetic practices* are ways of doing and making that intentionally increase experiential access, and thus change what is common and experientially shared. In his published writings, Rancière has focused on defining aesthetic practices within the domain of art and artistic production. We are interested in extending aesthetic practices (and their redistribution of the sensible) into all domains and arenas of life, particularly within landscapes.

Time and temporality also condition how sedimentary landscapes are experienced and engaged. This is a third aesthetic challenge. If we only focus on the experiential now and the short term, we miss understanding what was, and how a landscape has evolved into what it is. For example, shipping channels have been dredged progressively deeper in tandem with the design of larger and larger cargo vessels that require deeper drafts, in turn affecting tidal patterns, water quality, and habitat value. As its depth changes, a river becomes something qualitatively different. As another example, virtually all dredged landscapes in the US are colonial possessions, located on stolen land and shaped in ways never intended by the indigenous nations who should have sovereignty. How these nations experience a dredge disposal or reuse landscape can be radically different from the federal agencies regulating them and the ports using them. Similarly, experiential differences occur when we project forward. Our actions today (and over the past two hundred years) are radically affecting what the future can possibly be. Through the making of accelerated climate change and ecological degradation, we—a select *we* within the *we* of humanity—have colonized the future.[37] This is the reason for the rage, despondency, and activism of young people like Greta Thunberg. They are experiencing the onset of a foreclosed future. If we are able to feel into and imagine those futures in which sea levels are many feet higher, how does that affect the way we experience and value sedimentary landscapes today?

A fourth aesthetic challenge concerns who does and does not have access to design processes and decision-making in sedimentary landscapes. These terrains are created and destroyed by select, institutionalized expertise. Is engaging the public necessary or beneficial? Can't we avoid the delays and nimbyism of an open public process and just have groups of experts *efficiently* work on behalf of the public interest? This modernist and colonial ideal runs contrary to all efforts at democratizing design and limits the plurality of values and place-specific attributes that need to be integrated into the process of making these landscapes.[38]

As an example where affected publics were not initially included in a state-led design process, we can look to Franks Tract in the center of California's Sacramento-San Joaquin Delta. Franks Tract is a shallow, tidal lake (covering about 3,500 acres) and a peculiar California State Park.[39] The lake is an anthropogenic artifact, formed in the late 1930s after a breach in the ring of levees that *reclaimed* this subsided land. After repairs failed multiple times, the inundated tract was abandoned. Since then, this wetted place of infrastructural failure has continued to evolve along this formative pathway. Today it is a novel ecosystem, full of introduced plant species from around the globe and a world-class bass fishery, once introduced, now naturalized. Waterfront homes and

35 Illich, "Silence is a Commons."

36 Illich focuses on European grasslands (what was common pasture became fenced in private property), and roads in Mexico City that removed people, public markets and social activity for the sake of automobiles. According to Illich, who wrote in 1983, the enclosure of the commons has happened across the globe, through modes of capitalism, resourcism, and privatization, and exemplifies nothing less than a new ecological order that should be central to studies in political ecology. See also Singh, "Becoming a Commoner."

37 For more on futuring in contrast to defuturing (that is, expanding possibilities versus reducing them) and transition design see Escobar, *Designs for the Pluriverse* and Fry, *Design Futuring*, 71–77. For more on colonized futures, see Krznaric, *The Good Ancestor*.

38 Escobar, *Designs for the Pluriverse*.

39 "Franks Tract State Recreation Area," California State Parks.

large marinas line the edge of adjacent Bethel Island, catering to boaters and fishers. In fall and early winter, the lake is covered by an evenly spaced grid of hand-crafted hunting blinds, sheltering hunters pursuing waterfowl out on the open water.

For California's Department of Water Resources, Franks Tract is a challenging infrastructural problem. Its open water expanse, subject to the rhythms and flows of the tides, is the main source of salinity intrusion from the Pacific Ocean into the delta. This intrusion is expected to intensify through the twenty-first century, as droughts grow more frequent and sea levels rise. If the delta shifts from fresh to salty, its water would no longer be worth pumping and exporting to Bay Area cities and to the arid, southern half of California via the state's extensive water conveyance infrastructure. Meanwhile, for the state Department of Fish and Wildlife, the entire delta (including Franks Tract) is an ecological nightmare of accelerated species extinctions, induced by a suite of structural changes, including the eradication of tidal wetlands and floodplains that used to constitute the delta prior to reclamation's terraforming.

Around 2017, these two departments worked together to craft a technical solution to their problems. They proposed to fill in the northwest corner of Franks Tract with approximately twenty million cubic yards of dredged material, creating an earthen mass of tidal wetland 1,000 acres large. This mass would act as a hydrologic plug, discouraging salty water from entering Franks Tract and keeping the delta's water fresher. At the same time, the constructed wetlands would create much needed habitat and enhance food-web production to support endangered fish, such as salmon and smelt.[40] As large-scale, multifunctional blue-green infrastructure, the proposal has appeal, and modeling predicted that the design would be highly effective at attenuating salinity intrusion.[41]

But cultural and aesthetic values were left out of these design technics. When the design plan was revealed to local communities and regional recreationists, it was met with nearly uniform anger and vitriol. The earthen mass would be placed in front of the water entrances to most of the marinas, which would have cut off boating access, putting them all out of business and destroying a local economy. More broadly, the design would have disrupted and prevented boat travel in and around Franks Tract, which is the primary mode of travel for many people in the region. None of the homeowners on Bethel Island wanted their open-water views to be filled in with what they thought would be a stinky, mosquito-ridden mudflat (the project plan rendered the wetlands in brown, an aesthetic choice that persistently haunted the project). The design also failed to address just how much it would alter a much-loved public landscape, proposing no new amenities to compensate for potential impacts on current uses. Above all, local communities were outraged that they had not been consulted in the initial planning process, given how much they would be affected by it, and how they were being forced to deal with a state-level problem that involved extraction of local water. To many, the design effort seemed a waste of time and tax dollars, since it hadn't factored in any of these basic cultural and socio-economic factors. For the design to move forward, it would have to start anew with a more inclusive, equitable, and creative design approach.[42]

We will return to the saga of Franks Tract later in the chapter. Here, we can note that landscape values and valuations are divergently built upon

40 One of the main anthropogenic effects on delta ecology is severely reduced primary production based on solar energy inputs, such as phytoplankton, which is a foundation for aquatic food webs. Creating tidal wetlands is one way to bring back this essential productivity.
Vaughn, et al., *Delta Landscapes Primary Production.*

41 "Project Background, Franks Tract Futures," California Department of Fish and Wildlife.

42 Notably, this effort did not include any professionals who are trained to work with cultural and aesthetic values, like anthropologists, landscape architects, or cultural geographers. No single profession or professional is omnicompetent, and the next effort to redesign Franks Tract acknowledged this by assembling a more diverse range of disciplines on and alongside the design team.

each person's and each community's varied experiences.[43] Which aesthetic experiences and values are privileged and which are excluded or compromised is a foundational design parameter that should be made overt in any design process.[44] Achieving fair access to decision-making for all people and communities potentially affected by design decisions, particularly those who are the most vulnerable, is a core goal of environmental justice and equitable design.[45]

All four of the aesthetic challenges outlined above are not likely to change without intervention. Creative design, activism, and political pressure are needed to affect the status quo, and new sediment publics need to be part of that change.

3. Fostering New Aesthetic Sensibilities

Looking at an evenly curved seawall wrapping its way around a constructed waterfront, we can appreciate the crafting of its smooth, clean forms, and the geometric clarity with which it manifests as an object. In the moment of being there, observing, we might be impressed by its sheer mass and solidity, by the way the water is so mobile and fluid in comparison, by the way the waves break upon impact with its sheer bulk. But if we observe how the wall performs in relation to the coastal zone it occupies—particularly over time—we begin to find faults and defects. The shape and siting of the wall itself can augment and change erosional forces, potentially undermining its very purpose for being constructed.[46] The wall is also static, lacking capacity to adapt or change with changing conditions. These kinds of structures are engineered with eventual failure as a known outcome, requiring repair or replacement. We might say that the seawall (and many levees and flood control channels built over the past century) exhibits a *false* or misleading aesthetic. Its "infrastructural promise" of durability and strength, as expressed in its form and its smooth, opaque surfaces, cannot be sustained within the dynamic arena of landscape.[47] To sustain it even for a short time requires vast investments to maintain or rebuild a seemingly static form, and even as we do so, the landscape forces acting upon it change. Structural promises of safety can actually lead to harm when they generate a false sense of security that intensifies damage and loss, as in the systemic and catastrophic failures of New Orleans' civic infrastructure during Hurricane Katrina.[48]

These structures not only undermine themselves; they can also do the same to neighboring communities. Extrusions, sea walls, and other "hard" engineered coastal barriers (along outer coasts as well as within bays and estuaries), increase flood risk for neighbors by pushing more water toward them.[49] They also often generate increased wave heights. Waves that smash against a sea wall can rebound larger, increasing coastal erosion in neighboring areas.

So, when a sea wall or levee begins to fall apart, what should be done? Repair the structure, knowing that the process will repeat? Doing so is a known route to failure with known inequities including the off-site effects just described. Build it differently, so that some degree of adaptation, absorption, and ecological function might be achieved for some period of time? Or intentionally retreat, and let rising seas and riparian flooding exert their rewilding ways a bit sooner? This techno-political reversal—enabling landscape's collective agency to do what it is poised to do, rather than trying to control it—is often the most adaptive

43 For a deeper, empirical exploration of this for Franks Tract, see Milligan, Kraus-Polk, and Huang, "Park, Fish, Salt and Marshes."

44 This does not mean that designers' role is to ratify or materialize every intent expressed in a design process. (It is generally not possible to do that, particularly in complex and contested landscapes.) Rather, designers have a responsibility to engage those who our work affects in dialogue. The formation of publics can and should be a reciprocal activity, a back-and-forth between designers and those they work with.

45 Justice and equity also require that we consider not only people and communities who are able to speak now, but also the ways in which our designs affect future people and communities, who cannot speak for themselves. Failure to do this is a key cause of maladaptation, particularly when designing for climate change. A world in which everyone today is equally protected from rising seas by seawalls, tidal gates, and barriers is still an inequitable world, because those measures frequently destroy landscapes and value for future generations.

46 There is much scientific debate about the qualitative nature of these effects, and generalizations are highly complicated by site specific design factors (both for landscape and the wall design). Sea walls and armoring do affect local erosional and depositional patterns, and do so with high variability. Griggs, "The Impacts of Coastal Armoring."

47 Larkin, "Promising Forms."

48 Geographer Gilbert White famously observed this phenomenon in his influential 1945 paper, "Human Adjustment to Floods," which began as his doctoral dissertation at the University of Chicago. For contemporary examples and studies of levees actually increasing flood risk see Munoz, "Climatic Control of Mississippi River Flood Hazard" and Remo, Carlson, and Pinter, "Hydraulic and Flood-loss Modeling."

49 Hummel and Stacey, "Assessing the Influence of Shoreline Adaptation." Hummel et al., "Economic Evaluation of Sea-level Rise Adaptation."

50 "Google Plans To Expand Its Campus," NPR. "Who Will Pay To Protect Tech Giants," NPR.

51 For more on the aesthetic relationship between designer and landscape, and the implications of that relationship for design, see: Foxley, *Distance & Engagement.* Girot, "Four Trace Concepts in Landscape Architecture." Lutsky and Burkholder, "Curious Methods." Milligan, "Design Fieldwork." Milligan, "Making Terrains."

long-term response, but it presents many perceptual and ideological hurdles. This decision-making process itself, either for or against retreat, will create its own affective, politically charged events and situations, as it is doing in urbanized coastal areas, estuaries, and floodplains the world over.

Corporate-driven development proposals in the San Francisco area's vulnerable South Bay are an example. Wealthy tech companies like Facebook and Google are trying to expand their campuses at the bay edge. Google recently bought up over 70 properties in the area, and in 2021 the company got local approval to rezone, so that they could build high-density housing in a floodplain, with an agreement to help fund the construction of new and bigger levees along the bay edge, which could hold back three feet of sea-level rise.[50] Many have protested this move, arguing for retreat rather than intensifying development in a floodplain, given that climate risk is going to rise at an unknown rate, with no real accountability in the future for those who make decisions now.

To move beyond the deceptive aesthetics of sea walls and permanent levees, new aesthetic sensibilities are needed that accord with the kinds of adaptive landscapes that will sustain socio-ecological systems amid accelerated landscape change. We seek an aesthetics that is attuned to how infrastructures and landscapes *actually* perform over time in their place-based context; a sensory truthing to process and temporality, open to the relational quagmires of co-evolving landscapes.

Landscape, as medium, is a spatio-temporal field of encounters and events. It is a dynamic and complex field of affective becoming. It is an open, relational assemblage in which agents, including those of the landscape itself, take or select courses of action. It is thus molded toward one propensity (out of a multitude of possibilities) with greater or lesser efficacy, until other actions are taken and it reassembles itself in some other way. Because landscapes are complexly authored protagonists (rather than passive backdrops), they always exhibit their own collective ability to affect and to be affected, with and without intentional design intervention. Whatever state they may be in—polluted, desertified, dredged, feral, flooded, weedy, or covered in asphalt—landscapes perform and affect, both relationally within and external to themselves.

If we understand landscapes as fundamentally relational, and thus changeable based on those relations, it becomes important to place attention on those relationships and their performative qualities. Specifically, we can assess the relationships between designers and the landscapes they seek to shape, by examining the qualities of their encounters with landscape. Are designers' understandings of a place mostly conceptual and abstract? Are they embodied, material, or technologically augmented? Based upon these encounters and interpretations, what is learned and attended to? What is put forth as a design?[51] And more broadly, what are the aesthetic relationships between landscapes and publics? How might those relations be intentionally augmented and changed? All of these relations are open to aesthetic investigation and to design intent.

These questions emphasize the dynamic sets of relationships brought into play by a design intervention and that are fundamental to life itself. Performative aesthetic sensibilities, similarly, emphasize time, materiality, and how things continually change. To perform is to enact, to do, to work, to operate, to carry out. As described by theorist Maria Hellström Reimer, a performative landscape aesthetics emphasizes action; it entails a shift from landscape representation to actual agency, from landscape objects to landscape interactions.[52] Reimer states:

> *[a] performative aesthetics focuses on the establishment and exploration, through action, of a distributed field of reference...it is performative rather than representative; it works over time, through reiterations or contestations, repeatedly actualizing the relations, the tensions, or the field conditions of a historical or local setting.*[53]

In this way, "'life' does not unfold as a value 'above' or 'beyond,' but as conditioning 'within.'"[54] Life and sensation are enmeshed in a diverse collective, working things out in landscapes across scales, whether those things are plants, tides, sands, floods, or channel engineering. Within a performative aesthetic, we attune our senses to what is actually going on, and to the dynamic relationships that are constantly refashioning landscapes. Where and how does sediment accumulate in a flood control channel? What assemblies of plants and creatures live on these emergent formations? When and how often are maintenance and dredging teams deployed to remove them? What are the patterned refrains of this dredge cycle?

In between flood events, the 7,600-acre Bonnet Carré Spillway is a vibrant heterotopia. Large shoals of sediment deposited by each flood are commercially harvested, while many citizens hunt, fish, shrimp, boat, ride horses, off-road on ATVs and motorcycles, train dogs, fly model airplanes, observe wildlife, and engage in a wide assortment of covert and unsanctioned activities. The US Army Corps of Engineers has become the unlikely chaperone for all these feral social interactions now co-evolving with its tentative master plans.[55]

Drawing on Rancière, Reimer considers the aesthetic to be "a political site where the distribution of the sensuous is a dissensual matter of (landscape) concern."[56] This open, contested space can, like Bonnet Carré, give rise to a remarkable range of experiences, and not just the beautiful and pleasurable:

> *Aesthetics is not a sphere where you feel 'at ease.' On the contrary; following Jacques Rancière, it first and foremost constitutes a field of intervening forces, a disturbing micro-politics that actualizes and agitates the general and taken-for-granted "distribution of the sensible," opening up leeway, loopholes, and breathing space in an otherwise rigid world. Aesthetics thus forces us to face surroundings of discomfort and confusion in relation to which it takes on a new role as a "thinking of the new disorder," putting us in the critical position where we find ourselves at the same time "empowered and disempowered, knowing and unknowing"; thus forcing us to think and act.*[57]

In a landscape like Bonnet Carré, disorder and contesting of the sensible is obvious, intense, and feral. But some level of disorder and contestation always occurs in landscapes. Aesthetic experience can never be fully controlled, given experience's relational and affective nature.

In design, we can endeavor to deliberately sculpt, alter, and steward the distribution and experience of the sensible. New aesthetics can be forged and new publics called into being. But this will always remain an affective

52 In "Unsettling Eco-scapes," Reimer describes desires for both a shift within discourses about aesthetic experience ("aesthetics") and shifts in the aesthetic sensibilities expressed in aesthetic acts, such as the making of a landscape in a particular way ("an aesthetic"). Similarly, we advocate for both types of shift.
In developing this approach, Reimer draws from Nicolas Bourriaud's notion of relational aesthetics. For Bourriaud, relational aesthetics (and the open-authored, participatory artworks he uses to define them) operate at social interstices, opening up new avenues of exchange that exceed or break the confines of normative habits and cultural restrictions. "They actively produce a bundle of relations with the world, giving rise to other relations, and so on and so forth, ad infinitum." Bourriaud et al., *Relational Aesthetics*, 22.

53 Reimer, "Unsettling Eco-scapes," 32.

54 Ibid.

55 Bonnet Carré Spillway Masterplan, USACE.

56 Reimer, "Whose Goodness?," 76.

57 Ibid.

58 Meyer, "Beyond Sustaining Beauty," 48.

59 Graham, *Disrupted Cities*.

60 Star and Ruhleder, "Steps toward an Ecology."

61 Star, "The Ethnography of Infrastructure."

62 Carse, "Nature as Infrastructure," 543. Carse goes on to note that in science studies and actor-network theory, these inversions are called "opening the black box." "Black-boxing," according to Bruno Latour, is "the way scientific and technical work is made invisible by its own success. When a machine runs efficiently, when a matter of fact is settled, one need focus only on its inputs and outputs and not on its internal complexity. Thus, paradoxically, the more science and technology succeed, the more opaque and obscure they become." Opening the black box is an attempt to reveal and understand these operations. Latour, *Pandora's Hope*, 304.

63 Starr, "Ethnography of Infrastructure."

64 Ibid; for additional examples, see Kate Orff and SCAPE, *Toward an Urban Ecology* and Mattern, "Infrastructural Tourism."

65 "The Center for Land Use Interpretation," CLUI.

66 "Birdfoot: Where America's River Dissolves into the Sea," CLUI.

experiment. As Elizabeth Meyer states, "We chart out propensities without controlling outcomes. We design socio-ecological experiments in living with no promises. Something. Perhaps. It might. Who knows."[58] For us, it's in the *who knows*, or rather who senses what, that we can *collectively* leverage and agitate the "field of forces." This is where fostering publics becomes so important: the shared concerns, motivations, and desires of a public can provide orientations within aesthetic experiences.

* * *

Let's get back to that aging, problematic seawall, and what to do about it. How might we focus more attention on its performativity as a basis and criteria for design? Infrastructure typically works well when you don't have to think about it. Often, it's only when infrastructure fails, as in power outages, levee breaches, disruptions in consumer supply chains, or urban flooding, that infrastructure and its habitual work becomes apparent.[59] The background of infra inverts to foreground and "substrate becomes substance."[60] This kind of *infrastructural inversion* can also be performed intentionally, outside of crisis moments, by foregrounding the "truly backstage elements" of how infrastructure functions.[61] Publicly scrutinizing the sea wall—how it came to be, how it works, and how it is failing—sheds light into the black box of infrastructural invisibility.[62] Such practices call attention to habits and assumed ways of doing things, foregrounding not only the ways that infrastructure stealthily works, but also how it affects people in different ways, since what is infrastructure for one person might be a problem or a source of work for another person.[63]

So, how can we effectively reveal and explore infrastructure—sedimentary or otherwise—and make it sensible? How can sediment be brought into experience, so that it can be encountered and learned from? This experiential rousing is best achieved through creative, aesthetic means, using communicative media, events, and the design of landscapes themselves.[64]

Through bus tours, publications, exhibitions, and other events, the Center for Land Use Interpretation has, since 1994, unsettled the ways in which landscapes are ascribed particular "uses" within the United States, such as copper mining, waste disposal, or retail distribution.[65] They define themselves as "a research and education organization interested in understanding the nature and extent of human interaction with the earth's surface, and in finding new meanings in the intentional and incidental forms that we individually and collectively create," believing that "the manmade landscape is a cultural inscription, that can be read to better understand who we are, and what we are doing." Their geographic explorations typically reveal the contingency, vulnerability, absurdity, and shifting value of infrastructural uses over time. For example, the exhibit *Birdfoot: Where America's River Dissolves into the Sea* (2008) is a photographic atlas of the accelerating disappearance of the terrestrial end of the Mississippi River.[66]

The Center for Urban Pedagogy is another example. CUP makes urban policies, planning, and infrastructure legible to a wider public audience, through art and design, particularly for underrepresented communities. One of their early signature tools was a series of well-designed fold-out maps that explain how different infrastructures work. "The Cargo Chain," for instance, documented chokepoints within the global logistical system, with the aim of empowering longshore workers and their unions.[67] Similarly, *99% Invisible* is an audio podcast about the "unnoticed architecture and design that shape our world."[68] How did street and sidewalk curb cuts come to be shaped in the very specific way they are? What about standardized highway design? Episodes that range from bat bridges and culverted creeks to fire hydrants and see-through sidewalks describe how these artifacts came to be.

These examples operate largely at a distance from the landscapes they concern, much as this book does with respect to sedimentary landscapes. It is a mediatory technique between the landscapes we are discussing and you, the readers. This distance can be vast and abstract. Yet we can be attuned to both the limits and the affordances of media-based tactics. A podcast, a book, or a photo essay can enable affective shifts in landscape experience, which can in turn lead to transformative action. In the twentieth century, books like Aldo Leopold's *Sand County Almanac* and Rachel Carson's *Silent Spring* helped shape American environmentalism. Their words transformed innumerable landscapes via distanciated means. They created a two-fold redistribution of the sensible: first through calling collective imaginaries and concerned publics into being, followed by actual landscape change through land stewardship, environmental policies, and new designs.

Other aesthetic practices operate more directly and physically in landscapes. One of the most profound and clear examples of this was the artist collective Rebar's playful intervention that led to the creation of *Park(ing) Day*.[69] It began with a very simple experiment: pay for a metered on-street parking space, and repurpose it as a customizable D.I.Y. public space by rolling out turf grass, placing a bench, and adding a potted tree. Today, Park(ing) Day is an annual global event, and municipalities have been inspired to more permanently convert parking spaces into "parklets." This hack shows how influential a shift in the distribution of the sensible can be, and how the publics that grow and coalesce around that new experiential access can change worlds.

Then there are aesthetic practices that explicitly develop a critical and forensic relationship to landscapes, largely using the techno-media through which they attempt to sense it. Matthew Fuller and Eyal Weizman, of Forensic Architecture, call this type of design work *Investigative Aesthetics*, a type of sense-making activism.[70] Their work reconstructs past events, such as wars, conflicts, bombings, uprisings, and shootings, by carefully and arduously stitching together various ways through which those events can be sensed and aesthetically experienced: satellite imagery, Twitter feeds, forensic police reports, material evidence, public video footage, data scanning algorithms. They seek to create *aesthetic commons* within which fact and actuality can be collectively constructed and interrogated. Following in the footsteps of John Dewey and sociologist Noortje Marres, they understand publics as collective social-political entities that come into being through issues of shared concern and experience. Said another way, in today's complex and fractured social conditions, publics don't just exist. Rather, contested and complex issues drive public involvement in politics.[71] The resulting collective contestation of "what aesthetics might be, what its boundaries are, and how they might expand" can even constitute *aesthetic civil war*.[72]

Is this not at play in climate change debates and denials? Can't these be understood as an aesthetic civil war about how climate change is variably sensed, and intentionally hidden from sensibility? About how it is inequitably

67 "The Cargo Chain," Center for Urban Pedagogy.

68 "The Show," 99% Invisible.

69 Rebar's idea was inspired by an earlier creative urban hack: Gordon Matta-Clarke's Fake Estates Project. See "Park(ing) day: The Origin Story," Park(ing) Day.

70 Fuller and Weizman, *Investigative Aesthetics*, 6.

71 Marres, "Issues Spark a Public into Being".

72 Fuller and Weizman, *Investigative Aesthetics*, 25.

experienced? More specific to landscapes and sediment, we can look to Public Lab as a leader in these contestations. The lab got its start photographically exposing the impacts of the BP oil spill on the Gulf region's wetlands and beaches using homemade kites rigged with digital cameras. It has since branched out into a variety of public sensing tactics, pursuing environmental justice through community science and open-source technology.[73]

Across this selection of examples—from descriptive inversions to physical interventions—expanding and crafting aesthetic access has led to the transformation of landscapes and of our collective social existence.[74] Though there is no determinate *promise* or surety in outcomes when redistributing the sensible, it is clear that by instigating concerned and interested publics, we can create collective propensities for change.

Radical changes in how landscapes are understood, experienced, and valued can and do happen. Examples of intentional shifts can be found both in contemporary design culture and in historical precedents. The climate and ecological crisis, paired with sustainability mandates, has given rise to new landscape imaginaries in design and planning: *green infrastructure, Net zero, 350.org,* resilient shorelines, climate adaptation, the Green New Deal. Over the last few decades there has been a radical change in how urban stormwater is designed for. Many cities have shifted from highly centralized, subsurface infrastructures to more localized surface treatment, storage, and infiltration system. Stormwater gardens have, in many places, replaced concrete channels.

More broadly, one of the most poignant recent shifts in landscape perception concerns wetlands. During the reclamation era (in the United States, roughly the eighteenth and nineteenth centuries), wetlands were perceived as pestilent wastelands.[75] These wet landscapes were systematically drained and dried using networks of canals, levees, and ditches.[76] Over two hundred years, the lower 48 states lost about half of their wetlands—over 117 million acres from an original 221 million—due, in large part, to reclamation efforts.[77] We've since learned that reclamation had enormous costs: losses of species and richly diverse habitats, loss of indigenous sovereignty and land stewardship practices, loss in water storage and flood attenuation capacities, loss of water cleansing and cooling, loss of opportunities for recreational and cultural uses. This has been learned through experiencing the failures and unsustainability of reclamation infrastructure as well as through building a more expansive understanding of the value of wetlands.[78]

In a reversal of values, wetland publics are attempting to "reclaim reclamation,"[79] fostering ambitious efforts to undo reclamation era works in favor of coastal, estuarine, and freshwater wetlands.[80] This shift is happening because wetlands are now understood as being one of the most biodiverse and threatened environments on earth, serving a number of key environmental roles. This change manifests in myriad ways: the United Nations valuation of wetlands through "ecosystem services"; the US federal government's "no net loss" policy that has been place since 1989 (which, though not perfect, is a near-complete reversal of reclamation era policy); regional wetland restoration mandates (such as in California's Bay-Delta, which we will detail in "Technology, Pace, and Adaptation"); huge organizations, such as Ducks Unlimited that take a continental approach to wetland restoration via public constituency and

73 "About Public Lab," Public Lab.

74 When the Dredge Research Collaborative started investigating sediment, we found that access to this realm was remarkably limited, which is why we have so few examples specific to sediment to offer here, and why fostering sediment publics became the basis for much of our early work.

75 Vileisis, *Discovering the Unknown Landscape: A History of America's Wetlands.* Di Palma, *Wasteland: A History.*

76 Such infrastructural techniques might be considered "terrestrial strategies." Jensen, "Amphibious Worlds."

77 Dahl, *Wetlands Losses.*

78 For historic examples of hard lessons learned to arrive at this understanding and revaluation of wetlands, see Wilson, *Seeking Refuge.*

79 Milligan, "Landscape Migration." Relatedly, see Orff, "What is Design Now?" Specific to managed retreat, see Koslov, "The Case for Retreat."

80 However, this cultural shift is far from universal and is often displaced by both neoliberal ideology and the desire for short-term economic gains.

funding; as well as innumerable community-based wetland restoration groups, taking actions like the marsh-planting parties in Jamaica Bay.

Thus, both nationally and beyond, how human societies value wetlands has changed. Is it possible to foster a similar aesthetic shift for mud—for publics to value sediments and their distribution as an adaptive and vital infrastructure?

* * *

Mudflats are broad expanses of fine-grained sediment. When exposed, they look like a slick, wet strip of desert, often with a potent briny sea-smell. If you step out onto one, you may find that it is firm enough to walk on, or you might sink up your waist. It depends on the size and qualities of the sediments. Mudflats typically occur in the subtidal zones of bays and estuaries, and typically lie just below vegetated marshes, where daily tidal inundations bring sediment to them from both the ocean and upstream waterways. In both public perception and environmental advocacy, mudflats are typically overshadowed by other (vegetated) wetlands, yet they play a pivotal role in sustaining the larger "marsh profile" that they are part of.[81] Mudflats shallow water depths adjacent to the vegetated marshes by acting as dynamic sediment banks. This provides storm and wave protection and increases the resilience of upslope marshes. Mudflats are also a unique habitat for a range of organisms like invertebrates, fish, and migratory shorebirds.

If publics understood the functions, performance, and adaptive capacity of mudflats—combined with experiencing them as the unique places they are—mudflats might, like other wetlands, assume more value. Perhaps a *Muds Unlimited* could be for mudflats and their sources of sediment what Ducks Unlimited is for marshes and swamps. Like other wetlands, sensibilities could be changed through means ranging from the technical, conceptual, and rhetorical, to direct bodily immersion and the making of new landscapes. Aesthetically, mudflats offer experiences ranging from REI-esque adventures to the sublime. Mudflat hiking is an expanding form of recreation (popular in the Netherlands), which requires you to be both knowledgeable of the changing tides, so as not to get stranded by incoming waters, and attuned to the variable, squishy ground being traversed.

More quietly, imagine standing at the edge of a pier on a clear, still evening at Cooley Landing in San Francisco's South Bay. Vast mudflats stretch out over a thousand yards from the salt marsh. There's a slight, encompassing sound: the gentlest, intricate ensemble of ticking and tapping coming from an uncountable multitude of tiny crabs filtering through the exposed mud. A variety of shorebirds busily traverse the exposed land. They fly and walk, dipping their beaks into the soft mud to probe and feed on invertebrates. The birds satisfy themselves while providing pleasure and discovery for so many avid bay birders. Over an hour or so, the mudflats disappear under the advance of a thin mirror of water. The sounds of the crabs and the birds fade in tandem. This incoming tide reflects both a darkening sky and the screams of commercial jets passing overhead.

If you sit at the edge of these mudflats, perhaps upon the steep piers of old concrete demolition fill that you will find there, with your phone switched off and your senses tuned in, the landscape will work you over. Its vibrancy and eventful qualities—the curious sounds of the crabs, the calls of the birds, the dynamic

81 This overlooking of mudflats often extends to the way that "wetlands" are described and understood in design, planning, and policy, where swamps, marshes, and other vegetated wet areas are understood as wetlands, but mudflats (or, for instance, unconsolidated gravel shores) are not. However, mudflats do in fact meet technical definitions of "wetland," and are crucial parts of larger estuarine systems.
Federal Geographic Data Committee, "Classification of Wetlands."

82 Klein, "A New Formula."

83 "Roduin," Hehe.

84 "Maasvlakte 2," Van Oord.

85 Albrecht, "Solastalgia."

back and forth of the tides—will affect you. If you came here with a perception of mudflats as a dead zone, as a disposable, barren ecosystem, that perception will likely be transformed. Or, perhaps, none of these phenomena will affect you like this. Maybe these events and happenings are just noise to you, and the funky smell of drained sea and sediment will be just a bit too feral. In either case the landscape, as protagonist—as collective, affective agent in its own right—will make a bodily impression.

Dredge-made landscapes, whether wetlands, mudflats, beaches, dunes, confined disposal facilities, or golf courses, rarely call attention to the constructed nature of their terrain, or to the material and temporal processes that emerge out of their construction. Yet some wonderful examples do exist, such as those large, impromptu public gatherings that happen when the Bonnet Carre Spillway in Southern Louisiana is opened to pass sediment-laden floodwaters from the Mississippi River. We seriously doubt that the durational, technical means for opening this floodway was part of a deliberate aesthetic design. Opening the infrastructure requires the removal of up to 7,000 wooden "needles" (8-inch by 12-inch wooden beams) that span 350 linear bays. The number of needles removed depends on the magnitude of the flood. They are pulled out by rail-mounted gantry cranes perched atop the spillway; the process can take several days to complete. But there it all is, adding intrigue, anticipation, and rhythm to public events that were never really intended. Or consider the yearly sediment cleansing ritual of China's Xiaolangdi Reservoir on the Yellow River. Large tunnel gates on the face of the dam open to simulate flood events (and prevent flooding by passing sediment downstream), which produces enormous, explosive sprays of sediment and water that crowds and tourists come to observe and document.[82] At Xiaolangdi, the engineers were more aware of the aesthetic potential of the design, and they created public viewing platforms between the dam's sculptural flumes.

Artists have also successfully intervened in the blasé making of extraordinary contemporary landscapes to stage aesthetic happenings. In 2012, for instance, artists Helen Evans and Heiko Hansen navigated their way through Dutch institutional protocols to emplace 75 cubic meters of bright red sand on a constructed beach of the 2,000-hectare port of Maasvlakte 2, which had just been reclaimed from the sea near Rotterdam. The red sand formed "a large red stain that gradually dispersed by natural forces such as wind, rain and the sea," with the pattern of that dispersion being "dependent on the unpredictable weather conditions."[83] The 75 cubic meters were a tiny fraction of the 240 million cubic meters of dredged sand used to create Maasvlakte 2,[84] providing a material sense of the time and scale of both endeavors.

How might the designed restoration of destroyed marshes, using dredged sediment, creatively speak to the sublime induction or reversal of solastalgia—distress or sense of loss due to radical environmental change?[85] How is this approached in form? Can restoration be made legible? How is the nascent, vibrant establishment of the new landscape by a myriad of plants, birds, or invertebrates anticipated and structured? Where is human access encouraged, and what experiences are afforded? Very few dredged or reclaimed landscapes are designed with such a performative aesthetic sensibility in mind, or with an eye toward the socio-cultural, political, and ecological potentials they might afford. To be clear, we are not arguing that all designed or engineered landscapes should have some interpretive apparatus tacked onto them as a prescribed aesthetic overlay, or an artistic intervention to make them legible and valued by a larger constituency. Rather, we are advocating for a design sensibility that is more fundamentally attuned to political and performative dimensions of landscape experience and change, and to social experience generally, wherein social is conceived inclusively of human and other-than-human, of how all of these beings and things interact with and affect one another.[86] How can sedimentary and dredged landscapes be designed so that they are rich with aesthetic performative possibilities—ecologically, hydrologically, and recreationally? In their basic forms, and in how those forms evolve and are inhabited? As the Bonnet Carré Spillway, Xiaolangdi Reservoir, and Maasvlakte 2 demonstrate, aesthetic potentials are latent in working and dynamic landscapes themselves.

Performative aesthetic opportunities can be built into these designs, as has happened at the underwater fish windows at the Ballard Locks fish ladder in Seattle, and more grandly at the massive Bonneville Dam on the Columbia River between Oregon and Washington. At both infrastructures, an underground room has been made publicly available, with large aquarium-like windows on one side. The underwater view allows participants to see the different species of migrating salmon as they swim by. Those salmon have successfully made it up and over the fish ladders that allow them to surmount the otherwise impassable infrastructure. These observation points are directly tied into the technical, scientific monitoring of the salmon, using sensors and cameras near the window space, and are integral to the design. Both dam and locks have been retrofitted in two ways, allowing physical and aesthetic passage to fish and people, which in turn will pay its way forward in increased ecological awareness, advocacy, and adaptation.

What about integral windows on mud? We humans are now inseparably enmeshed in the co-evolution of the earth's upper crust.[87] Recognizing this should not lead to pride. Rather, it should produce a sense of responsibility, owing to our collective participation in processes and events that we cannot fully control, and which are far from being fully desirable or equitable. Sediment design should be deeply cognizant of these dimensions of experience and how they can be leveraged to affect landscapes—from inception through construction to stewarding their co-evolution. Such efforts can lend more value to sediments, both for what they can do—like making marshes and increasing the adaptive capacity of shorelines—and simply for what they are.

4. Building Sediment Publics

Over the past decade, as members of the Dredge Research Collaborative, we have tried to broaden public and professional engagement with sediment through a variety of design tactics that have evolved over time. We began with DredgeFest, an event series held in four coastal regions around the US—New York, Louisiana, the Great Lakes, and California—between 2012 and 2016. Each DredgeFest was "particular to its host territory, explicitly working to uncover that region's distinctive qualities, processes, and networks of actors"[88] through a combination of events and actions.

In New York, DredgeFest was a one-day public symposium and exhibition in Manhattan (hosted by Columbia University's Studio-X NYC), followed the next day by a boat tour of the New York and New Jersey harbor. We didn't know if

86 A more integrative and diverse understanding of the social has been advocated for and advanced in a number of fields, disciplines, and philosophies, such as actor-network theory, assemblage theory, posthumanism, environmental humanities, new materialism, and anthropology. Accessible entry points can be found in Tsing, "More-than-human Sociality" and Ash Amin, "Collective Culture and Urban Public Space."

87 Church, "The Trajectory of Geomorphology." Wilkinson, "Humans as Geologic Agents." Harden, "The Human-landscape System."

88 Holmes and Milligan, "DredgeFest," 3.

89 This work also formed the foundation of the four visual chapters in this volume.

90 See "Choreographing Sediment" for more discussion of these challenges, both in terms of data collection and computational modeling.

a public audience would be interested in talking about sediment and design. The answer was yes: Studio-X was packed full, and the boat tour sold well over a hundred tickets. For Louisiana and the Great Lakes, we expanded the symposium to two full days, retained the public tours and exhibition, and added multi-day design workshops that grappled with issues raised in the symposium and the tours. By the time we reached California, the design workshops had become the driving element around which we structured tours, the symposium, and an exhibition. This evolution reflects our efforts in first trying to access, experience, and understand the complex worlds of sediment design, and then moving toward greater emphasis on designing and effecting change in these worlds. Participants and audiences were diverse, ranging from interested citizens to experts and industry representatives. We promoted the events widely and publicly, reaching out to local stakeholders, public agencies, universities, and others. Generally, the symposiums and the public tours were attended by larger audiences, up to about 250 people. The design workshops were intentionally smaller and more intensive, engaging as many as 60 participants.

Building up to each DredgeFest, we worked to understand dredged and sedimentary landscapes across scales through fieldwork, mapping, and visualization. We sought to reveal the magnitude, qualitative features, and effects of human sedimentary manipulation within a particular region.[89] This work was first for ourselves, as a method to engage and understand these networked landscapes. It was a starting point for considering where and how design intervention might be most needed. Then, these mappings were exhibited and shared with others for discussion, debate, and design speculation.

This investigative work was often challenging, in ways that are tied directly to the aesthetic challenges inherent in making sediment public. With sediment, cohesive and consistent geospatial data is rarely available. Different sedimentsheds overlap with different Corps districts, between whom record-keeping practices have historically varied. Often there is no coordinated data between private ports and federal projects. There are also gaps and differences in data over time within one organization, like a single Corps district. There is a broad lack of data on sediment movement and dispersal generally—because it is so difficult to sense and document, or to track change over time, or to compare trajectories within and across regions.[90] But by using what information we could find and assemble (online data, technical reports, direct consultation with state and federal agencies and corporate actors, and interviews) we developed a synthetic and practical understanding of these sedimentary geographies that we could begin to work with. Where we had enough dredging and sediment management data to identify trends and spatial patterns, we critically assessed what was working and what was not. Over longer time scales (like comparing historical and current data), we were able to observe how sediment supply and distribution was affected by built infrastructure—such as dams, flood control channels, and levees—and the downstream effects of land subsidence and altered deposition patterns. Where data was missing, we spoke with scientists, industry experts, and public officials to understand why, what the implications might be, and to lobby for more research.

These regional, integrative mappings were aesthetically generative. They provoked new and more informed questions and speculation as much as they addressed the initial research questions our team started with. They served as the foundation for each DredgeFest's symposium and design workshops, wherein designers, engineers, scientists, regulators, public agency representatives, corporate practitioners, and students came together to critically discuss current sediment management and geographic conditions (the symposium) as a basis for design propositions for potential futures (the design workshops).

DredgeFest helped ally professionals around sediment by creating a new and different space for interdisciplinary interaction. The gatherings' emphasis on discovery, curiosity, and design helped foster a more diverse and creative environment.[91] Modes and levels of participation were largely up for each person to invent, alter, and decide for themselves. After all, they were Fests, rather than structured decision-making consortiums. Acting independently and outside of the institutional arenas and governmental pulpits in which sediment management is typically worked out, each DredgeFest performed as a "play upon a play"—a space where people could perform their habitual roles in new, improvisational ways.[92] The workshops permitted institutional actors, such as state and federal agency personnel, regulators, and corporate representatives, to break out of their working habits and protocols, to experiment with very different ideas and approaches to designing with sediment.[93] In turn, those same actors provided useful checks on radical or ambitious schemes, since they had firsthand experience with the political and institutional impediments these schemes would have to work through to ever manifest.

Each DredgeFest included guided public tours of local sedimentary landscapes. By boat and bus, these tours traversed a range of sites, including sediment disposal facilities, constructed wetlands, successful beneficial reuse projects, shipping channels, ports, active dredging operations, and vulnerable or damaged landscapes in need of intervention. The tours emphasized locations that illustrated different stages in the cycling of sediment, sites where new and innovative practices were being prototyped, and places where such prototyping was strongly needed. Tours were narrated by both members of the DRC and outside experts. Many of the landscapes we visited are typically closed to the public, and so the tours provided unique, immersive opportunities to experience them, while interacting with the people who own and manage them.

The happenings and outcomes of DredgeFest were further distributed through social media, exhibitions, online reports, articles, and websites. For example, the *Hidden Sediment Reserve* is an interactive online mapping interface where users can visually explore surpluses and deficits in sediment distribution patterns through time across California's entire Bay-Delta Watershed (covering approximately 100,660 square miles), based on climatic conditions and the construction of extensive dam infrastructure.[94]

DredgeFest was enacted as an iterative series of social experiments.[95] Operating on the premise that most complex and wicked problems are (in their most immediate, stubborn, and experiential dimensions) social problems, DredgeFest created encounters and an affective, uncertain space for innovation and work-play. The most significant impacts of DredgeFest were likely the connections and commonalities forged by the diverse sets of actors participating in them,[96] which is a form of social innovation.

One of the biggest questions coming out of each DredgeFest was how some of the promising design ideas they generated could be actualized, propagated, and carried forward, particularly the more activist and community-engaged projects.

91 All DredgeFest symposiums and design workshops were held on university campuses, which created a more neutral and equally new space for most of the participants (in contrast to hosting these events in the proprietary domains of regulators, public agencies, or corporate contractors).

92 Holmes and Milligan, "DredgeFest," 4.

93 The workshop format varied substantially from DredgeFest to DredgeFest, and there were no workshops at the first DredgeFest in NYC.

94 "The Hidden Sediment Reserve," Landscape Metrics.
This website was built by Landscape Metrics in collaboration with the Dredge Research Collaborative and was the outcome of extensive mapping efforts prior to *DredgeFest California*, as well as the design research produced by the *Regional Choreography* workshop at the same event.

95 Holmes and Milligan, "DredgeFest."

96 Ibid, 9.

Ezio Manzini broadly defines *design for social innovation* as "everything that expert design can do to activate, sustain, and orient processes of social change toward sustainability." Manzini, *Design, When Everybody Designs*, 62.

Based on Corps participation and collaboration in the DredgeFests, the DRC helped organize a two-day *Engineering With Nature* workshop at the US Army Engineer Research and Development Center in Vicksburg, Mississippi, which focused on ways landscape architects could collaborate with ERDC's Engineering With Nature research program. Since that time, we have been working as design research consultants with several Corps districts and EWN on a variety of Corps infrastructure projects across the continental US, in states including New Jersey, Louisiana, Texas, and Alaska.[97] *Healthy Port Futures* is a research project funded by the Great Lakes Protection Fund involving a wide range of agencies and disciplinary expertise. This grant has supported in-depth investigation of how small ports can operate more sustainably in the Great Lakes region and grew directly out of conversations and relationships forged through DredgeFest Great Lakes. The project is piloting new strategies for beneficial placement of dredged sediments that increase shoreline resiliency and habitat value.[98]

Franks Tract Futures followed the flawed feasibility study described earlier in this chapter.[99] Unlike the feasibility study, FTF was a co-design effort that from the very beginning included a broad range of stakeholders and publics, such as marina and property owners, hunters, farmers, ranchers, anglers, boaters, local government, and reclamation districts.[100] The FTF process began with a public meeting, where we launched an online survey asking members of the public to geospatially map their boating routes through Franks Tract and the larger region, where (and how) they engage in recreation, where improvements could be made, and where tidal marsh might be located that would have the least impact on current use values. These crowd-sourced maps created a foundation for place-based design investigation, revealing an *uncommons* where the public space of the tract was perceived, experienced, and valued in radically different ways.[101]

The initial design for Franks Tract was largely reviled by members of the public, with the predominant survey response being to "leave it alone." But these mappings showed that, when asked about their own experience and what could be improved, citizens pointed to many things they didn't like about the existing tract, like the many dangerous submerged boating hazards and sharp turns in boating routes, eroded beaches and levees, declining sport fisheries, lack of sheltered areas, and the copious amounts of invasive aquatic weeds that make it difficult to boat across the tract. And perhaps most of all they disliked the "temporary" salinity barrier that has been installed by the Department of Water Resources with increasing frequency as droughts have increased in severity and frequency. The barrier is installed by placing rock across the false river channel near Franks Tract, a crude and costly importation that blocks boat passage and alters flow conditions.

We also collectively envisioned what, based on scientifically observed trends, a "no-action" scenario would look and feel like over the next few decades, which showed that keeping Franks Tract as-is will not be possible. Left alone, the tract was changing in many ways that local residents did not want or like.

With this work as a shared foundation, we began collaborating with state agencies and a large advisory committee composed of local stakeholders representing a broad mix of interests and perspectives. Through a series of in-person design workshops, we focused on ways that the tract might be molded and recontoured (through extensive cut and fill dredging) to meet collectively

97 DRC members Sean Burkholder, Brian Davis, Rob Holmes, Justine Holzman, Brett Milligan, and Gena Wirth have participated in this work, along with O-RISE fellow Tess Ruswick, supported by an array of staff and research assistants at Auburn University, the University of Pennsylvania, the University of Toronto, and the University of Virginia.

98 "Healthy Port Futures," Great Lakes Protection Fund.
Healthy Port Futures has been led by Sean Burkholder, Brian Davis, and Tess Ruswick. See "Inland Chokepoint" for examples of these pilot projects.

99 "Franks Tract Futures," California Department of Fish and Wildlife.
Brett Milligan led the public engagement and co-design efforts, and co-led overall project conceptual design.

100 "Franks Tract Futures Final Report," California Department of Fish and Wildlife.

101 Milligan, Kraus-Polk, and Huang, "Park, Fish, Salt and Marshes."

agreed-upon objectives that included improving recreation and the local economy alongside the earlier, institutionalized aims of ecological recovery and attenuation of salinity intrusion in the central delta.

We started with a broad range of possible design scenarios and whittled them down to three design options that we repeatedly refined to meet all project objectives. At each stage, the designs were evaluated by representatives of each involved community and agency. Working closely with the Department of Water Resources, we developed an iterative, stakeholder-led hydrological modeling process, whereby preferred landform configurations were tested to see if they could meet water quality objectives. A key parameter was the width and depth of boating access channels. If they were too wide, they let too much salt in; if too narrow, they wouldn't meet user needs.

Aesthetic and experiential values were a core element of the design process. We conducted visual surveys of tidal marsh and tidal marsh channel preferences to see what the advisory committee liked and didn't like. University of California, Davis undergraduate landscape architecture students designed recreational features—such as cove and beach landforms, overlooks and rest areas, camping areas, docks, trails, and water access—and presented their concepts at one of the project workshops with the advisory committee for review and refinement.

Upon completion of the design workshops, we invited citizens to review and rank the three final design concepts and the no-action alternative. Overall, public preference for each design concept was higher than for the no-action alternative, and thus in sum, overwhelmingly in favor of the designs over the status quo. This outcome was consistent for local and regional survey respondents.[102] This represents a remarkable change in public values. The project was able to build a sediment public around an issue of concern, and many local community members continue to advocate for the project to move forward. It's also quite clear from the results of the effort that public input into the design greatly improved it. Rather than just a wetland salinity plug, the preferred design concept from FTF would create a multi-functional recreational landscape in the Sacramento-San Joaquin Delta that is befitting of its 2019 National Heritage Area designation.[103] Further, the preferred design didn't seek consensus in values per se. Rather, it allowed for a diversity of values and experience—the uncommons that Franks Tract is—by diversifying the sedimentary form of the landscape itself.[104]

Downstream from Franks Tract and the delta, *Public Sediment* is a transdisciplinary and participatory design proposal for multi-benefit adaptation of creeks and sedimentsheds in the San Francisco Bay Area. DredgeFest California seeded conceptual momentum for Public Sediment, and we discuss the project in detail in the final chapter of this book, "Why Design?"

Much of our work since DredgeFest seeks to test and develop new design techniques for making, managing, and guiding the evolution of sedimentary landscapes, based on what we learned. One of the central lessons from DredgeFest, as well as our applied design work, is that the testing and prototyping of such techniques and landscapes is highly dependent on the complex, overlapping policies and protocols that govern a region. Learning, evolution, and adaptation often occur through these instruments. This critical relationship between technologies and adaptation, particularly in the context of accelerated change, is the focus of "Technology, Pace, and Adaptation."

102 Ibid.

103 "Sacramento-San Joaquin Delta National Heritage Area," NPS.

104 Milligan, Kraus-Polk, and Huang, "Park, Fish, Salt, and Marshes."

Bibliography:

"The Show," 99% Invisible, https://99percentinvisible.org/about/the-show/.

Ash Amin, "Collective Culture and Urban Public Space," *City* 12, no. 1 (2008): 5-24.

Nikhil Anand, Akhil Gupta, and Hannah Appel, eds., *The Promise of Infrastructure* (Duke University Press, 2018).

Glenn Albrecht, "'Solastalgia': A new Concept in Health and Identity," *PAN: Philosophy Activism Nature* 3 (2005): 41-55.

Nicolas Bourriaud, Simon Pleasance, Fronza Woods, and Mathieu Copeland, *Relational Aesthetics* (Dijon: Les presses du réel, 2002).

"Franks Tract State Recreation Area," California State Parks, https://www.parks.ca.gov/?page_id=490.

Ashley Carse, "Nature as Infrastructure: Making and Managing the Panama Canal Watershed," *Social Studies of Science* 42, no. 4 (2012).

Ashley Carse and Joshua A. Lewis, "New Horizons for Dredging Research: The Ecology and Politics of Harbor Deepening in the Southeastern United States," *Wiley Interdisciplinary Reviews: Water* 7, no. 6 (2020): e1,485.

"Birdfoot: Where America's River Dissolves into the Sea," The Center for Land Use Interpretation, https://clui.org/section/birdfoot-where-americas-river-dissolves-sea.

"The Center for Land Use Interpretation," The Center for Land Use Interpretation, http://www.clui.org/.

"The Cargo Chain," Center for Urban Pedagogy, http://welcometocup.org/Projects/MakingPolicyPublic/CargoChain.

Michael Church, "The Trajectory of Geomorphology," *Progress in Physical Geography* 34, no. 3 (2010): 265–86.

Zachary A. Collier, Matthew E. Bates, Matthew D. Wood, and Igor Linkov, "Stakeholder Engagement in Dredged Material Management Decisions," *Science of the Total Environment* 496 (2014): 248–56.

The Critical Infrastructure Studies Collective, "Critical Infrastructure Studies.org," https://cistudies.org/.

Thomas E. Dahl, *Wetlands Losses in the United States, 1780s to 1980s* (US Department of the Interior, Fish and Wildlife Service, 1990).

Brian Davis, "Public Sediment," in *Toward an Urban Ecology*, eds. Kate Orff and SCAPE Landscape Architecture (The Monacelli Press, 2016).

Brian Davis, Rob Holmes, and Brett Milligan, "Isthmus," *Places Journal* (2015).

Manuel DeLanda, *A New Philosophy of Society: Assemblage Theory and Social Complexity* (Bloomsbury Publishing, 2006).

Vittoria Di Palma, *Wasteland: A History* (Yale University Press, 2014).

Arturo Escobar, *Designs for the Pluriverse* (Duke University Press, 2018).

Federal Geographic Data Committee, "Classification of Wetlands and Deepwater Habitats of the United States," FGDC-STD-004-2013, Second Edition (Washington, DC: Wetlands Subcommittee, Federal Geographic Data Committee and U.S. Fish and Wildlife Service, 2013).

Billy Fleming, "Design and the Green New Deal," *Places Journal* (2019).

Alice Foxley, *Distance & Engagement: Walking, Thinking, and Making Landscape*, ed. Günther Vogt (Lars Müller Publishers, 2010).

Richard M. Frank, "The Public Trust Doctrine: Assessing its Recent Past & Charting its Future," *UCDL Rev.* 45 (2011): 665.

Tony Fry, *Design Futuring* (Sydney: University of New South Wales Press, 2009).

Matthew Fuller and Eyal Weizman, *Investigative Aesthetics: Conflicts and Commons in the Politics of Truth* (Verso Books, 2021).

Matthew Gandy, *The Fabric of Space: Water, Modernity, and the Urban Imagination* (MIT Press, 2014).

Paul H. Gobster, Joan I. Nassauer, Terry C. Daniel, and Gary Fry, "The Shared Landscape: What does Aesthetics have to do with Ecology?," *Landscape Ecology* 22, no. 7 (2007): 959–72.

Christophe Girot, "Four Trace Concepts in Landscape Architecture," in *Recovering Landscape: Essays in Contemporary Landscape Architecture*, ed. James Corner (Princeton Architectural Press, 1999), 59–68.

Stephen Graham, ed., *Disrupted Cities: When Infrastructure Fails* (Routledge, 2009).

Stephen Graham and Simon Marvin, *Splintering Urbanism: Networked Infrastructures, Technological Mobilities, and the Urban Condition* (Routledge, 2002).

"Healthy Port Futures," Great Lakes Protection Fund, https://glpf.org/funded-projects/healthy-port-futures/.

Gary B. Griggs, "The Impacts of Coastal Armoring," *Shore and Beach* 73, no. 1 (2005): 13–22.

Carol P. Harden, "The Human-landscape System: Challenges for Geomorphologists," *Physical Geography* 35, no. 1 (2014): 76–89.

Penelope Harvey, Casper Bruun Jensen, and Atsuro Morita, eds., *Infrastructures and Social Complexity* (New York: Routledge, 2016).

"Roduin," Hehe, http://hehe.org.free.fr/hehe/roduin/index.html.

Kregg Hetherington, ed., *Infrastructure, Environment, and Life in the Anthropocene* (Duke University Press, 2018).

Rob Holmes and Brett Milligan, "DredgeFest: Social Experiments in Sedimentary Landscapes," in *The Expanding Periphery and the Migrating Center: Proceedings of the 103rd ACSA Annual Meeting*, eds. Lola Sheppard and David Ruy, 2015.

Michelle A. Hummel and Mark T. Stacey, "Assessing the Influence of Shoreline Adaptation on Tidal Hydrodynamics: The Role of Shoreline Typologies," *Journal of Geophysical Research: Oceans* 126, no. 2 (2021): e2020JC016705.

Michelle A. Hummel, Robert Griffin, Katie Arkema, and Anne D. Guerry, "Economic Evaluation of Sea-level Rise Adaptation Strongly Influenced by Hydrodynamic Feedbacks," *Proceedings of the National Academy of Sciences* 118, no. 29 (2021).

Ivan Illich, "Silence is a Commons," *CoEvolution Quarterly* 40 (1983): 5–9.

Jeffrey Jenkins, Brett Milligan, and Yiwei Huang, "Seeing the Forest for more than the Trees: Aesthetic and Contextual Malleability of Preferences for Climate Change Adaptation Strategies," *Ecology and Society* 25, no. 4 (2020).

Casper Bruun Jensen, "Amphibious Worlds: Environments, Infrastructures, Ontologies," *Engaging Science, Technology, and Society* 3 (2017): 224–34.

JoAnna Klein, "A New Formula to Help Tame China's Yellow River," *The New York Times,* June 2, 2017.

Eduardo Kohn, *How Forests Think* (University of California Press, 2013).

Liz Koslov, "The Case for Retreat," *Public Culture* 28, no. 2 (2016): 359–87.

Marianne E. Krasny and Keith G. Tidball, *Civic Ecology: Adaptation and Transformation from the Ground Up* (MIT Press, 2015).

Roman Krznaric, *The Good Ancestor: A Radical Prescription for Long-Term Thinking* (The Experiment, 2020).

"The Hidden Sediment Reserve," Landscape Metrics, http://dredge3.s3-website-us-west-2.amazonaws.com/.

Brian Larkin, "Promising Forms: The Political Aesthetics of Infrastructure," in *The Promise of Infrastructure,* eds. Nikhil Anand, Akhil Gupta, and Hannah Appel (Duke University Press, 2018), 175–202.

Bruno Latour, *Aramis, or the Love of Technology* (Harvard University Press, 1996).

Bruno Latour, *Pandora's Hope: Essays on the Reality of Science Studies* (Harvard University Press, 1999).

Karen Lutsky and Sean Burkholder, "Curious Methods," *Places Journal* (2017).

David Maddox and Keith G. Tidball, *Grassroots to Global: Broader Impacts of Civic Ecology* (Cornell University Press, 2018).

Erin Manning and Brian Massumi, *Thought in the Act: Passages in the Ecology of Experience* (University of Minnesota Press, 2014).

Ezio Manzini, *Design, When Everybody Designs: An Introduction to Design for Social Innovation* (The MIT Press, 2015).

Noortje Marres, "Issues Spark a Public into Being: A Key but often Forgotten Point of the Lippmann-Dewey Debate," *Making Things Public: Atmospheres of Democracy*, eds. Bruno Latour and Peter Weibel (The MIT Press, 2005).

Brian Massumi, *Politics of Affect* (John Wiley & Sons, 2015).

Shannon Mattern, "Infrastructural Tourism," *Places Journal* (2013).

Elizabeth Meyer, "Beyond Sustaining Beauty: Musings on a Manifesto," in *Values in Landscape Architecture and Environmental Design: Finding Center in Theory and Practice*, ed. M. Elen Deming (LSU Press, 2015).

Elizabeth Meyer, "Sustaining Beauty: The Performance of Appearance," *Journal of Landscape Architecture* 3, no. 1 (2008): 6–23.

Elizabeth Meyer, "Uncertain Parks: Disturbed Sites, Citizens, and Risk Society," *Large Parks*, eds. Julia Czerniak and George Hargreaves (Princeton Architectural Press, 2007): 59–86.

Brett Milligan, "Design Fieldwork: Reclaiming Affect and Experience as a Primary Locus of Design Knowledge and Expertise," *Landscape Review* 18, no. 1 (2018).

Brett Milligan, "Making Terrains: Surveying, Drones and Media Ecology," *Journal of Landscape Architecture* 14, no. 2 (2019): 20–35.

Brett Milligan, Alejo Kraus-Polk, and Yiwei Huang, "Park, Fish, Salt and Marshes: Participatory Mapping and Design in a Watery Uncommons," *Land* 9, no. 11 (2020).

Samuel E. Munoz, "Climatic Control of Mississippi River Flood Hazard Amplified by River Engineering," *Nature* 556.7699 (2018): 95–98

"Code of Federal Regulations," National Archives, https://www.ecfr.gov/current/title-33/chapter-II/part-329.

"National Beach Nourishment Database," National Oceanic and Atmospheric Administration, https://coast.noaa.gov/digitalcoast/tools/beach-nourishment.html.

"Sacramento-San Joaquin Delta National Heritage Area," National Park Service, https://www.nps.gov/places/sacramento-san-joaquin-delta-national-heritage-area.htm.

"Google Plans To Expand Its Campus—Which Might Become Unsafe As Sea Levels Rise," National Public Radio, https://www.npr.org/2021/07/27/1021373017/google-plans-to-expand-its-campus-which-might-become-unsafe-as-sea-levels-rise

"Who Will Pay To Protect Tech Giants From Rising Seas?," National Public Radio, https://apps.npr.org/sea-level-rise-silicon-valley/.

Kate Orff, "What is Design Now? Unmaking the Landscape," *Architectural Design* 90, no. 1 (2020): 94–99.

Kate Orff and SCAPE Landscape Architecture, *Toward an Urban Ecology* (Monacelli Press, 2016).

"Park(ing) day: The Origin Story," Park(ing) Day, https://www.myparkingday.org/about.

Orrin H. Pilkey and Linda Pilkey-Jarvis, *Useless Arithmetic: Why Environmental Scientists can't Predict the Future* (Columbia University Press, 2007).

"Bayou St. John Urban Marsh," Pontchartrain Conservancy, https://scienceforourcoast.org/pc-programs/coastal/coastal-projects/bayou-st-john-urban-marsh/.

"Bayou St. John Urban Marsh Monitoring Report 2014–2019," Pontchartrain Conservancy, https://scienceforourcoast.org/wp-content/uploads/2020/08/BSJ_UrbanMarshMonitoringReport2014-19.pdf.

"About Public Lab," Public Lab, https://publiclab.org/about.

Jacques Rancière, *The Politics of Aesthetics* (Bloomsbury Publishing, 2013)

Anne Raver, "Where the Water Was," *Landscape Architecture Magazine,* October 2018.

Maria Hellström Reimer, "Unsettling Eco-scapes: Aesthetic Performances for Sustainable Futures," *Journal of Landscape Architecture* 5, no. 1 (2010): 24–37.

Maria Hellström Reimer, "Whose Goodness? Ethics and Aesthetics in Landscapes of Dissensus," *Journal of Landscape Architecture* 7, no. 2 (2012).

Jonathan Remo, Megan Carlson, and Nicholas Pinter, "Hydraulic and Flood-loss Modeling of Levee, Floodplain, and River Management Strategies, Middle Mississippi River, USA," *Natural Hazards* 61.2 (2012): 551–75.

Stephen Sheppard, "Bridging the Sustainability Gap with Landscape Visualisation in Community Visioning Hubs," *Integrated Assessment* 6, no. 4 (2006).

Neera Singh, "Becoming a Commoner: The Commons as Sites for Affective Socio-nature Encounters and Co-becomings," *ephemera: theory & politics in organization* 17, no. 4 (2017).

Susan Leigh Star, "The Ethnography of Infrastructure," *American Behavioral Scientist* 43, no. 3 (1999): 377–91.

Susan Leigh Star and Karen Ruhleder, "Steps toward an Ecology of Infrastructure: Design and Access for Large Information Spaces," *Information Systems Research* 7, no. 1 (1996): 111–34.

Julie Sze, *Environmental Justice in a Moment of Danger* (University of California Press, 2020).

"Franks Tract Futures Final Report—Public and Stakeholder Engagement Appendix," California Department of Fish and Wildlife, https://ucdavis.app.box.com/s/hl3qpglcu9ibf919sfby1txeb8qu6unl.

"Project Background, Franks Tract Futures," California Department of Fish and Wildlife, https://franks-tract-futures-ucdavis.hub.arcgis.com/pages/project-background.

"Bonnet Carré Spillway Masterplan," US Army Corps of Engineers, https://www.mvn.usace.army.mil/Portals/56/docs/Recreation/BCS/Docs/Master%20Plan_2010.pdf.

Anna Tsing, "More-than-human Sociality: A Call for Critical Description," in *Anthropology and Nature*, ed. Kirsten Hastrup (Routledge, 2014), 37-52.

US Army Corps of Engineers, "Fact Sheet—Jamaica Bay Marsh Islands, NY," last modified March 24, 2021, https://www.nan.usace.army.mil/Media/Fact-Sheets/Fact-Sheet-Article-View/Article/487604/fact-sheet-jamaica-bay-marsh-islands-ny/.

Mon Vanderostyne and Marsha Cohen, "From Hand-Drag to Jumbo: A Millennium of Dredging," *Terra et Aqua* 77 (1999): 1–47.

"Maasvlakte 2 Expansion Project Port of Rotterdam, the Netherlands," Van Oord, https://vimeo.com/82379896.

L. Smith Vaughn, et al., *Delta Landscapes Primary Production: Past, Present, Future*, SFEI Contribution No. 988 (Richmond, CA: San Francisco Estuary Institute, 2020).

Ann Vileisis, *Discovering the Unknown Landscape: A History of America's Wetlands* (Island Press, 1999).

Bruce H. Wilkinson, "Humans as Geologic Agents: A Deep-time Perspective," *Geology* 33, no. 3 (2005): 161–64.

Robert M. Wilson, *Seeking Refuge: Birds and Landscapes of the Pacific Flyway* (University of Washington Press, 2010).

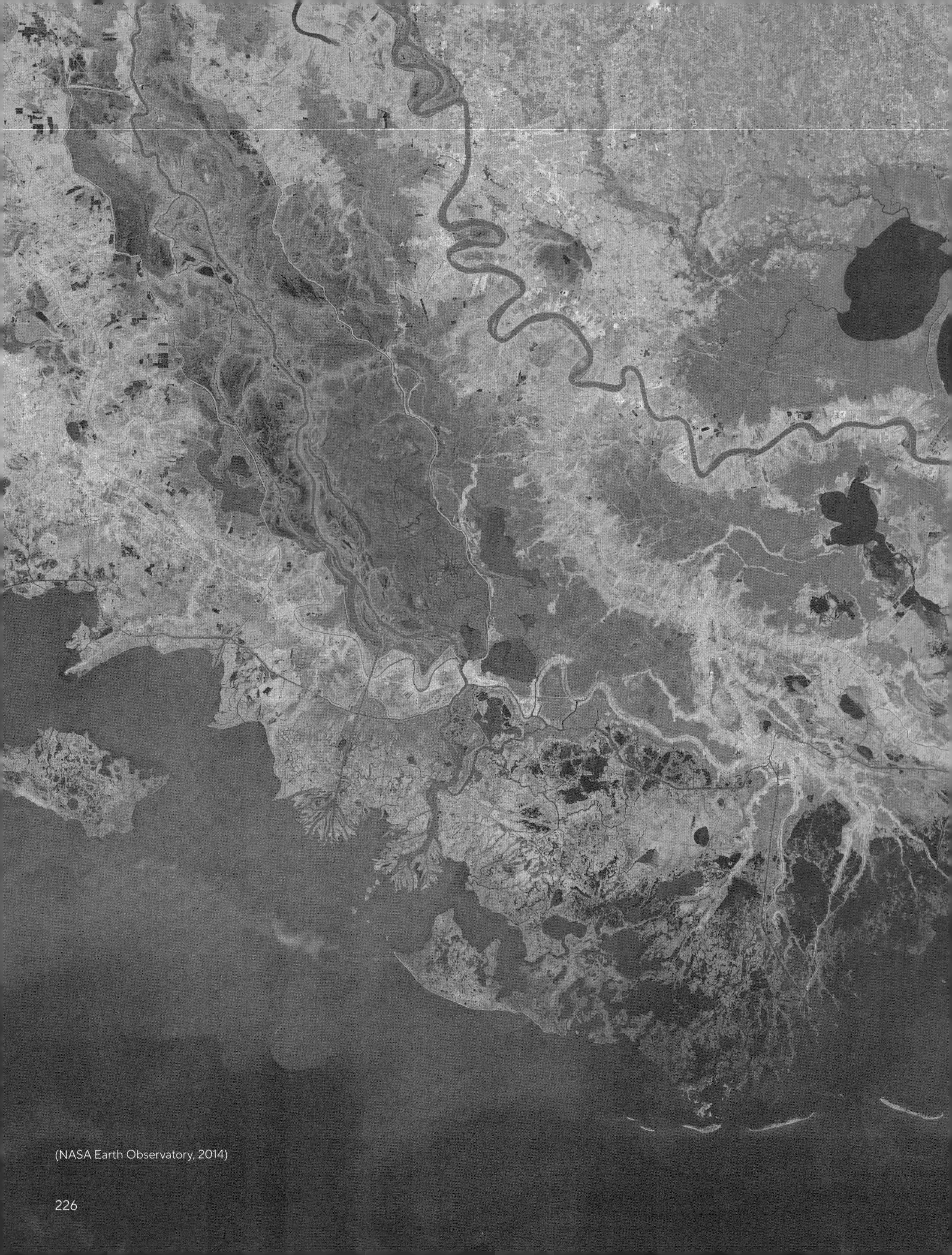

(NASA Earth Observatory, 2014)

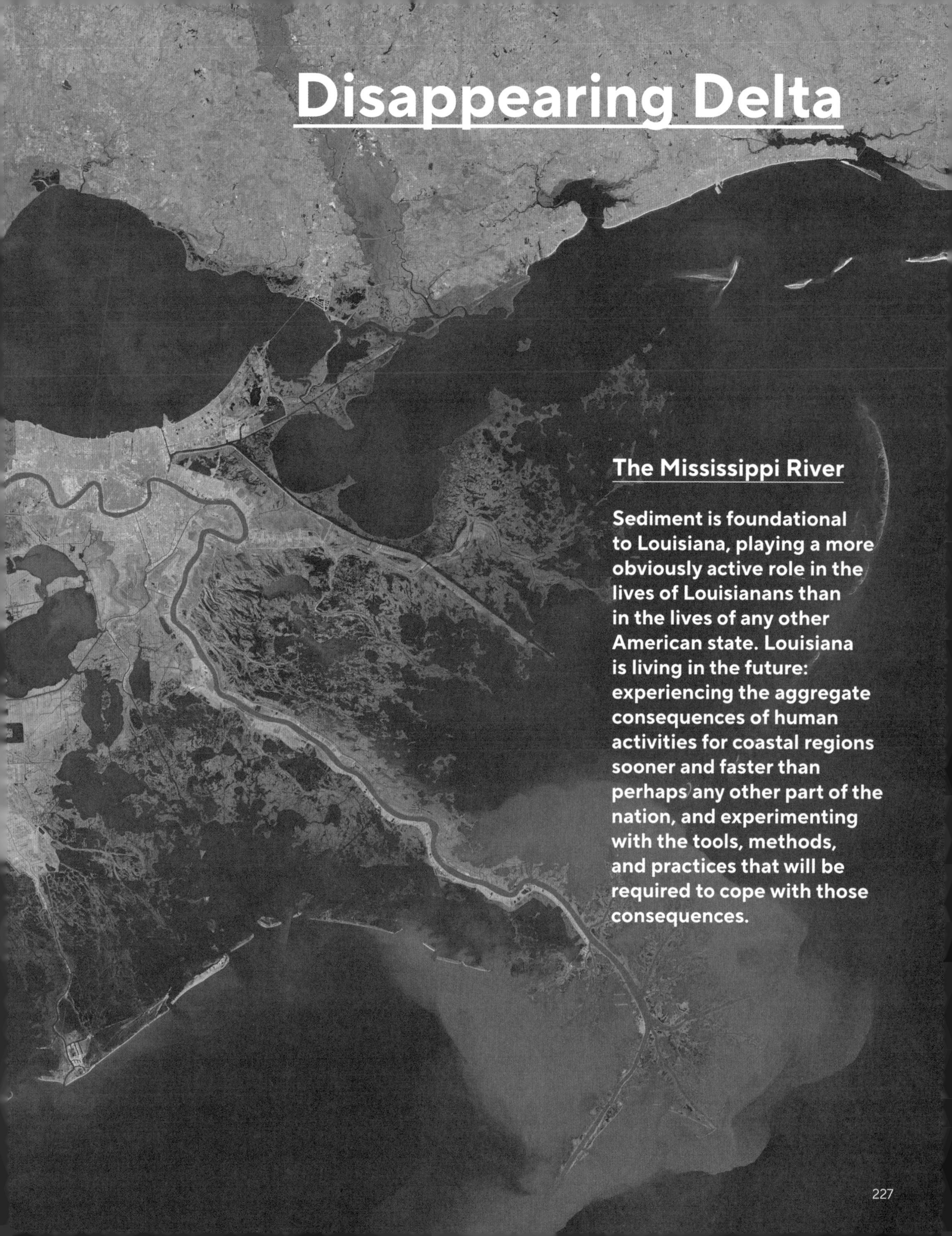

Disappearing Delta

The Mississippi River

Sediment is foundational to Louisiana, playing a more obviously active role in the lives of Louisianans than in the lives of any other American state. Louisiana is living in the future: experiencing the aggregate consequences of human activities for coastal regions sooner and faster than perhaps any other part of the nation, and experimenting with the tools, methods, and practices that will be required to cope with those consequences.

The Land-Making Machine

The Mississippi River Delta was built over thousands of years by deposition of sediment from the river's basin.

For the entirety of its geologically brief existence, the Mississippi River Delta has been a dynamic landscape, simultaneously growing through the deposition of sediments and shrinking due to erosion and subsidence. The delta is the outlet of the Mississippi/Atchafalaya River Basin, the largest drainage area in the United States. More than 1.2 million square miles of runoff and sediment flows through the system, meandering its way to the Gulf Coast.[1] The Mississippi meets the coast at the delta, where fertile sediment is deposited as the river floods, breaches natural levees, and meets the open ocean. The river feeds the delta's expanding coastal marshes, swamps, and cheniers. As sediment builds over thousands of years, the river shifts to find a new path, forming new delta lobes and reshaping the coastline of the gulf.[2]

Barges along the Mississippi River, 2016

Sediment rich water of the Mississippi River, 2016

The reservoir behind Gavins Point Dam is expected to lose 50% of its capacity due to sediment build up by 2045.
(US Army Corps of Engineers, Trisha Dorsey)

Dams impounding the Missouri River are considered a major factor in the decline of sediment transported to the Mississippi delta.
(Maj. Gen. David Sprynczynatyk, North Dakota National Guard)

A Supply Limited System

Flood control infrastructure along the Mississippi River rivals the length of the Great Wall of China.[3] These modifications have led to radical changes in the downstream flow of water and sediment.

Over 3,500 linear miles of embankments and floodwalls, thousands of upstream dams, and hundreds of miles of bank stabilization devices control the river and its sediment.[4] This system, overseen by the US Army Corps of Engineers, controls water levels, enables navigation, and reduces flood risk in low-lying vulnerable communities along the river and its tributaries. Yet these fortifications, including the construction of the largest dams in the 1950s, have had the unintended consequence of trapping sediment upstream and reducing the amount of land-building sediment that transports to the delta's marshes and mudflats.[5] Prior to large-scale river control, the Mississippi annually transported an average of 400 million metric tons of sediment to the delta; in recent decades, this average has dropped to 145 million metric tons.[6]

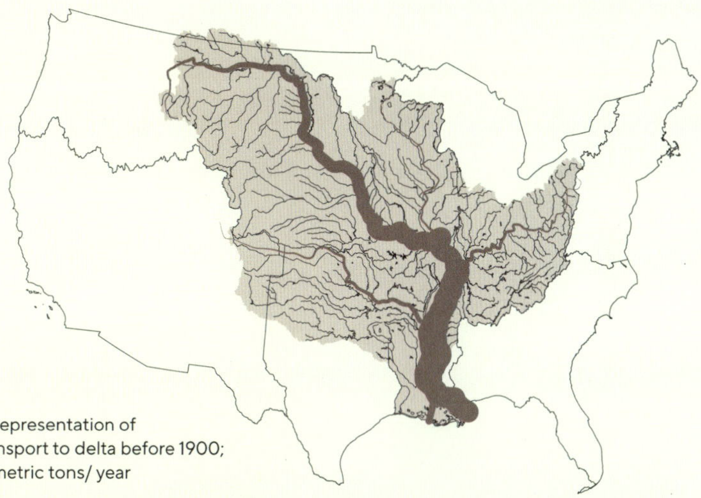

Conceptual representation of sediment transport to delta before 1900; 400 million metric tons/ year

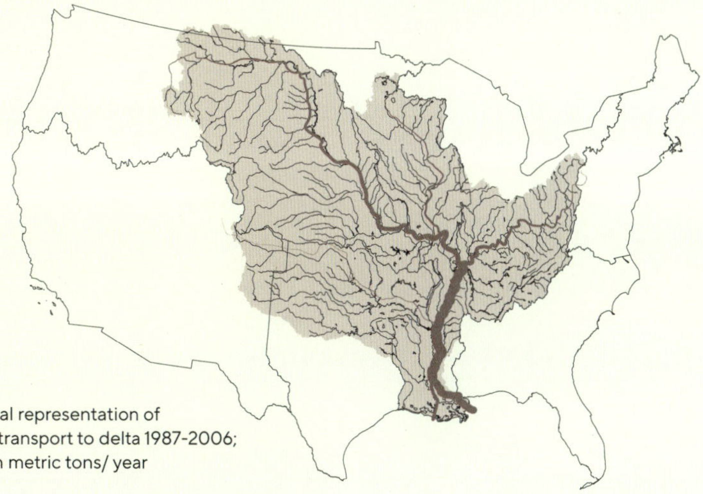

Conceptual representation of sediment transport to delta 1987-2006; 145 million metric tons/ year

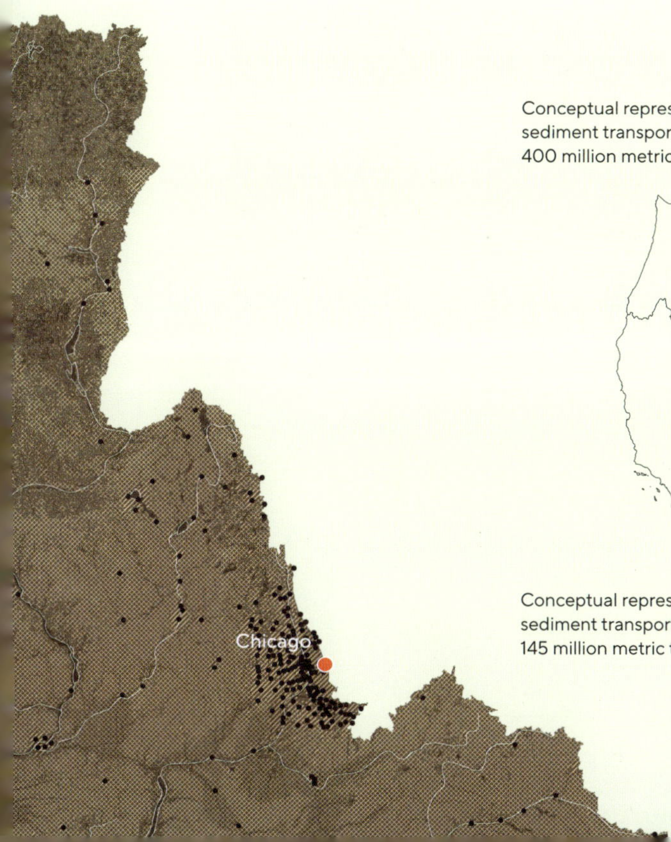

Levees and Land Loss

Levees along the river are being raised to protect communities against a changing climate, yet their very existence accelerates a longer-term threat—the rapid loss of the delta's land.

Subsidence, or the sinking of land, is a natural process that has been occurring in southern Louisiana for thousands of years. Human interventions, like levee building, have accelerated this process by preventing annual floods from depositing sediment that rebuilds land and encourages wetland accretion. While processes of subsidence and erosion (loss) and land building (gain) are constantly reshaping the coastline, today the region faces a net loss of land. From 1932 to 2010, coastal Louisiana lost nearly 2,000 square miles—roughly a quarter of the delta's land—at a pace equivalent to a football field every 100 minutes.[7] Much of this loss has been coastal wetlands, which serve essential roles in managing inland flooding for delta communities, providing habitats for a broad range of species, and supporting 30% of the commercial fishing catch in the United States.[8]

New Orleans
Delacroix Island
Port Sulphur
Boothville - Venice
Grand Isle
Bird's foot
Southwest Pass

Levees and Labor

The enormous earth-moving project of controlling the river is rooted in a legacy of slavery and oppression.

Slavery and forced labor were used to construct and maintain levee systems over generations, starting with the construction of New Orleans' first levee and drainage structure by enslaved people.[9] Enslaved people also built and maintained plantation levees, as landowners were obligated to build earthworks to hold back the river.[10] During the Civil War the levee systems deteriorated, which resulted in the establishment of the Mississippi River Commission (MRC) in 1879[11] and increased authority for the US Army Corps of Engineers. Federal regulation of the river intensified after the Great Flood of 1927, the most destructive river flood in the history of the nation, which displaced over 200,000 African Americans from their homes and killed 250 people.[12] From the late 1800s to 1940s, the MRC and the Corps organized temporary settlements known as levee camps to provide the labor for levee construction, repair, and enlargement.[13] Workers at the levee camps were primarily black male sharecroppers, and the camps were notorious for "long work hours, ... exploitive pay practices, (and) harsh forms of white disciplinary violence."[14] Levees built by camp workers after the Great Flood are the foundation of the modern infrastructure system we see today.

Fascine Willow Mat Construction at Deer Park Levee (US Army Corps of Engineers)

Plaquemine Parish levee workers, 1935 (US Army Corps of Enginers)

People imprisoned at the notoriously brutal Angola Prison working to close Morganza Crevasse, 1890 (US Army Corps of Enginers)

Controlling the Banks

While dams and levees are often considered the river's primary infrastructure, there are many techniques used to manipulate the river's flow. Over 600 miles of the Mississippi's banks are stabilized with underwater concrete mattresses, invisible from the surface.[15]

Revetments control the river flow and prevent channel migration, locking a river's path in time. Historically made from wooden mats, today's revetments are often made of concrete tiles. Manufactured at select sites along the river, the tiles are sewn together into a continuous mattress on a specialized launching barge and draped over the bank. Every year, approximately 50 full-time and 220 seasonal employees work continuously throughout the low water (summer) season, living on barges that travel up and down the river, laying the underwater mattresses that stabilize the river's edge.[16] Sometimes the mattresses are placed in a new location. Sometimes they replace old mattresses that were worn away by the river. The strategy is successful at preventing the bank from caving during floods, yet it ignores the role that destabilizing floods once played in the river's alluvial ecology as a generator of valuable sediment that supported downstream ecosystems.

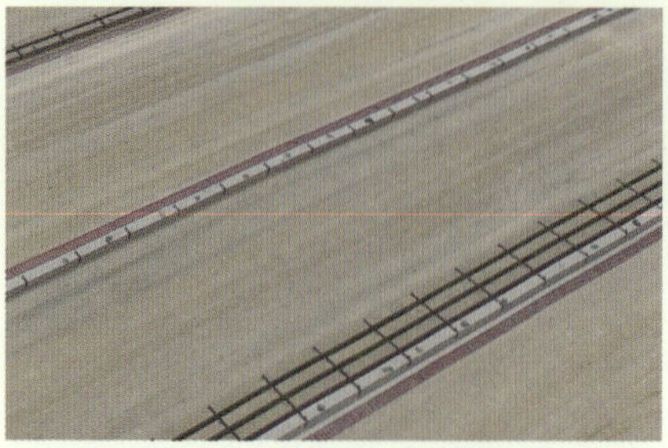

The St. Francisville mat casting fields

US Army Corps of Engineers mat sinking unit along the Mississippi (Team New Orleans, US Army Corps of Enginers, 2006)

Controlling the Depth

The New Orleans district of the USACE dredges an annual average of 77 million cubic yards of sediment from federal channels, dwarfing the volumes of other districts (5 MCY in the New York District).[17]

Sediment that historically would deposit within the river's floodplain now is channeled downstream and accumulates in the delta's riverbeds, including Southwest Pass, the primary navigation channel for river industry. The result is a never-ending race between dredgers and the river.

Dredge placement in the Bird's Foot (Ben Mendelsohn and Alex Chohlas-Wood, 2013)

Controlling Resources

River control and resource extraction are jointly responsible for the fragility of the region. Oil and gas operations have impacted Louisiana's coastline through the digging of canals and widespread construction of pipelines and oil platforms throughout the delta.[18]

Coastal Louisiana hosts one of the largest concentrations of oil and gas extraction infrastructure in the world, much of it enabled through thousands of miles of intersecting dredged canals that allow boat traffic, pipelines, and material transport through extensive marshlands. Canals act as conduits for salt water in some areas and blockades to water flow patterns in others, changing the salinity gradient and weakening the marshes.[19] As the wetlands recede, this infrastructure of extraction is increasingly exposed to the open water of the Gulf and presents ecological and national security threats as sea levels rise and the intensity and frequency of storms increase.[20]

Canal cuts in the Mississippi River Delta (2016)

Louisiana is Sinking

The consequences of river control are accelerated subsidence and rapid land loss, now understood as the new normal for Louisiana residents.

The infrastructure investments that protected river communities and enabled current economic conditions have made the region more fragile over the long term.

The construction of upstream dams that deprive the river of sediment, the construction of levees that constrain the river's floodplain, and the digging of canals for resource extraction have altered the hydro-sedimentary calculus of the region, causing the delta to shrink radically and rapidly. About two million people, or 70% of Louisiana residents, live within 50 miles of the receding coast.[21]

They are increasingly susceptible to land loss, subsidence, rising sea levels, and hurricane damage exacerbated by climate change.

Gulf of Mexico
Port Fourchon
Grand Isle
Barataria Bay
Golden Meadow
Isle de Jean Charles →
Lafitte
Lake Salvador
Davis Pond
Diversion
Paradis
St Rose
Lac des Allemands

A Coastal Masterplan

The challenges facing the coast gained widespread attention after Hurricanes Katrina and Rita, leading to the establishment of the Coastal Restoration and Protection Authority (CPRA) and the development of the Louisiana Coastal Master Plan, the nation's most ambitious plan for addressing flood risk, sediment scarcity, and restoration at the scale of the state's coast.

The 2017 Coastal Masterplan covers 124 projects, including large structural protection infrastructure (e.g., levees, seawalls, and storm surge barriers), non-structural measures (retreat and home elevation), and restoration projects (wetland and barrier island building, ridge restoration, hydrologic connection, and sediment diversion projects).[22] Together these projects aim to build or maintain more than 800 square miles of land, protect and restore habitat, and reduce community risk from coastal floods and sea level rise. The plan is updated every five years, and it shifts along with the developing science and community input. The 2012 Coastal Master Plan predicted "no net loss" of land by 2035 with implementation,[23] yet the 2017 update acknowledges that land loss remains a likely outcome even with the plan's proposed investments of $50 billion over 50 years.[24] The *Deepwater Horizon* oil spill recovery funding has played a significant role in recent implementation of the plan, yet the scale, ambition, and timeline of the necessary work will require substantially more investment.

PROJECT TYPES
Structural protection
Non-structural risk reduction
Oyster barrier reef
Ridge restoration
Shoreline protection
Barrier island restoration
Marsh creation
Sediment diversion
Hydrologic restoration
Infrastructure
Other

Designing with Sediment

The land loss crisis is driving Louisianans to extraordinary lengths, as they experiment with new methods of large-scale sediment choreography and land building.

This dredged material placement project, next to the one-road, water-based community of Delacroix Island, aims to reduce flood risk with thicker, sediment-based living systems rather than perimeter seawalls or levees. Projects like this often take different forms and extents than the natural systems they aim to restore, due to the costs and logistics of moving sediment and the speed of implementation. A new hybridized coastline is emerging, one shaped more intentionally by human action. The following pages examine different types of land-building and restoration techniques being tested through the framework of the Coastal Master Plan.

Coastal protection with living shoreline, Delacroix Island (2016)

Moving Sediment

The Long Distance Sediment Transfer Pipeline (LDSTP) project sucks sediment from the Mississippi River bottom and pumps it over 12 miles to replenish subsiding marshlands in Barataria Basin.[25]

The LDSTP was used to construct some 300 to 400 acres of marsh and ridge habitat over three years at Bayou Dupont at a cost of $37 million. Nothing is permanent in the delta; in twenty years it is anticipated that only 190 of these acres will remain, due to hydrologic processes and subsidence.[26] With over 2.6 million acres of coastal land at risk over the next fifty years, the scale and pace of the problem is clear.[27]

▲ Placement of sediment for wetland building

◀ LDSTP

Movement of sediment by pipeline to Barataria Bay / Bayou Dupont for wetland building

▼ LDSTP

Dredging of sediment in the Mississippi River for wetland building (2016)

Channeling Sediment

In the struggle to combat land loss, freshwater diversions have been designed to artificially mimic historic spring floods, puncturing levee walls to allow valuable river water into adjacent wetlands.

Filled with sediment and nutrients, river water that enters the freshwater diversion slows saltwater intrusion and reestablishes the salinity gradient that Louisiana wetlands require for sustainable growth. The Davis Pond diversion was designed to move freshwater, but has ended up moving more fine sediments than expected, leading to higher accumulations of land. Some have retroactively understood this project as a prototype for larger diversions being planned along the Mississippi River,[28] though scientific debate continues on the ability of these systems to sustainably build land.[29]

Water and sediment move to 9,200 acre ponding area, reducing saline intrusion into mashlands and building land

Davis Pond freshwater diversion project (2016)

Diverting Sediment

Sediment diversions harness the power of high-velocity river mud, diverting it to places where land building is key to ecological and human survival.

The Coastal Master Plan calls for eight mid-basin sediment diversions, rerouting small portions of the river into engineered structures that will shoot sediment-laden water into subsiding marshlands. Although they take decades to permit and construct, the mid-basin sediment diversions are a step toward designing *with* the river system of sediment flows, albeit in a highly controlled and modified system. The Mid-Barataria Sediment Diversion was identified in the Coastal Master Plan as a priority project for implementation and is funded in part with *Deepwater Horizon* settlement funds, as Barataria Basin experiences one of the fastest rates of land loss in Louisiana and was heavily impacted by the oil spill. The project is expected to build 20 to 40 square miles of land and may begin construction in 2023.[30] Yet there is much indeterminacy about how these infrastructures will actually perform, and physical and computational modeling are under way to test, visualize, and communicate the potential impacts of this $1.4 billion project.

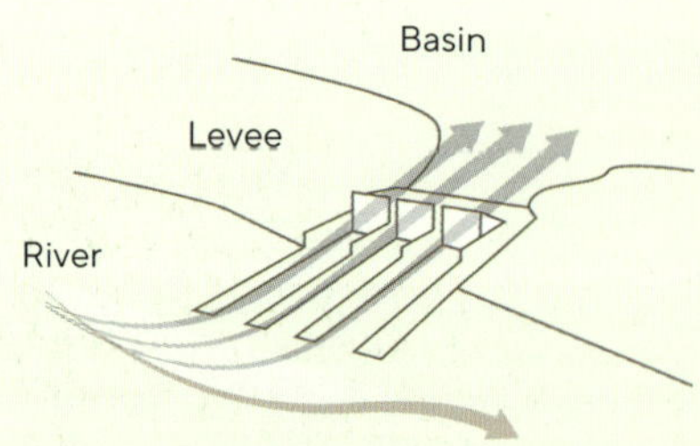

Sediment diversion structure

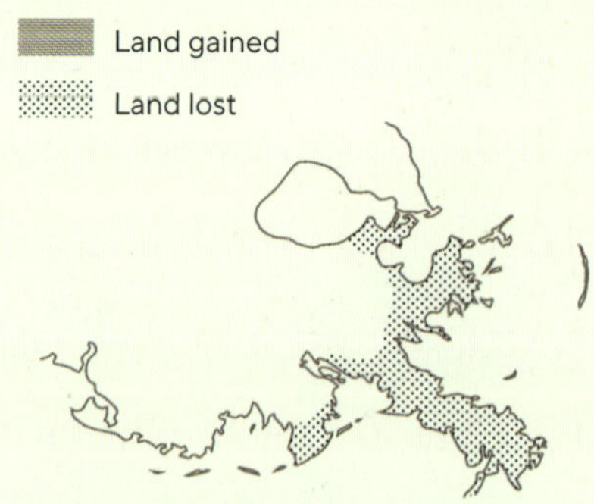

Projected 50 year land loss without four diversion structures

Projected 50 year land loss and gain with four diversion structures

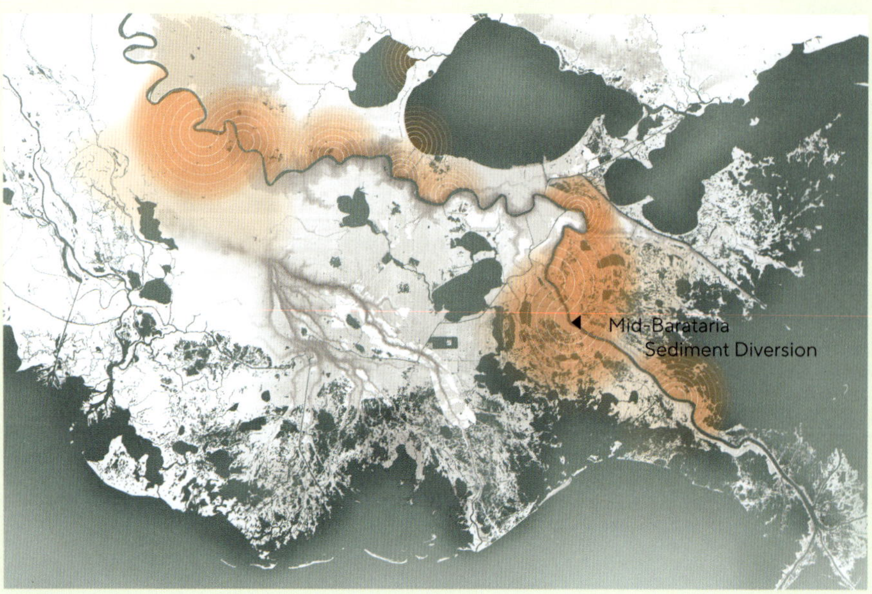

Planned sediment diversion projects from the Coastal Masterplan 2017

Physical hydraulic model of diversions at the Louisiana State University Center for Water Studies

Sediment Politics

Technical, design, and engineering questions cannot be separated from the social and political effects of sediment management in Louisiana. Climate change exacerbates complicated problems and magnifies existing inequalities.

Conventional engineering projects have intentionally and unintentionally reinforced inequality in communities. New sediment management practices bear these same risks, and have the potential to differentially impact community safety, individual property values, local ecosystems, and livelihoods, particularly in the short term. There are long-term inequalities that will be produced by this infrastructure, as diversions are very likely to negatively impact the local fishing and seafood economies if sediment-laden freshwater is introduced into today's saltwater environment. While the proposed shift to freshwater more closely mimics the ecosystems of the past, people who depend upon saltwater species like brown shrimp and oyster populations for their livelihoods will be negatively affected in the near term.[31] These social tradeoffs illustrate the charged political nature of sediment design in Louisiana, with CPRA scientists and engineers recommending large-scale diversions and many fishery advocacy groups advocating for the continuation of pipeline and dredging projects and against diversions.[32] The figure of the future Louisiana coastline and the long-term fate of vulnerable communities will be shaped by the outcomes of these debates.

(above) (Monica Barra 2015)
(right) Flooding of vulnerable communities after Hurricane Katrina
(Lieut. Commander Mark Moran, NOAA Corps, NMAO/AOC, 2005)

Endnotes:

1. "The Mississippi/Atchafalaya River Basin (MARB)." EPA. Environmental Protection Agency. Accessed April 16, 2022. https://www.epa.gov/ms-htf/mississippiatchafalaya-river-basin-marb.

2. "How Mississippi River Delta Formed." How the Delta formed. Restore the Mississippi River Delta, February 1, 2017. https://mississippiriverdelta.org/our-coastal-crisis/how-the-delta-formed/.

3. "Home." Levee Systems. Mississippi Valley Division, US Army Corps of Engineers. Accessed April 16, 2022. https://www.mvd.usace.army.mil/About/Mississippi-River-Commission-MRC/.

4. US Army Corps of Engineers, Mississippi Valley Division, Memphis District, Vicksburg District, New Orleans District, Levee System Evaluation Report for the National Flood Insurance Program, April 2014. https://www.mvd.usace.army.mil/Portals/52/docs/MRC/LeveeSystem/MRT_levee_system_eval_report_for_NFIP.pdf.

5. US Geological Survey and Department of State, A Brief History and Summary of the Effects of River Engineering and Dams on the Mississippi River System and Delta, Circular 1375, Reston, VA: 2012. https://pubs.usgs.gov/circ/1375/C1375.pdf.

6. Ibid.

7. Harvey, Chelsea. "Scientists Say the Rapid Sinking of Louisiana's Coast Already Counts as a 'Worst Case Scenario.'" The Washington Post. WP Company, October 27, 2021. https://www.washingtonpost.com/news/energy-environment/wp/2017/06/15/scientists-say-the-rapid-sinking-of-louisianas-coast-already-counts-as-a-worst-case-scenario/.
McCoyer, Michael. "'Rough Mens' in 'the Toughest Places I Ever Seen': The Construction and Ramifications of Black Masculine Identity in the Mississippi Delta's Levee Camps, 1900–1935." *International Labor and Working-Class History*, no. 69 (2006): 57–80. http://www.jstor.org/stable/27673022.

8. "Louisiana's Rate of Coastal Wetland Loss Continues to Slow: US Geological Survey." Louisiana's Changing Coastal Wetlands: Lack of Major Hurricanes Since 2008 Is Likely the Main Reason. US Geological Survey Communications and Publishing. July 12,2017, Accessed April 16, 2022. https://www.usgs.gov/news/national-news-release/usgs-louisianas-rate-coastal-wetland-loss-continues-slow.

9. Organization of American Historian. "New Orleans Levee System Timeline: Through the Eye of Katrina: The Past as Prologue?" *Journal of American History*, 94 (December 2007), 693–876. http://archive.oah.org/special-issues/katrina/resources/levee.html

10. Davis, John Dean. "Levees, Slavery, and Maintenance." *Technology's Stories*. vol. 6, no. 3 - DOI: 10.15763/jou.ts.2018.08.20.01. Accessed April 16, 2022. https://www.technologystories.org/levees-slavery-and-maintenance/.

11. "Historical Vignette 088—Congress Established the Mississippi River Commission." US Army Corps of Engineers. Headquarters US Army Corps of Engineers. Accessed April 16, 2022. https://www.usace.army.mil/About/History/Historical-Vignettes/Civil-Engineering/088-Mississippi-River-Commission/.

12. Hohle, Randolph. *Racism in the Neoliberal Era: A Meta History of Elite White Power*. New York, NY: Routledge, 2018.

13. McCoyer, Michael. "Levee Camps." Mississippi Encyclopedia. Center for Study of Southern Culture, July 11, 2017. https://mississippiencyclopedia.org/entries/levee-camps/.

14. Ibid.

15. Hearings before a subcommittee of the committee on appropriations, House of Representatives, 92nd Congress, First session. p 278 (1978). (Major General J. P. Rollins, J. R., Division Engineer, US Army Corps of Engineers). https://books.google.com/books?id=irMdAAAAMAAJ&pg=PA278&lpg=PA278&dq=600+miles+of+revetment+mississippi+power&source=bl&ots=INruEvSpxZ&sig=ACfU3U3rF5hG91vj816pQUxqoCuOeRiQ2Q&hl=en&sa=X&ved=2ahUKEwjG8v2No8_wAhWMiOAKHTwsDhIQ6AEwCXoECAoQAw#v=onepage&q=600%20revetment&f=false

16. Schuler, Mike. "Mississippi River's Specialized Mat Sinking Unit to Get Modern Upgrade." gCaptain, December 12, 2019. https://gcaptain.com/mississippi-rivers-specialized-mat-sinking-unit-to-get-modern-upgrade/.

17. "Beneficial Use of Dredge Material." Operations. New Orleans District, US Army Corps of Engineers. Accessed April 18, 2022. https://www.mvn.usace.army.mil/About/Offices/Operations/Beneficial-Use-of-Dredged-Material/.

18. Scaife, William W., R. Eugene Turner, and Robert Costanza. "Coastal Louisiana Recent Land Loss and Canal Impacts." *Environmental Management* 7, no. 5 (1983): 433–42. https://doi.org/10.1007/bf01867123.

19. Ibid.

20. Carey, John. "Louisiana Wetlands Tattered by Industrial Canals, Not Just River Levees." Scientific American. Scientific American, December 1, 2013. https://www.scientificamerican.com/article/carey-louisiana-wetlands-tattered-by-industrial-canals/.

21. "Louisiana's Oil." Wetlands and Coastal Erosion. Accessed April 18, 2022. https://www2.southeastern.edu/orgs/oilspill/wetlands.html.

22. "2017 Masterplan Projects." Coastal Protection And Restoration Authority. State of Louisiana. Accessed April 18, 2022. https://coastal.la.gov/our-plan/2017-coastal-master-plan/planning-process/projects/.

23. Peyronnin, Natalie, Mandy Green, Carol Parsons Richards, Alaina Owens, Denise Reed, Joanne Chamberlain, David G. Groves, William K. Rhinehart, and Karim Belhadjali. "Louisiana's 2012 Coastal Master Plan: Overview of a Science-Based and Publicly Informed Decision-Making Process." *Journal of Coastal Research* 67 (2013): 1–15. https://doi.org/10.2112/si_67_1.1.

24. Coastal Protection and Restoration Authority of Louisiana. Louisiana's Comprehensive Master Plan for a Sustainable Coast. State of Lousiana., 2017. https://coastal.la.gov/our-plan/2017-coastal-master-plan/

25. Schleifstein, Mark. "Pipeline Moves Mississippi River Sand to Create 12-Mile Land Bridge across Three Parishes." NOLA.com. The Times-Picayune, May 16, 2015. https://www.nola.com/news/environment/article_619cb14a-05ab-58b7-8d27-8ae4aaf381dd.html#:~:text=A%20pipeline%2C%20in%20yellow%2C%20carries,in%20white%2C%20red%20and%20purple.

26. "Bayou Dupont Ridge Creation and Marsh Restoration (BA-48)." Coastal Wetlands Planning, Protection, and Restoration Act. Louisiana Coastal Wetlands Conservation and Restoration Task Force. Accessed April 18, 2022. https://lacoast.gov/reports/gpfs/BA-48.pdf.

27. "A Changing Landscape." Coastal Protection And Restoration Authority. Accessed April 18, 2022. https://coastal.la.gov/whats-at-stake/a-changing-landscape/.

28. Mesehle, Ehab. 2016. The impact of the Davis Pond freshwater diversion on water and sediment exchange in Barataria Basin. In Conference: State of the Coast At: New Orleans, Louisiana, USA.

29. Turner, R. Eugene, Michael Layne, Yu Mo, and Erick M. Swenson. "Net Land Gain or Loss for Two Mississippi River Diversions: Caernarvon and Davis Pond." Restoration Ecology 27, no. 6 (2019): 1,231–40. https://doi.org/10.1111/rec.13024.

30. Mid-Barataria Sediment Diversion Deis Virtual Open House—faqs. Coastal Protection and Restoration Authority. Accessed April 18, 2022. https://coastal.la.gov/midbarataria/faqs.html.

31. Crcl. "The Mid-Barataria Sediment Diversion: Fact vs. Fiction." CRCL. CRCL, September 20, 2021. https://www.crcl.org/post/the-mid-barataria-sediment-diversion-fact-vs-fiction.

32. Schleifstein, Mark. "$2 Billion Mid-Barataria Sediment Diversion Attracts More Opposition." The Courier / Houma Today. Times-Picayune | New Orleans Advocate, April 22, 2021. https://www.houmatoday.com/story/news/2021/04/22/2-billion-mid-barataria-sediment-diversion-attracts-more-opposition/7335311002/.

Image Credits:

Spread 1:
Mississippi OLI 2014. Digital Image. NASA Earth Observatory, 2014. https://eoimages.gsfc.nasa.gov/images/imagerecords/85000/85519/mississippi_oli_2014_lrg.jpg

Spread 2:
(left)
US Geological Survey. Mississippi River Basin. Shapefile. July 2012. https://www.sciencebase.gov/catalog/item/55de04d5e4b0518e354dfcf8
US Department of Transportation/ Bureau of Transportation Statistics. Major Ports. Shapefile. 2019 https://hifld-geoplatform.opendata.arcgis.com/datasets/geoplatform::major-us-port-facilities/about
Office of the Assistant Secretary for Research and Technology/ Bureau of Transportation Statistics. Dams (National). Shapefile. 2016. https://catalog.data.gov/dataset/dams-national
(right)
Dredge Research Collaborative. "Barges along the Mississippi." (Photograph). 2016.
Dredge Research Collaborative. "Sediment rich water of the Mississippi River." (Photograph). 2016.

Spread 3:
(left)
Map- same as spread 2.
Sprynczynatyk, Maj. Gen. David. North Dakota National Guard. "Garrison Dam upstream of the Bismarck-Mandan." (photograph). Public domain. https://www.flickr.com/photos/usacehq/5815611688/in/photolist-9RUwK3-9SaTc5-9S835M-aCYByR-aB1fKJ-aD3t7w-axvucP-ch6GVq-bfWgqx-cDAjf9-bfWgva-bfWgB8-9CCyrp-btJZ96-KLaR99-cCoxk9-eaM8WA-Eaiztn-h9RtAn-eaFuQH-ars7bw-KLaUUq-9Wo9xg-eaM8Kh-eaFuZV-e1wL4x-aHBwcK-hyVacA-kNgwbh-btJZ7T-8s1iQ5-ptLmQu-mbSGqp-9RRAzz-aB1rQC-aD3t91-cCoyfb-LxFfso-cnWeJh-aD3ta7-fAdhg-T-9WogTg-8wEHYg-e1wLut-e1wL64-e1Crqf-8UzUB2-e1wLmt-bmvnuh-e1Cr2y
Dorsey, Trisha. US Army Corps of Engineers. "Lewis and Clark Lake, Gavins Point Dam." (photograph) 2011. Public domain. https://www.flickr.com/photos/usacehq/6022208529/in/photolist-c1eNAs-c1eNGh-abaoKZ-abaoMx-9RSdDa-abdfBC-c1eNy5-9TKPPg-c1eNDq-abaoPx-c1eNwY-abaoS8-adQLLn-adTxcy-9TNRj3-abaoRg-9TL4A2-9TNKzq-9TNMvY-abdfws-9TNNMC-abaoRD-abdfxW-abaoKn-abaoQ8-abdfyu
(right)
Mapping Adapted from:
Meade, Robert H., and John A. Moody. "Causes for the Decline of Suspended-Sediment Discharge in the Mississippi River System, 1940–2007." Hydrological Processes, 2009. https://doi.org/10.1002/hyp.7477.

Spread 4:
US Fish and Wildlife Service. National Wetlands Inventory. Shapefile. 2017. https://www.fws.gov/node/264586
Barras, John & Hartley, Steven & Johnston, J. B. & Jones, W. R. 100+ years of land change for coastal Louisiana (map). 2004
Louisiana Oil Spill Coordinator's Office (LOSCO). Bathymetry for Louisiana, Geographic NAD83. Shapefile. 1999. http://lagic.lsu.edu/data/losco/bathymetry_NOAA_1994.zip
United States Army Corps of Engineers (USACE), US Department of Transportation (USDOT)/Bureau of Transportation Statistics (BTS). Navigable Waterway Lines. Shapefile. 2006. https://data-usdot.opendata.arcgis.com/datasets/c91bc22014744f9389488f8ca203b4d4_0/about
US Army Corps of Engineers (USACE), Federal Emergency Management Agency (FEMA). National Levee Database. Shapefile. 2006.
United States Census Bureau. TIGER/Line Shapefiles—Urban Census. Shapefile. https://www.census.gov/geographies/mapping-files/time-series/geo/tiger-line-file.2000.html

Spread 5:
(left)
Team New Orleans. US Army Corps of Engineers. "1929 09-19 Fascine willow mat construction at Deer Park MRm728R" (photograph) 1929. https://www.flickr.com/photos/teamneworleans/16777410532/in/album-72157650824210777/
(right)
Team New Orleans. US Army Corps of Engineers. "1935—Plaquemine Parish MR levee workers" (photograph) 1935. https://www.flickr.com/photos/teamneworleans/16156184354/in/photolist-58pHLu-8yZmXB-qBEEJE-7MBNmH-7MFMPo-8z3Gbs-8yZnbt-8yZnqD-8yZnkV-4XqPK-rwroey-9muU3B-89CEYf-89zrrZ-2iYeAt8-8fHGTr-89JEFL-8f1rkd-87LAQf-87LACh-87HoCc
Team New Orleans. US Army Corps of Engineers. "Morganza Crevasse Closure using prison labor 1890" (photograph) 1890. https://www.flickr.com/photos/teamneworleans/16592407779/in/album-72157650824210777/

Spread 6:
(left)
Dredge Research Collaborative. "St. Francisville Casting Field Survey." (Photograph series). 2016.
(right)
Team New Orleans, United States Army Corps of Engineers. "Mat Sinking Unit." (Photograph). 2006.

Spread 7:
Mendelsohn, Ben and Chohlas-Wood, Alex. "Dredging in the Bird's Foot." Still image from video footage. 2013.

Spread 8:
Dredge Research Collaborative. "Canal cuts in the Mississippi River Delta." (Photograph). 2016.

Spread 9:
US Fish and Wildlife Service. National Wetlands Inventory. Shapefile. 2017. https://www.fws.gov/node/264586
United States Army Corps of Engineers (USACE), US Department of Transportation (USDOT)/Bureau of Transportation Statistics (BTS). Navigable

Waterway Lines. Shapefile. 2006. https://data-usdot.opendata.arcgis.com/datasets/c91bc22014744f9389488f8ca203b4d4_0/about
US Army Corps of Engineers (USACE), Federal Emergency Management Agency (FEMA). National Levee Database. Shapefile. 2006.
United States Census Bureau. TIGER/Line Shapefiles- Urban Census. Shapefile. https://www.census.gov/geographies/mapping-files/time-series/geo/tiger-line-file.2000.html

Spread 10:
The data depicted here was provided by the Louisiana Coastal Protection and Restoration Authority and were originally produced to inform the development of the 2017 Coastal Master Plan.
Louisiana Coastal Protection and Restoration Authority (CPRA). 2017 Coastal Masterplan: Restoration Projects. Shapefile. 2017. https://cims.coastal.louisiana.gov/masterplan/GISDownload/
Louisiana Coastal Protection and Restoration Authority (CPRA). 2017 Coastal Masterplan: Structural Projects. Shapefile. 2017. https://cims.coastal.louisiana.gov/masterplan/GISDownload/
Louisiana Coastal Protection and Restoration Authority (CPRA). 2017 Coastal Masterplan: Nonstructural Projects. Shapefile. 2017. https://cims.coastal.louisiana.gov/masterplan/GISDownload/

Spread 11:
Dredge Research Collaborative. "Delacroix." (Photograph). 2016.

Spread 12:
(left)
Dredge Research Collaborative. "Movement of sediment by pipeline to Barataria Bay." (Photograph series). 2016.
(right)
Dredge Research Collaborative. "Dredging of sediment in the Mississippi River for wetland building." (Photograph). 2016.

Spread 13:
Dredge Research Collaborative. "Davis Pond freshwater diversion project." (Photograph). 2016.

Spread 14:
(left)
The diversion location data depicted here was provided by the Louisiana Coastal Protection and Restoration Authority and were originally produced to inform the development of the 2017 Coastal Master Plan.
Louisiana Coastal Protection and Restoration Authority (CPRA). 2017 Coastal Masterplan: Restoration Projects. Shapefile. 2017. https://cims.coastal.louisiana.gov/masterplan/GISDownload/
Land Loss / Land Gain maps abstracted from:
Haase, Bren. "Presentation to the Expert Panel on Diversion Planning and Implementation, Meeting #6. MS River Sediment Diversions: Fall 2015 Recommendation." PDF presentation, Coastal Protection and Restoration Authority, October 27, 2015.
(right)
Dredge Research Collaborative. "Physical model of diversions at the LSU Center for Water Studies." (Photograph). 2018.

Spread 15:
(left)
Dredge Research Collaborative. "Physical model of diversions at the LSU Center for Water Studies." (Photograph). 2018.
Barra, Monica. "10_stop-diversions-now." (Photograph) 2015. https://aesengagement.files.wordpress.com/2016/03/10_stop-diversions-now.jpg
(right)
Moran, Lieut. Commander Mark. NOAA Corps, NMAO/AOC. "Wea02630." (photograph) Flickr Creative Commons. 2010. https://www.flickr.com/photos/noaaphotolib/5033310659/in/photostream/. License: https://creativecommons.org/licenses/by/2.0/

Todd Bridges: Engineering by, for, and with Nature

Todd Bridges is the US Army's Senior Research Scientist for Environmental Science and is the National Lead for the US Army Corps of Engineers' Engineering With Nature® Program.

DRC: We've heard you talk often about the importance of "going big." What do you mean by that?

TB: The scale of intervention that is needed to develop more sustainable and more resilient systems with respect to climate change—it's tens and hundreds of thousands, millions of acres. Not an acre here, two acres there... I don't want to diminish that localized value of small projects, but it's not on the scale that society needs, ultimately.

DRC: Do you think that a larger framework is needed to make those decisions?

TB: You need to develop a collaborative vision for what the system should look like. So, you have to start with the vision. If there's no vision, you're left with a collection of piecemeal projects. With no vision, that's all you're going to get, ever.

The large-scale, long-term vision is missing—sometimes at an individual organization level, but also when you start thinking across organizations or even the "whole of government." Where's the vision?

You need to start by talking about how the agencies that relate to the system at hand collaborate on developing a vision, and then you can figure out how to actually work with each other to achieve that vision and all agency mandates. There's no agency that owns every aspect of the work. And the private sector and private land holders, they have to be brought into this process.

I don't want to disparage action by individual landowners, for example, an action someone might take on their hundred-foot living shoreline project. There's value to be achieved there. But we're not going to satisfy our needs fully across the entirety of the country's landscape one hundred feet at a time.

DRC: Are there pathways that you've seen toward this scale and vision?

TB: Yes, I'll give two examples. One would be the partnership and collaboration between the Corps and the National Oceanic and Atmospheric Administration on Engineering With Nature. This collaboration and partnership has been underway for several years now. We have several projects that we're working on together in the same place, and I think that's where it starts. One agency works with another agency and they co-develop and co-learn a new arrangement. We did that at a place called Mordecai Island in the Back Bays of New Jersey. We're doing that right now at Swann Island in the Chesapeake Bay. We're going to be doing that in other places.

The other level would be engagement between state and federal levels. We have a partnership that's been developing over the last few years with the California Department of Water Resources. They're a multi-billion dollar water infrastructure agency, and we've identified several things that we're working on together because we recognize this desire to achieve integration.

Whether you're talking about a state agency or you're talking about the Corps, you can't keep doing things the way that you've always done them and expect to make progress. There's no logic to that. You've got to change to make progress.

DRC: What would you say are the biggest methods or entrenched practices that prevent progress?

TB: I think the biggest hurdle is what you might call the "siloization" of work. We use this word a lot within the Corps about the silos that are created within an organization, but you can see that between organizations as well. Siloization is where I'm only concerning myself with what my organization's mandates or authorities are, and other objectives or considerations are immaterial to me. We can't make real progress if siloization is the way we do business.

DRC: What does real, demonstrable progress look like?

TB: There are many ways to describe what needs to happen to bring about transformation. In fact, we've been using the term "revolution" in the Corps to describe this need. Our Chief of Engineers has stimulated a whole range of activities that are focused on revolutionizing and innovating how the Corps of Engineers delivers civil works projects. As we think about how to deliver a new arrangement between humans and nature, every organization in government, whether it's an infrastructure agency like the Corps or a regulatory agency or a resource agency that's focused on conservation... all organizations have to commit themselves to innovation.

DRC: Why is "nature-based infrastructure" important to this vision?

TB: It certainly is good engineering to consider what the landscape is affording you as a starting condition. Let's say I have a blank canvas. There is no flood risk management infrastructure out there, it's just, "here's the landscape." I think it's consistent with good design and good engineering practice to say, "okay, what is nature offering me here?"

What nature offers is the palette that I have to work from. "Okay, if it's coastline I may have barrier islands, dunes, wetlands, oyster reefs, maritime forests... hmm... how could I make investments in conserving and maybe enhancing those features as a part of an integrated system of flood risk management, where we may also have some 'conventional engineering,' whether it's in the form of levees or flood walls or breakwaters?" Or... how do I integrate the natural with that system to get truly multifunctional, multipurpose, multilayered resilient flood risk management? It's a natural part of good design or engineering to see what nature's offering you as a starting condition and to figure out how you can integrate with that.

The other way to approach the situation doesn't make as much sense to me. There, you start with "okay, well our design philosophy here is that we're going to create an entirely artificial system and figure out what green stuff we can add to it." I think that approach is more likely to lead to conflicts with nature, where you're working to impose human will upon nature.

And I think mother nature is powerful and persistent, and she will always win, ultimately. So, you either start with what nature's offering you and figure out how to join that with human engineering intervention, or you start with a design concept to impose an artificial system and then adding some form of 'nature' to it to satisfy 'green' interests. I think the former is the better, lasting way to develop infrastructure.

DRC: Why did Engineering With Nature, which deals with so many kinds of nature-based infrastructure, emerge from research at ERDC related to dredging operations?

TB: As you know, the Corps moves a lot of sediment in the US to support navigation and other infrastructure—two hundred million plus cubic yards a year, and that represents a lot of what I might call "natural infrastructure opportunity." Sediment is, in a sense, the currency of natural infrastructure. It's so fundamental because it's the material that you use to build natural infrastructure connected to water and waterways. Of course, there's the important biological component. You can't just pile up a bunch of mud and call it "natural infrastructure." There has to be something living associated with it—that's fundamental. Combining the living and the physical together makes the natural infrastructure hold together—it makes it congeal into something that is, by its nature, natural.

DRC: Could you talk a bit about how you understand design and how you see its value?

TB: There are at least two forms: there's "little-d" design and "big-D" design. An example of little-d design would be determining the size of rock you need to stabilize a shoreline with riprap. You go to this page in this guideline and look-up an answer in a table or, maybe calculate it with an equation or run a numerical model to get the design answer. That's little-d design. The equation might be very complicated, it might even require calculus, but it's equation-solving design. Now, big-D design is an entirely different undertaking where you're asking questions like what should this system be? What values are we trying to derive from it? What does the solution set for that system include? What are the nature-based interventions that we would put into the landscape? How are people going to interact with the interventions? How long is it going to last? How can we integrate the natural with the engineered? That's big-D design. We haven't gotten to equations yet... that's not to say that math isn't involved there, it is, but those are the big-D design questions.

And I would further argue that I think some of the weaknesses that we have in our current process are that we rush through big-D design to get to little-d design. The process of solving equations is not the most important step. The most important part is figuring out what you should be doing.

DRC: What do you think it would take to put more emphasis on the "big-D" design?

TB: This is something that we've undertaken with Engineering With Nature—putting more focus on the total design process. We want to be more intentional about introducing big-D design questions into the early stages of a project to help reveal the broader value propositions within a project.

That involves bringing people with big-D design skills into projects early, before people start working on equations. We want to think about the big opportunities to create multi-purpose value. What project should we be building? Let's focus on that before we move on to figuring out how to build the project. Vision before calculation.

DRC: If you were to take what you've learned over the last decade or so and go back and speak to yourself in 2010, what would you want your younger self to know?

TB: We have to spend more time outside, in the landscape, visiting projects and talking to people. There's so much value to be gained by doing those things. I've spent a lot of time on the road in the last year, doing that and writing and talking about what've I seen, heard, and thought about. We've called it "Engineering With Nature On the Road." Just talking about the experience now brings forward the positive emotions I've experienced from this immersion in our landscapes. Your thinking evolves as a part of being there, seeing, being struck by what you're seeing and the landscape, and the concept of it. You put yourself in the place and just let your eyes and your mind go. The words still elude me to describe the impact it's had on my thinking.

The Science of Sediment

Justine Holzman

Sediments have so much to tell us, more so than scientists say. In *The Sea Around Us*, first published in 1950, Rachel Carson describes sediments as an "epic poem of the earth," imagining that, with a bit more wisdom and scientific knowledge, "perhaps we can read in them all of past history."[1] For Carson's subject, the vast and hard-to-scientifically-know ocean, sediments—the small particles of sand, silt, and clay—moving from continents to the ocean floor were a book just beginning to open. During that time, US military scientific projects to make oceans visible and navigable sites of warfare shaped how the ocean was and would be known.[2] Carson's excitement to read this new text for scientific insights came with deep concern for the impacts nuclear waste and nuclear fallout would have on oceans. In the postwar period, practices of irradiation and the tracing of radionuclides were promoted as significant tools for scientific inquiry, heavily supported with federal funding and projects. The purposeful irradiation of environments and the observation of their effects influenced the ecosystem concept, contributed to the perception that ecosystems are vulnerable and can be lost to human disturbance, and set the foundation for ecological restoration science and approaches.[3] Radioactive sources that traveled well in water and adhered to sediments were utilized for their ability to move through biological material in ecosystems—tracing their interrelations and entanglements. Novel methods of tracking sediments were established by releasing irradiated sediments into watersheds and rivers, providing insight into the movement of dynamic landscapes. The adoption of nuclear scientific methods transformed the scientific study of sediments. During this time, new scientific disciplines and practices emerged under the heading of environmental sciences that focused on environmental management, and with this dramatic shift, sediment-related research began to expand.

Before the bottoms of lakes and oceans could be grasped, the science of sediment emerged in shallower waters along rivers, coasts, and their meetings in estuaries. Scientists have long poured time, energy, and resources into interpreting and making meaning out of sediment. As I write, practitioners of varied scientific backgrounds are sampling shallows, trapping underwater sediments, examining cores, sifting sizes, observing fluid interactions in clear containers, leaning over giant map-like models of synthetic sediments, and delving into droves of sensor-detected data. But these practices are hardly new or novel, they have histories: layered, sedimented, and murky, some made into stone and others eroding away. Sediment is used to reveal the furthest stories of our geologic past, while studies of sediment-transport—the movement of sediment from one place to another—have prompted scientists to develop some of the most challenging dynamical models about our geologic future. Sediment is at once archival, in motion, and new ground.

Scientific knowledge is not *found* in nature, it is *constructed* by human beings. It is social, political, cultural, and contextual—made through scientific and cultural practices. Scientific knowledge is produced historically and contingently, and while it is often presented as a sequential series of truths, that is *not* actually the case.[4] The science of sediment is composed of the many scientific pursuits that look to physical and abstracted forms of sediment to answer questions about sediment and sediment-related matters. The focus and output in the science of sediment has been dynamic in response to major physical and sediment-related environmental changes, and heavily swayed by changing cultural attitudes toward these environmental changes. Sediment-related knowledge is not devoid of politics, culture, and context; it is constituted by them. The study of sediments crosses a number of scientific disciplines, has prompted scientific fields, and inspired new research areas. At its broadest, the science of sediment can be found within cartography, the earth sciences, and environmental sciences. Through history, sediment has appealed to natural historians and scientists working in physical and life sciences, surfacing in stratigraphy (the study of strata over geological time), sedimentology (the study of modern sediments), potamology (the study of rivers), limnology (the study of lakes), and oceanography (the study of oceans). A fundamental concern within studies of sediment is its movement, which produced the interdisciplinary research field of sediment dynamics through the combined disciplines of physical oceanography, geology, biology, and chemistry. And most recently, sediment and the contaminants it holds have become a critical subject of inquiry in toxicology. Practices in the science of sediment include varied methods of observation, measurement, quantification, valuing, modeling, and prediction. These practices are executed across a number of places and spaces from field to laboratory to archive—in estuaries, on boats, in warehouses, over tanks, on experimental sites, and through digital models. This chapter begins to reveal the science of sediment and the landscapes they compose by drawing from the works of historians of science, scientific publications, scientists themselves, popular pieces, and landscape studies. Small windows are opened onto places, instruments, people, and laboratories that figure into practices of knowledge production and landscape-making with sediment.[5]

Recent works in the humanities have grappled with the history and context of the science of sediment. This scholarship shows how deeply intertwined sediment, and our knowledge of sediment, is with economic industries of resource extraction, commerce and global transportation networks, European/Euro-colonial settlement, and global development regimes. In an article on the history of eighteenth-century European littorals, Wilko Graf von Hardenberg, an environmental historian of modern Europe, reminds us that an established global understanding of relative sea level is a recent development, which arose from a set of differing and specific local understandings of sea level across disparate coasts with varied practices related to local conditions and needs. The movement from marking exceptional or catastrophic events toward recording the daily fluctuations of the sea took place over centuries as Europeans worked to settle in coastal areas: "it was, in other words, the product of an infrastructural turn that radically transformed certain European coastal landscapes and economies after the Middle Ages and throughout early modernity."[6] Debjani Bhattacharyya looks to the silts and sediments of the Bay of Bengal (the Sundarbans) to understand "the material and social processes of making property in a swamp" by the East India Company in eighteenth- and nineteenth-century Calcutta.[7] There, the material movements of sediments resisted colonial surveys for claims to land. Bhattacharya argues that scientific explanation for fluvial and deltaic movement were notably absent.[8] The Company's merchants and the Army of Bengal engineers failed to execute their engineering projects, and instead, turned to the courts to establish property rights. By reading the legal archive, Bhattacharya reveals how the mobile landscapes of the delta "threatened the legal architecture of property, undoing the stability of leases, one of the central mediating paper technologies of extracting rent for the expanding British Empire."[9]

The sediments and sediment-laden landscapes of India have inspired other thoughtful ruminations on the role of sediments through history and contemporary life. Lindsay Bremner's piece "Sedimentary Ways" looks to the materiality of sediment and its attendant politics. Moving from the Himalayas to the Bay of Bengal, Bremner traces sediment through local environments while thinking through the actual and metaphorical possibilities of the geomorphological processes sediment might transform through: "weathering," "saltating," "alluviating," and "lithifying." Bremner expands upon these scientific terms through ethnographic work, in order to reveal bodily entanglements with lively matter that constitute the conditions for health, labor, and political struggles. For Bremner, sediment is both connective and constitutive of relations and politics: "for sediments mobilize surfaces and materials, unsettle dry, grounded notions of place and undercut binary notions of geo-physical and geo-political worlds."[10] Both Bremner and Bhattacharya look to Rohingya refugees living within the char landscapes of the Bengal Delta to pose the question: what does it mean for your livelihood to depend on land that will disappear and reappear with monsoons?[11]

Through their body of works within rivers and estuaries in post-colonial landscapes, Anuradha Mathur and Dilip da Cunha have called attention to the colonial misappropriation of *the line* within maps and surveys, drawn to alter the rivers and lands of colonized territories across the world.[12] They falsify "the line" by articulating landscape in moments of rain and flood, when the line is nowhere to be found and water takes entirely different forms. Their reading extends to sediment, which moves with water, and similarly reveals colonial-scientific cartographic practices for delineating "land" and "water" as constructs invented for the appropriation and commodification of land. Their collective drawings and design work aims to unravel this distinction and propose new forms of knowing the landscape.

Analyses and histories such as these are needed to understand the contemporary landscape of sediment-related research, to situate, critique, and expand our knowledge of sediment. Although there is much more study of sediment in today's scientific discourse, sediment still takes a backseat to the study of water. As a scientist and hydrologist, Sean Fleming's 2017 narrative and illustrative text, *Where the River Flows: Scientific Reflections on Earth's Waterways*, communicates the science of rivers and how they have shaped our world.[13] Given the importance of sediment to understanding watersheds, it is surprising that "sediment" did not make it into the index as a term worth searching. Although the text is primarily concerned with water, hydrology and the science of rivers is deeply connected to many areas of sediment-related science. Fleming's reflections illuminate aspects influencing sediment science while also exposing the need for more attention to and integration of sediment. As a crucial factor for watershed planning and assessing watershed health, sediment is only beginning to figure more prominently in such studies.

A bibliometric analysis of sediment-related research published in scientific journals revealed a steady growth of sediment-related research across the twentieth century, gaining velocity in the 1990s around topics with the keywords of "sediment transport," "dynamics," and "heavy-metals."[14] This rise coincided with scientific acceptance and new disciplinary methods associated with the emergence of complexity theory and access to computing, which helped scientists model complex systems like weather and water. Spatial clusters of scientific authorship appear in the United States, Western Europe, and Eastern Asia, with the US producing the largest number of both single-country and collaborative articles. Institutionally, the Chinese Academy of Sciences was most productive, followed by the US

Geological Survey and the Russian Academy of Sciences. (These institutions produced more than double the number of articles of other institutions.) The most productive "single-site" institutions were Utrecht University in the Netherlands, the Woods Hole Oceanographic Institution on the coast of Massachusetts, and Bremen University in Germany. Across this network of international collaborations, the US emerges at the center.[15] Woods Hole Coastal and Marine Science Center supports the USGS Coastal and Marine Hazards and Resources Program, which is "the primary Federal marine geology and physical science research program responsible for the Nation's entire coastal and marine landscape."[16] The US Army Corps of Engineers collaborates with the USGS to understand areas of navigation, water control, and flood control in the US. As the largest mover and influencer of sediment in the US, they hold an incredibly important knowledge-base about sediment and sediment management. Studies of water and sediment directly inform the USACE's Dredged Material Management Plans, which guide where sediment is dredged and placed.[17]

What becomes clear is that knowledge of sediment is highly uneven across different geographies and contexts. A salient point of Fleming's is how dependent contemporary models and predictive efforts are on past observations and records. These records are relatively limited in scope, fidelity, and duration. Although the technologies for river science have improved and expanded dramatically, these models and predictions rely on physical infrastructures of sensing and monitoring devices that measure certain attributes of water systems. The USGS monitoring stations that measure river stages (the depth of water within a river at a particular point) have often required small check dams to increase the stability and reliability of the data. It is telling that efforts to scientifically *know* rivers have required methods of control and engineering. Unsurprisingly, sediment is tricky to measure within water systems. Monitoring stations will sometimes capture measurements of turbidity—the amount of suspended sediment in the water—by measuring how light passes through the water. These measurements need to be supplemented by physical samples, where the sediments collected can be analyzed in a laboratory. The difficulty of measuring sediment buildup behind dams has prompted new methods of acoustic monitoring, where the volume of sediment is configured through the specifics of audio waves. Monitoring and calculating sediment flux within estuaries and coastal landscapes has become increasingly important as scientists seek to understand how historic and restored wetlands might withstand the impacts of sea level rise and increased storms across different coastal areas.

The climate crisis has caused scientists to revisit sediment in a number of other ways. A subglacial sediment core extracted from the site of the Camp Century military project in 1966, now warns of the impossibly fast glacial melt to come.[18] As salt marshes are lost around the coasts of North America, sediments are examined for their role in releasing greenhouse gasses into the atmosphere.[19] And, a giant physical hydraulic model of simulated engineered sediment diversions at the LSU Center for River Studies visualizes over a thousand cumulative years of Mississippi River Delta growth and decay for scientists' and the public's eyes. Physical hydraulic modeling is the use of scaled physical models of hydrologic environments to evaluate the design and engineering of hydraulic structures and machines. It is also used in geomorphology to simulate river history and predict possible futures. Practices using sediments began in the nineteenth century, when engineers noticed the similarities between the movement of sediments at very small and very large scales. Over time, through experimentation and failure, mathematical relationships were established that make the practice of physical and numerical modeling valuable today. For scientists and engineers working on coastal Louisiana's urgent land loss crisis, the sediment diversions modeled at the Center are an extraordinary measure to sustain existing wetlands and rebuild some of the loss.[20]

Macro-environmental engineering projects of this scale are not confined to Louisiana. Plans for the Florida Everglades and the San Francisco Bay-Delta similarly apply converging practices of restoration ecology and engineering at the watershed scale in highly disturbed and engineered environments—each with unique sediment challenges and opportunities. In the 2019 IPCC Special Report on the Ocean and Cryosphere in a Changing Climate, sediment and "sediment-based protection"—the restoration and creation of dunes, beaches, wetlands and other coastal ecologies as coastal infrastructure—figured prominently in the chapter on sea-level rise and adaptation of low lying coastal areas and communities.[21] The futures of such regions will be shaped by cultures of science and engineering, the political environment, corporate forces, and funding; while the impacts on residents will be highly uneven due to intersecting geographic and social factors.

Sediments *do* tell many stories. We need stories and histories about the science of sediment to put the many ways of knowing with sediment into context, to break path-dependencies, to include more perspectives, and to choose how to enter into knowledge- and landscape-making with sediments. The study, design, and management of sediment landscapes are increasingly framed as *experimental*, as objects for knowing, and as places to better understand how we might design and adapt to environmental degradation and the climate crisis. These designations shift authority and discourse into the realm of science. The science of sediment is only expanding, in its intellectual scope and ambition, as well as its physical footprint. There is more work to be done to understand its history and contribute to its future.

1 Rachel Carson, *The Sea Around Us*, rev. ed. (New York: The New American Library, Inc., 1961 [1950]), 78.

2 Naomi Oreskes, *Science on a Mission: How Military Funding Shaped What We Do and Don't Know about the Ocean* (Chicago and London: Chicago University Press, 2021).

3 Laura J. Martin, *Wild by Design: The Rise of Ecological Restoration* (Cambridge: Harvard University Press, 2022).

4 This has been the core work of science studies, the history of science, the sociology of science, and science and technology studies.

5 Parrinello and Kondolf argued that rivers and water history are composed of more than water—they contain sediment—and asked for the inclusion of sediment in water histories, histories concerned with water, and hydrology, which occupy a large historiography in environmental history. See Giacomo Parrinello and G. Mathias Kondolf, "The Social Life of Sediment," *Water History* 13, no. 1 (2021): 1–12. As this paper demonstrates, the inclusion of sediment should extend to histories of science, which has begun to overlap more significantly with histories of the environment and environmental history.

6 Wilko Graf von Hardenberg, "Making a Stable Sea: The Littorals of Eighteenth-Century Europe and the Origins of a Spatial Concept," *Isis* 112, no. 1 (2021): 135.

7 Debjani Bhattacharyya, Empire and Ecology in the Bengal Delta: The Making of Calcutta (Cambridge: Cambridge University Press, 2018), 45.

8 Debjani Bhattacharyya, "A River Is Not a Pendulum: Sediments of Science in the World of Tides," Isis 112, no. 1 (2021): 141–49.

9 Bhattacharyya, "A River Is Not a Pendulum," 142–43.

10 Lindsay Bremner, "Sedimentary Ways," *GeoHumanities* 7, no. 1 (2021): 24–43.

11 See, Lindsay Bremner, "Sedimentary Logics and the Rohingya Refugee Camps in Bangladesh," *Political Geography* 77 (2020): 102–9 and Bhattacharyya, "A River Is Not a Pendulum."

12 Anuradha Mathur and Dilip da Cunha, *Mississippi Floods: Designing a Shifting Landscape* (New Haven: Yale University Press, 2001).
Anuradha Mathur and Dilip da Cunha, *SOAK: Mumbai in an Estuary* (New Delhi: Rupa & Co., 2009).
Dilip da Cunha, *The Invention of Rivers: Alexander's Eye and Ganga's Descent*. Penn Studies in Landscape Architecture (Philadelphia: University of Pennsylvania Press, 2019).

13 Fleming, Sean W., *Where the River Flows: Scientific Reflections on Earth's Waterways* (Princeton and Oxford: Princeton University Press, 2017).

14 Beibei Niu, et al., "Global Trends in Sediment-related Research in Earth Science During 1992–2011: a Bibliometric Analysis," *Scientometrics* 98 (2014): 511–79.

15 Niu, et al., "Global Trends."

16 United States Geological Service, "Woods Hole Coastal and Marine Science Center," https://www.usgs.gov/centers/whcmsc.

17 Data from these efforts informs many of the drawings produced for the visual chapters of this book.

18 Andrew J. Christ, Paul R. Bierman, Joerg M. Schaefer, Dorthe Dahl-Jensen, Jørgen P. Steffensen, Lee B. Corbett, Dorothy M. Peteet, et al., "A Multimillion-Year-Old Record of Greenland Vegetation and Glacial History Preserved in Sediment beneath 1.4 Km of Ice at Camp Century," *Proceedings of the National Academy of Sciences* 118, no. 13 (2021): e2021442118.

19 Nathan D. McTigue, Quentin A. Walker, and Carolyn A. Currin, "Refining Estimates of Greenhouse Gas Emissions from Salt Marsh 'Blue Carbon' Erosion and Decomposition," *Frontiers in Marine Science* 8 (2021): 431.

20 I learned of this project working as a research fellow at the LSU Coastal Sustainability Studio from the principal investigator, Clint Willson, and the Louisiana Coastal Protection and Restoration Authority.

21 M. Oppenheimer, B. C. Glavovic , J. Hinkel, R. van de Wal, A. K. Magnan, A. Abd-Elgawad, R. Cai, M. Cifuentes Jara, R. M. DeConto, T. Ghosh, J. Hay, F. Isla, B. Marzeion, B. Meyssignac, and Z. Sebesvari, "Sea Level Rise and Implications for Low-Lying Islands, Coasts and Communities" in IPCC Special Report on the Ocean and Cryosphere in a Changing Climate, H. O. Pörtner, D. C. Roberts, V. Masson-Delmotte, P. Zhai, M. Tignor, E. Poloczanska, K. Mintenbeck, A. Alegría, M. Nicolai, A. Okem, J. Petzold, B. Rama, N. M. Weyer, eds., (Intergovernmental Panel on Climate Change, 2019).

Technology, Pace, and Adaptation

1 Thompson and Dutra, *The Tule Breakers*.

2 Easterling, *Extrastatecraft*, 73.

3 Ibid, 92.

4 Thompson, "Early Reclamation and Abandonment."

1. Stacking Earth

The Gregorian calendar year is 1892. A towering clamshell dredger lowers its open steel mandibles to the bottom depths of a murky channel within California's vast Sacramento-San Joaquin Delta to scoop up a load of sand. This steam-powered machine is a ship-like platform assembly of timber and steel. Its extremely long boom is customized for moving and transforming the landforms of this place.[1] This dredger, and others like it, will "reclaim" the delta by dredging sediments from within its many sinuous channels, and placing the granular material on the outer banks to form earthen berms for the purpose of separating land from water. Through the linear dredging and piling of sediment, the verdant fluidities of tidal marsh and riparian forests will give way to arduously manufactured levees. The mechanized process is challenging, costly, and monotonous. Scoop; raise; pivot; dump; pivot; lower; scoop. Repeat; repeat; repeat...

The emergence of these machines in the delta was neither arbitrary nor predestined. Rather, they were made to be and act like this through a particular set of values, policies, political regimens, ideologies, and attitudes toward the landscape. The dredgers, levees, and policies enabling them are expressions of a certain disposition.

A disposition is a specific tendency, inclination, or temperament. It defines the manner in which something tends to act across various situations. For example, a dog or a landlord can be said to have a friendly, cheerful, or aggressive disposition. Keller Easterling has applied the term to infrastructure, and by infrastructural disposition she means "the accidental, covert, or stubborn forms of power—political chemistries and temperaments of aggression, submission or violence—hiding in the folds of infrastructure space."[2] For Easterling, disposition describes *disclosed or undisclosed* political bearing in infrastructure space, bearing produced by the actors, bureaucracy, governance, social systems, and technologies assembling it.[3]

According to the predominant sensibilities of nineteenth century settlers, wetlands were "barren" wastelands[4] and harbors of pestilence, which fostered the infrastructural disposition that motivated land reclamation. *Swamp lands* came to be understood as impediments to cultivation and settlement of these territories. They presented a challenge to fulfilling the American ideology of Manifest Destiny, and thus "draining and reclaiming such lands, perfecting

5 Garone, "Managing the Garden," 18.

6 Matthew, *Down by the Bay*, 79.

7 Garone, "Managing the Garden."

8 Ibid; Booker, *Down by the Bay*.

9 Thompson and Dutra, *The Tule Breakers*.

10 Thompson, "The Settlement Geography."

what was perceived as an imperfect nature, was a national imperative."[5] Swamps fit into a set of landscapes that, in the representative phrasing of Thomas Jefferson, required the "finishing" of nature to be productive. "Finishing" also implies an optimizable and static understanding of landscapes.[6] "Reclamation," as a national objective, was enabled and incentivized through a string of experimental policies, including the Swamp Land Act (1849) and the Swamplands and Overflowed Lands Act (1850). California created a state swampland fund, the State Lands Office, and formed the State Board of Swamp Land Act Commissioners to oversee local reclamation districts.[7] But most of these centralized state efforts to regulate and incentivize reclamation in California failed to achieve their intent, and it wasn't until the passage of the Green Act (1868) that reclamation really got going.

The Green Act had two key provisions. First, it allowed reclamation projects to be managed locally. Second, it removed restrictions on the number of acres of *swamp and overflowed lands* that an individual could purchase. With these scalar provisions, the pace of delta reclamation greatly accelerated. The Green Act facilitated the transfer of public lands over to unrestricted private land speculation, enabling large, capital-intensive investments to come into play.[8] The act took authority to review reclamation plans away from county boards of supervisors and placed authority within the local reclamation districts themselves. Thus, large corporations owned by wealthy investors and land owners—such as the Tide Land Reclamation Company (established 1869)—now had the motivation, property, and capital to pursue large-scale reclamation efforts.

This evolution in California's reclamation policy shows an underlying tendency away from centralized planning toward localism and unfettered capitalism. Authority shifted downward in scale from federal governance to local reclamation districts, enabling corporately financed production and profit making. Working together, these policies, machines, and corporations took about 80 years to terraform the delta. They transformed 12,000 square miles of dynamic floodplain into 57 engineered diked tracts destined for agricultural production. It could be said that the delta is California's netherlands.

The transformation of the delta led to the eradication of over 95% of its tidal marshes, and the razing of their 10-foot-tall thickets of *Tule* sedges. The delta dredgers—known as the *Tule Breakers* for the work they did[9]—were not the first stage of delta reclamation efforts. Rather, they were designed to replace exploited Chinese immigrant wage laborers, who, using shovels and wheelbarrows, had dug, moved, and stacked layers of the local peat soil (of the former wetlands) into earthen berms encircling reclaimed tracts. These dikes were particularly hard-won, both in terms of the back-breaking labor and the difficult, unpredictable qualities of the endemic peat soil. Being largely composed of decayed organic material, the stacked peat would often slump, shrink, crack, catch fire, or float away on a high tide.[10]

So later, when the steam-powered dredgers came on-line, rather than building with peat, those Tule Breakers could scoop up the clays and sands down in the delta's channels that had previously been far out of reach of spades and pitchforks. And the clamshells' massive steel buckets could scoop up vastly more sediment. Much of that dredged sand—a superior levee-building material compared to wetland peat—was itself an unintended artifact of upstream technological doings.

Hydraulic gold mining is a Gold Rush-era practice invented in California. It entailed blasting apart the Sierra Nevada foothills using high-pressure water cannons, which turned entire hillsides into a slurry that could be sieved to extract gold. This anthro-geologic event created an unprecedented pulse of eroded sediment flowing through montane creeks and rivers down into the delta. This sediment filled in channels, leading to widespread flooding in the mid- to late-1800s in the Central Valley and the delta. Between 1860 and 1914, more than 800 million cubic yards of mining debris passed through the delta.[11] Unexpectedly, this colossal mass of waste from gold mining became an optimal material (far more durable and solid than the local peat soils) for constructing a more robust and extensive network of dikes, and the sheer number of workers rushing to California to mine significantly increased demand for produce from delta farms, incentivizing the conversion of wetlands into farmland. Gold mining was thus inadvertently a key driver of delta reclamation.[12]

But prior to the techno-colonial enterprises of delta reclamation, other ways of being and making within the region's landscapes prevailed. For at least six millenia, many groups of native Californians—the Plains and Bay Miwok, the Nisenan, and the Yokuts—lived within its immense labyrinth of wetlands, riparian forests, and sinuous channels.[13] As indigenous geographer and land steward Don Hankins notes, "California, particularly the delta, supported one of the highest human population densities in pre-Columbian North America," sustained for thousands of years through the tending of grasslands, oak woodlands, vernal pools, riparian woodlands, scrub, and other ecosystems that constituted the pre-colonial delta.[14] Of an estimated 300,000 indigenous people in California at the time of European contact, 10,000 lived in the delta region.[15]

Unlike the nineteenth-century settlers who re-engineered the delta, native peoples nurtured its existing forms and dwelled within the dynamic landscape processes that sustained it. These people also changed the delta, but in very different and more subtle ways—ways that enriched biodiversity rather than diminishing it. Their cultural practices and lifeways involved a co-making of landscape through extensive tending, seeding, hunting, gathering, and the skillful use of fire; these co-evolutionary practices were embodied and refined over millennia.[16] The landscapes of pre-colonial California were gardened at the grandest of scales, and the design was so seamless and integral to the land that it wasn't visible as such to most colonial settlers. Kat Anderson, an ethnobotanist and leading scholar on California's traditional ecological knowledge, describes this *tended wilderness*:

> *California Indians protected and tended favored plant species and habitats, harvested plant and animal products at carefully worked out frequencies and intensities, and practiced an array of horticultural techniques... During the course of their long history in California, Indians so exhaustively explored the plant kingdom for its uses and so thoroughly tested nature's responses to human harvesting and tending that they discovered how to use nature in a way that provided them with a relatively secure existence while allowing for the maximum diversity of other species...*
>
> *Through twelve thousand or more years of existence in what is now California, humans knit themselves to nature through their vast knowledge*

11 Lund et al., *Envisioning Futures*.

12 Milligan and Holmes, "California Delta, USA."

13 Garone, "Managing the Garden."

14 Hankins, "Ecocultural Equality in the Miwko? Waali?," 6.

15 Whipple et al., *Sacramento-San Joaquin Delta Historical Ecology Investigation*.

16 Anderson, *Tending the Wild*.

17 Ibid, 1–2.

18 Zedler and Stevens, "Western and Traditional Ecological Knowledge."

19 Ibid.

20 Ibid; Whipple et al., *Sacramento-San Joaquin Delta Historical Ecology Investigation*.

21 Anderson, *Tending the Wild*.

22 Garone, "Managing the Garden," 11–15.

23 Hankins, "Ecocultural Equality in the Miwko? Waali?"
Kimmerer, "Restoration and Reciprocity."

24 Hankins, "Ecocultural Equality in the Miwko? Waali?"

25 Zedler and Stevens, "Western and Traditional Ecological Knowledge."

26 Zedler and Stevens, "Western and Traditional Ecological Knowledge," 11.
Kimmerer, *Braiding Sweetgrass*.

27 Zedler and Stevens.

base and practical experience. In the process, they maintained, enhanced, and in part created a fertility that was eventually to be exploited by European and Asian farmers, ranchers, and entrepreneurs, who imagined themselves to have built civilization out of an unpeopled wilderness.[17]

Joy Zedler and Michelle Stevens note that native Californians harvested up to 500 indigenous species of plants, and their landscape management practices encouraged the presence and expansion of some species—such as the delta's native fishes—and deliberately decreased others.[18] For example, the open, gallery-like structure of California's riparian woodlands that Spanish colonizers encountered and admired was a result of intensive indigenous tending and harvesting of white root sedges for basket weaving.[19] Native people crafted successional disturbance processes through selectively timed, deliberate burns. Their lifeways were responsive to annual seasons and the delta's transformative cycles of flooding and drought; they shifted their tending and migration accordingly. They moved from place to place, fashioning a constellation of elevated earthen and midden mounds as they inhabited the floodplain.[20] Over thousands of years, many California plants and other organisms were evolutionarily dependent on, and thrived because of, human tending of the land.[21] But native peoples and their landscape stewardship practices were largely displaced by European colonization through direct physical violence, introduced diseases, and forced emigration to distant reservations. Most native Californians had already been pushed out of the delta by the time the *tule breakers* got to work.[22]

Delta traditional ecological knowledge, as an integrative set of spatial techniques, was radically different from "reclamation" knowledge. For the Miwok, the Nisenan, the Yokut, and others, this knowledge was inherently place-specific and adaptive, meaning that it emerged from dwelling within a specific landscape and by paying attention to how that landscape responded to human action. Their stewardship practices were guided by a belief system of gratitude and care-giving,[23] developed through intimate interaction with landscape as a dynamic spatial assemblage. These technologies operated on entirely different time scales from reclamation. They were learned and empirically developed over generations, and intended to be sustained across multiple future generations.[24] Indigenous practices and knowledge are distinct from "reclamation" in their *ethic of reciprocity*, wherein "people tend the land so the land can sustain the people."[25] As Zedler and Stevens note, "to build sustainability and resiliency of socio-ecological systems, people need to know how their land-care actions affect their own well-being and that of future generations."[26] Traditional ecological knowledge accomplishes this, but its foundational understanding of human-landscape interdependency and co-evolution is often unrecognized.[27]

Today, most of the delta's reclamation-era dikes are still in place, forming a network over 1,000 miles long. Since their initial construction, some of these levees have failed and been repaired. Both the US Army Corps of Engineers and the local reclamation districts have built upon these earthworks, augmenting them with additional earth, and diversifying them in form and function. Through state policies, incentives, and mandates, their design criteria have been expanded to include the creation of aquatic and riparian habitat, in order to mitigate the destruction they caused. And these dikes, along with the delta's deepened channels, have been co-opted into a new role in the state's massive water conveyance infrastructure, the Central Valley and State Water projects. The Central Valley Project, which is operated by the Bureau of Reclamation, consists of 20 dams and reservoirs, 11 power plants, and 500 miles of major canals, conduits, tunnels, and related facilities. Annually, it delivers about 7 million acre-feet of water for agricultural, urban, and wildlife use.[28] The larger State Water project, built and operated by the California Department of Water Resources, spans more than 705 miles north to south across the state and includes 36 storage facilities, 21 pumping plants, five hydroelectric power plants, four pumping-generating plants, and approximately 700 miles of canals, tunnels, and pipelines that transport and deliver water to 27 million Californians and 750,000 acres of farmland.[29] These two infrastructure projects overlap and share waterways, and together constitute the largest such infrastructure in the Western Hemisphere, if not the world.

The delta is central to both of these water projects, as the logistical switching yard for conveying water from the wetter northern half of the state to urban centers and the arid south. This centrality has come at a significant cost ecologically. Since the late 1950s, when delta water exports were maximized, native fish populations have plummeted. The largest deltaic system on the west coast of the Americas has been radically transformed by technology: its flows, salinity, habitats, ecology, and earthen forms have all been altered. Once a shape-shifting, complex, absorptive sponge, it is now a dredged expressway of water with disconnected floodplains. Because former floodplains are ringed by levees, they are no longer nourished by sediments carried into the delta by the Sacramento and San Joaquin Rivers, and they are made ever-more vulnerable by the subsidence of their peat soils. Reclaimed wetland soils that are dried out and exposed to air decompose and atrophy (releasing greenhouse gasses), compressing and sinking over time. This has created an *anthropogenic delta air space* of more than 2.5 billion cubic meters.[30] Thus the topographic lay of the delta has been systemically inverted. Water channels are, like earthen aqueducts, at higher elevations than many of the polders, which have subsided to as much as 25 feet below sea level. Nearly all tidal and riparian habitat has been eradicated. The once highly variable salinity, which favored native species, is now kept constant by calculated flow releases from upstream reservoirs, favoring introduced and undesirable species. Climate change is expected to increase the intensity of flood events in the delta (further threatening the levees), the severity of droughts (further stressing water supplies), and water temperature (further threatening fish). A strong earthquake (which is expected) could create multiple levee breaches, bringing the water projects to a standstill and creating a crisis for the state's lucrative agricultural industry. Economic effects would ripple across the globe. Clearly the delta of today will not be the delta of the future. Change is inevitable, be it accidental or deliberate. The pivotal moment in this history, which set the delta on its current technological trajectory, is the creation of the reclamation-era earthworks. So much for "finishing" nature.

The delta is a rapidly evolving landscape, constituted by multiple emplacements and displacements, shaped by varying intentions, and realized

28 "About the Central Valley Project," Bureau of Reclamation.

29 "State Water Project," California DWR.

30 Deverel and Leighton, "Historic, Recent, and Future Subsidence."
Mount and Twiss, "Subsidence, Sea-level Rise, and Seismicity."

31 Milligan and Kraus-Polk, "Evolving the Evolving."

32 Renaud et al., "Tipping from the Holocene to the Anthropocene."

33 These discourses include the philosophy of technology as well as science and technology studies. A few examples of this broad way of approaching technology include:
Arendt, *The Human Condition*.
Ellul, *The Technological Society*.
Latour, "Morality and Technology."

through changing technologies.[31] The region's colonial, earthwork-enabled urbanization is a technological experiment. It is now vulnerable, failing, and teetering on thresholds of unwanted transformation.[32] More broadly, deltas altered by earthmoving collectively constitute a relatively recent global experiment. This experiment exceeds the goals, predictions, and control of its designers, who have made a plethora of undesirable conditions and risks that current and future generations will have to contend with. Reclamation techniques are teaching hard lessons.

In the pre-reclamation delta, knowledge and skills were acquired through embodied learning, through physically dwelling within the landscape over long durations. In contrast, reclamation worked through extraction, abstraction, and subjugation, carelessly remaking land in accordance with colonial desires. Settlers did not know or understand the delta before they started re-plumbing it, and thus could not know, with any real depth, what they were doing and what the longer impacts might be. Kat Anderson and other scholars have documented the ways through which native Californians learned and created as they tended wildness. But what about the reclamation technologies that largely replaced them? If it is through technologies that we learn, what have we learned, and how does that inform future design? Given the many unpredicted technological mishaps of the past and present, how do we move forward with greater humility? How do we assess the situated values of landscape technologies? How can we effectively apply them, particularly in this time of accelerated change and crisis?

2. Technologies of Making, Becoming, and Learning

Technology is etymologically rooted in the craft involved in making and doing. Like most thinkers who study technology, we understand technology broadly and inclusively, encompassing both tools (or, machines, devices, and other technical artifacts) and techniques (or, actions, skills, and particular ways of doing things).[33] Tools and techniques are technology when they are used to achieve an objective or to effect change. Using a coal-powered clamshell dredger to scoop up huge buckets of sediment and stacking them into the form of a levee is an application of technological tools, using a set of specific techniques. Likewise, planning protocols that allow for unlimited purchase and ownership of public lands by private individuals are a technology that works through codifying politically enabled techniques. Technological forms and systems vary substantially. They can be embedded in machines, computers, documents, devices, factories, and landscapes, which are often managed and operated by individuals without detailed knowledge of the workings of such things. But they can also take the form of embodied knowledge and action.

To simplify, we can say that technologies are the ways in which things get done, and how things get done changes over time, shaped by access to resources and power. In the delta, native Californians and their land-tending practices were largely eradicated by colonial settlers, and succeeded by an extractive gold-mining economy and the exploitation of immigrants for their manual labor. This early reclamation was followed by corporately financed shovel-machines, which today have been replaced by bulldozers, environmental regulations, barges, neoliberal financial instruments, and GPS-guided hydraulic dredgers. Humans use technologies to make change within an environment, and those changes lead to the emergence of new landscapes, which generates both intended and unintended reformattings of space, which in turn leads to further applications and adaptations of technologies. Thus, relationships between technologies, beings, and landscapes are reciprocal and co-evolutionary. As Barbara Adam and Chris Groves put it, technologies are "quasi-social things that stand in an interactive and transformative relation to their creators and users as well as to their environments."[34] Technologies are the dominant way we encounter, learn, and adapt to different environments, rendering these co-evolutionary pathways as specific, contextual, and alterable. Technologies are incredibly diverse, each with their own affordances and limitations.

Many people understand technology to be singularly driven by science; technology, in this understanding, is the application of scientific knowledge. This understanding is deceptive. It is stripped of the politics and values that factor heavily in the production and representation of scientific knowledge, and it omits other ways of knowing.[35] As exemplified in the California Delta, technologies are never value neutral. They are instruments and artifacts of socio-political orders and geographic contexts, bearing the marks of the motivation behind their invention and application, as well as the technical arrangements that precede them.[36] Technologies enable and disable particular things, conditions, patterns of order, worldviews, prejudices, and disparities, and do so both deliberately and beyond intention. Technologies nearly always surprise. When enacted, their benefits and damages are unequally and broadly distributed, for humans, and for multitudes of other beings and things. Anthropogenic climate change is produced through machines that function by incinerating the buried remains of our predecessors and projecting their carbon into the atmosphere. But not all of "us" engaged in that effort. Some have access to and working knowledge of technologies. Others don't. Some are subjected to their damaging effects and others are shielded. And, as we will discuss, technologies can create "lock in," wherein the paths of becoming and evolution they foster can be difficult or costly to escape, due to habit, dependency, and entrenched power dynamics, even when those technologies are outmoded and the conditions under which they evolved no longer exist.[37]

Thus, pivotal and pressing questions for how we inhabit and design landscapes concern how we learn and evolve with technology. What ways of doing and being should we choose? Why? Based on what values, knowledge, and projective thinking? What is emplaced and displaced as techniques of making and dwelling change? What is the nature of the feedback loops between the design of technologies and the actual changes they help manifest in place and time? If technology is critical to the futures we are making, how do we intentionally refine how we make, remake, and care for landscapes across increasingly complex and enmeshed technical, material, social, political, economic, and ecological domains?

3. Landscape Complexity and Shifting Baselines

One pervasive and problematic trend in recent landscape history is oversimplification of the landscape medium itself.[38] Oversimplification, often motivated by desired uses and forms of extraction, motivates technologically driven change to landscapes, which then creates unplanned and undesirable

34 Adam and Groves, *Future Matters*, 113.

35 Latour and Woolgar, *Laboratory Life*. Latour, *Science in Action*.

36 Winner, "Do Artifacts Have Politics?"

37 van Staveren and van Tatenhove, "Hydraulic Engineering in the Social-Ecological Delta."

38 Recent here meaning roughly since the emergence of human-designed phenomena of intensive farming and agriculture, sedentary lifestyles (in contrast to nomadic and semi-nomadic), cities, and the industrial revolution.

effects. Human societies have tended to invent technological means to intensify the production and extraction of desired things, leading to growth and consumption that destroys the landscapes that once supported them.[39] In response to phenomena associated with intensive extraction, such as deforestation, landscapes elastically remake themselves in a chaordic cascade of material transformations.[40] For example, on a logged hillside stripped of trees and other vegetative cover, unprotected topsoil loses fertility and organic matter, and thus changes in texture. This ground is then easily mobilized by rain storms, resulting in catastrophic gully erosion, which further displaces nutrients and topsoil in an aggregating, unwanted feedback cycle that is difficult to reverse. At that point, re-establishing the former forest is almost unimaginable.

Be it trees, fish, or fertile soil, there are many accounts of human societies unable to anticipate the effects of their selective landscape consumption. Or, if societies are aware of emergent detrimental effects, they can't shift course or quell their entrenched desires. In the American dust bowl of the 1930s, for instance, prairies and their soils were lost; transformed to dust that was blown across the continent by wind, and carried by creeks and rivers as far as the Gulf of Mexico. Today, many coastal deltas around the world are atrophying due to sediment starvation caused by upstream dams. We see this problem far more extensively in the global ecological crisis, loss of biodiversity, and accelerating climate change. In the phrasing of Amitav Ghosh, we are trapped in the *inertia of habitual motion*.[41]

Histories of colonial and industrial technologies are full of these patterns, wherein physical transformation is built upon a conceptual abstraction of landscape for instrumental purposes.[42] For the settlers of California, the delta was a reviled space to be refashioned (or *finished*) in a manner that could make extractable products to create profit and wealth. In Marxist terms, the delta became an abstracted object of *exchange value*—the monetary value that something has on the market in relation to other commodities. The delta's exchange value was detached from how it came to be, as capital and labor transformed its landscapes.[43] The Gold Rush was the instrumental event that led to the commodity abstraction of both California as a whole and the delta in particular—"setting the tone, mind-set, fervor, and conditions for the exploitation of other resources and the mistreatment of minorities."[44] But the nature of such abstractions (and the actions they lead to) is that they are always exceeded by the dynamic complexity and physical reality of landscapes themselves. A landscape is never *finished, completed,* or *solved*.[45] Regardless of the perception of techno-scientific control, landscape always continues to evolve based on the diverse, changing cast of forces and actors at play within it, a cast far wider than those trying to control the overall action.[46]

Like California's Bay-Delta, most American landscapes today are vastly different from what they were like prior to European colonization. Even after the massive, tectonic changes of *reclamation* efforts, the delta's watery environments have continued to rapidly change due to human influence, and the picture we see shifts based on when we look. If, for instance, you were to visit the delta just after reclamation, in, say, the 1920s, you would see the massive tracts of drained, ditched marsh, recently turned into farmland. The main water channels between the tracts are still full of tiny pelagic fish called Delta Smelt, who thrive in the delta's still-turbid, murky water. Salinity levels fluctuate in

39 See for example Scott, *Seeing Like a State*.

40 Holling and Gunderson, *Panarchy*.

41 Ghosh, *The Great Derangement*, 54.

42 In addition to Scott, *Seeing Like a State*:
Da Cunha, *The Invention of Rivers*.
Sunseri, *Wielding the Ax*.
Candiani, *Dreaming of Dry Land*.
Elvin, *The Retreat of the Elephants*.
Ash, *The Draining of the Fens*.
Worster, *Rivers of Empire*.
White, *The Organic Machine*.

43 Exchange value as monetary abstraction contrasts with the concept of use value—or what a commodity's actual utility is. So, for example, asparagus and Bartlett pears—both of which grew exceptionally well on reclaimed Delta land, when harvested, would have a certain monetary value on the market—their exchange value—that is abstracted from all that went to making them, including the land and labor, as well as their use value. The use value of these products is nutritional, as food. See Karl Marx, *Capital Volume* 1 (first published Hamburg, 1867).

44 Anderson, *Tending the Wild*, 91.

45 Holmes, "The Problem with Solutions."

46 Milligan, "Accelerated and Decelerated Landscapes."

the delta, in accordance with season and rainfall. During storms, large flows of suspended and bed load sediment from the Sierra Nevada Mountains make their way through the delta's channels, and much of those flows continue downstream to San Francisco Bay and, ultimately, the Pacific Ocean.

Fast forward to the early 1960s and you would experience a much different landscape. Volumes of water flowing through the delta have decreased significantly due to upstream diversions, as well as high volume pumping of delta water to southern California from the recently completed Banks Pumping Plant. The pumps are so strong and effective at what they do that they alter the currents and flows in the delta's channels, causing migratory fish like salmon and smelt to get confused and head toward the pumps, only to get shredded by turbine blades when they do. The salinity gradient in the delta is now statically maintained *at bay* (literally at the meeting of bay and delta), in order to facilitate those freshwater exports. All of this is closely monitored and controlled through freshwater releases from the network of dams now built in the Sierra Nevada Foothills. These dams catch and retain all the upstream sediment that flows into their reservoirs, dramatically decreasing the amount of sediment flowing into the Bay-Delta that would support the growth and maintenance of marshes and mudflats. Delta waters become more clear and less turbid due to the lack of sediment, making the smelt more susceptible to predators. Their presence begins to decline, while introduced sport fish, such as striped bass (introduced to California in 1879 with amazing success), continue to thrive.

In the 2020s, the water in the delta would be even clearer due to the pervasiveness of a small Asian clam, *Corbicula fluminea*, which was introduced in the 1940s. It voraciously filters out and consumes much of the phytoplankton in the water column, undermining the foundation of the native aquatic food web. The Bay-Delta has received the moniker of *world's most invaded estuary*, due to all the organisms that cargo ships, travelers, and others have brought into it.[47] During periodic droughts, which are becoming more common and more intense as the climate changes, thick mats of aquatic weeds choke sloughs and waterways of the delta, making navigation impossible. The California Division of Boating and Waterways sprays copious amounts of herbicidal chemicals into delta waters in a Sisyphean effort to control undesired plants. Delta Smelt—suffering from changes in just about every habitat condition they need, from turbid, plankton-rich cool water to variable salinity—are considered by many to be functionally extinct. Exhaustive annual monitoring efforts turn up very few of them, and stories of how monitoring teams used to habitually catch and fry up smelt for a tasty snack—since they were so plentiful and easy to catch—seem like unlikely, yet compelling, lore of a former landscape. The only significant number of smelt that can be seen are raised in an experimental hatchery just for smelt, appropriately housed near the Banks Pumping Plant.[48] This hatchery produces about 20,000 smelt a year, largely for research aimed at better understanding them. On the heels of much scientific debate, these *cultured smelt* will be released into the wild in an emergency effort to try to revive the species.[49] And it is not only the Delta Smelt. Ecological declines are broad across aquatic and terrestrial delta communities—birds, mammals, even introduced sportfish are not doing well. Thus, in a relatively short span of time, just about every landform, physical

47 Cohen, "Nonindigenous Aquatic Species."

48 "Fish Conservation and Culture Laboratory," Fish Conservation and Culture Laboratory.

49 Hobbs, Moyle, and Fangue, "Is Extinction Inevitable."
Moyle et al., "Futures for Delta Smelt."

50 Pauly, "Anecdotes and the Shifting Baseline Syndrome of Fisheries."

51 Soga and Gaston, "Shifting Baseline Syndrome."

52 Ibid.

53 Ghosh, *The Great Derangement*. Ghosh, *The Nutmeg's Curse*.

54 Jasanoff, "Technologies of Humility," 224.

process, materiality, and ecological composition of this delta has fundamentally changed. The delta is radically different from what it was. It cannot go back.

For a moment, imagine if settler society had not made the very particular choice to reclaim and replumb the delta. Imagine if the delta was still akin to what it had been for thousands of years prior. Imagine just how different that evolutionary track could have been, and how different our collective Californian experience and knowledge base would be today.

The term shifting baseline syndrome was coined by marine biologist Daniel Pauly in 1995.[50] It refers to gradual changes in accepted norms for conditions of landscapes and environments due to a lack of experience, memory, and/or knowledge of past conditions.[51] *Gradual* in this definition is highly relative, and implies accelerated rates of environmental change that can be experienced both within the short time frame of a human life and between generations. The *environmental generational amnesia* of shifting baseline syndrome is caused by landscape change and migration occurring at such a fast rate that two seemingly identical conservationists born thirty years apart will experience the same geographic space as two qualitatively different landscapes. Pauly initially described the syndrome through declines in fisheries, which were correlated with shifting cultural expectations and loosening conservation standards. Thus, shifting baselines are typically understood negatively, in terms of diminished resources, reduced populations of various species, damaged ecosystems, and increasing pollution. The syndrome describes a tendency to aesthetically accept degraded landscapes as "natural"—since they are what we are used to experiencing—rather than sensing a trajectory of progressive degradation and loss. This cultural plasticity has been observed in multiple studies that compare young people's attitudes to anthropogenic landscape change to those of people who have lived in it much longer. Older people recognize and feel changes; younger ones typically much less so.[52]

The obvious danger with shifting baseline syndrome is the acceptance of progressive environmental damage over time; cultural changes continually ratchet conservation standards downward. But there is another critical implication to consider: the very existence of shifting baselines contradicts the modernist, positivist understanding of technology and assumed assurance of linear progress. Over the last three centuries, far more landscapes have been destroyed and damaged than have been improved, and systemic societal risk has been expanded and manufactured, not reduced. The techno-artifact of global warming firmly obliterates the comforting hubris of most human-made technology. The uneven effects of global warming allude to concealed differences in values, power, and impacts that were never equitably or intentionally worked through. All the false *externalities* of technological advancement resurface as more challenging internalities for all.[53] If we are going to try to learn from these hard lessons, where is the humility, the inclusivity, and the adaptive means to do so?

The human ability to create and operate vast technological systems now far outpaces the capacity for prediction and control of their effects. Theorist Sheila Jasanoff suggests that the corresponding problem we urgently face is "how to live democratically and at peace with the knowledge that our societies are inevitably 'at risk,'" wherein risk is "part of the modern human condition, woven into the very fabric of progress."[54] She recommends that we develop and deploy *technologies of humility*, rather than the *technologies of hubris*—the predictive methods used by science, industry, and governments (like risk assessment, cost-benefit analysis, climate and other environmental modeling) that "are designed, on the whole, to facilitate management and control, even in areas of high uncertainty"—that our societies have become accustomed to relying on. As Jasonoff describes technologies of humility:

> *These are methods, or better yet institutionalized habits of thought, that try to come to grips with the ragged fringes of human understanding—the unknown, the uncertain, the ambiguous, and the uncontrollable. Acknowledging the limits of prediction and control, technologies of humility confront "head-on" the normative implications of our lack of perfect foresight. They call for different expert capabilities and different forms of engagement between experts, decision-makers, and the public than were considered needful in the governance structures of high modernity. They require not only the formal mechanisms of participation but also an intellectual environment in which citizens are encouraged to bring their knowledge and skills to bear on the resolution of common problems.*[55]

People operating within technologies of hubris tend to focus on what is known and make claims of objectivity, while downplaying risks and unknowns outside their unavoidably limited purview. In so doing, they preempt political discussion and hide their normative presuppositions.[56] This hubris lacks tools to address issues beyond an assumed technical and methodological framing of mastery and control. In contrast, Jasanoff's technologies of humility call attention to the normative and situated qualities of technological knowledge, giving representation to an inclusive plurality of viewpoints and realities, and thus instigating "richer deliberation on the substance of decision making."[57] She cites governmental and institutional efforts at increasing transparency, accountability, and contextualization of biotechnology over the past couple of decades as an example. Jasanoff calls for critical and public debate on the normative intentions and purposes of technology to improve upon how we learn and adapt from collective experiences over time.

The goal of Jasanoff's recommendations is to "integrate the 'can do' orientation of science and engineering with the 'should do' questions of ethical and political analysis."[58] She speaks to four domains of such questions: the *framing* of issues of concern, assessing *vulnerability*, assessing the potential *distribution* of impacts, and the capacity for *learning*. If we put this inquiry into action in delta landscapes, what are the hard lessons we have learned from reclamation-era technology and its colonial values? Beyond deltas, what about the systemic construction of levees along rivers that eradicated floodplains and facilitated development in harm's way? What about technological lock in and path dependency in general? When should we augment and refine existing landscape technologies, like building wider and ever taller levees along shorelines, versus when should we change course to something far more adaptive in the long term, like strategically retreating and giving some terraformed landscapes back to water?[59]

55 Ibid, 238.

56 Ibid. See also Pilkey and Pilkey-Jarvis, *Useless Arithmetic*.

57 Jasanoff, "Technologies of Humility," 240.

58 Ibid, 243.

59 Koslov, "The Case for Retreat."

60 Milligan, "Landscape Migration."

61 Milligan, "Accelerated and Decelerated Landscapes."

62 Steffen, Broadgate, and Deutsch, "The Trajectory of the Anthropocene," 82.

4. Living and Designing in the Milieu of Accelerated Change

Anthropogenic climate change is a global techno-crisis. If our human ability to change landscapes using technologies far exceeds our ability to anticipate the effects of those actions, and if, consequently, we have to accept that the world is rifely animated by technological mishaps and unplanned effects, what then?

Climate change is a predominant force of landscape migration.[60] A landscape migrates when the diverse material and relational assembly of things that constitute it changes. As assemblages, landscapes have always changed in this way (have always moved in patterns across space and time), but landscapes are migrating now at unprecedented, accelerated speeds.[61] Within this *great acceleration*—defined as "the holistic, comprehensive and interlinked nature of the post-1950 changes simultaneously sweeping across the socio-economic and biophysical spheres of the Earth System, encompassing far more than climate change"[62]—the past is not a solid or realistic picture for the present or the future. But, historic baselines can serve as temporal instruments to inform how far we have collectively moved, what has been lost, what we might mourn, and what we might attempt to reclaim.

A danger, though, is that those baselines can also become weird, unachievable socio-technical abstractions cast upon novel, complex, and somewhat wrecked landscapes. Many contemporary landscapes are now occupied by a globally scrambled milieu of species, and all landscapes are moving into new and more volatile climates while still geographically tethered in place. In these circumstances, untethered baselines—datums constructed from patchy, incomplete memories—are deceptive tools for understanding new environments that lack historic analogues. How are notions of the past to be reconciled with designing for an ever more chaotic, ever more different future?

For another example of a radically altered delta, let's move from California to the Mississippi River Delta. The Mississippi River Basin covers more than half of the continental United States, and the river system currently conveys less than half the sediment supply it did before the era of dam-building. In the delta itself, petrochemical industry canals accelerate the erosion of the remaining, malnourished marshes, while accelerated sea-level rise further converts drowning land to open water. Under this compounded trajectory of rapid loss, what should be kept, restored, and fought for? What should be let go? Since these forces are too powerful to reverse, how hard should we try to retain and restore? What is the measure of success or failure? Based on what values and political schema? By what specific techno-sedimentary means? Who benefits and who loses according to those means?

Given the sheer magnitude of human sedimentary manipulation—the globalized shift in the patterned movement of sediment and of sedimentary systems—there is no going back to what was, nor, given the trajectory of anthropogenic climate change, will there be a return migration to earlier climate regimes. We are faced with overwhelming novelty, and the pressing design question of how and what we can choreograph within it. The challenges and transformations facing the Mississippi River Delta are place-specific in terms of their landscapes and techno-political histories, but the overarching trends of ecological transformation, reduced sediment supply, and vulnerability to sea-level rise are now common. Human action and technologies have transformed the key factors that shape most deltas, including sediment and water flux, structural flow pathways, and shoreline position.[63] The largest anthropogenic impact affecting global deltas today is the upstream alteration of water flow,[64] which is tightly coupled with sediment supply. Overall, human actions have globally increased rates of soil erosion in upland areas due to various land uses by approximately 2.3 billion metric tons per year. This is partially counterbalanced by retention of sediment within dam reservoirs, which has reduced the flux of sediment reaching the world's deltas by approximately 1.4 billion metric tons per year.[65] Projecting forward with climate change and subsidence, most of the world's deltas are likely to drown. Marine inundation is expected to exceed the potential for growth or sustainment in all deltas larger than 10,000 square kilometers and most of those from 1,000 to 10,000 square kilometers.[66]

How then might we, as designers, planners, engineers, and publics, effectively and affectively work within what is becoming? "Shifting" does not need to be a passive process. Baselines—as measures for locating ourselves in reference to the present and to desired futures—can be shifted with care, inclusive deliberation, and intent. To do this collectively we need technologies of humility and effective forms of public participation. This is the double bind of the present: we need to humbly, creatively deal with the vast technological problems and assorted mishap "externalities" stemming from the past, while adapting to landscapes, technologies, and climates that are changing faster than ever before. Not only do we have to get better at learning from past mistakes, we also have to learn and adapt more quickly.

The Louisiana Coastal Master Plan is an example of a large-scale design for a drowning delta that engages with its own indeterminacy and shifting baselines. Its projective maps and diagrams of the region reveal that all the planned efforts at sedimentary restoration of this deltaic plain will come nowhere near recreating its former extent or even reversing the overall vector of marsh loss.[67] Such a recreation is not feasible, given the set of forces, infrastructures, and political regimes acting on the region. The upstream dams that retain over half of the delta's sediment supply are not going anywhere soon. The suite of technologies that are strategically sited in the overall plan, such as the sediment diversions (those cuts in the levees designed to route sediment laden storm flows into the delta rather than the Gulf), though ambitious, will only be able to create targeted patches of marsh within a currently expanding sea. How much sediment may accumulate in these areas, and thus how much marsh may be created, has been studied by leading-edge computational modeling, but great uncertainties still exist. The landscape technologies employed are understood as experiments. This evolving plan transparently states its goals and humbly admits what it does not know. It implies that what the coast is now is different than what the coast will become, and that what has been done cannot be fully undone. Its optimism consists in holding that, despite these truths, the coast's future is still worth investing in. Its wisdom is in understanding that this investment requires learning through doing, and being flexible enough to adapt to what is being learned.

63 Syvitski and Saito, "Morphodynamics of Deltas Under the Influence of Humans."

64 Syvitski et al., "Impact of Humans on the Flux."

65 Ibid.

66 Giosan et al., "Climate Change," 31.

67 CPRA, Louisiana's Comprehensive Master Plan (2017), 94-107.

68 Matthew Coolidge and the Center for Land Use Interpretation offer a photographic tour of these landscapes in *Around the Bay*.

69 Netzer, "Future development of the San Francisco Bay Area 1960-2020," 394.

5. Events, Institutions, and Policies

Environmental policies and regulations are designed to direct, control, and codify modes of action within landscapes. As protocols for decision making, policies are typically based on assessments of current conditions, and the actions and processes that have led to those conditions. Policies frequently come about to address the emergence of undesired trends and situations. (These are precisely the conditions under which baselines tend to shift.) Such policies are usually coupled with regulations aimed at enforcing corrective actions to minimize or reverse current trends, by dictating what actions are acceptable and incentivized, and what actions are unacceptable and punishable. Thus, regulations operate as technologies for fostering more desired futures according to the conditions and socio-political values of the moment.

Let's now go back to California's Bay-Delta. The Baylands—the wide, tidally influenced edges of the San Francisco Bay—were privatized and reclaimed in a similar fashion to the upstream Delta, though not as extensively or uniformly. For nearly a century, from around 1870 (when the bay's marshes and mudflats were, like delta tracts, transferred from state to private ownership) through the early 1960s, large portions of the bay were diked for farmland and salt production. Other Baylands were indiscriminately filled in with cast-off materials, including construction fill, demolition debris, and garbage. By the late 1950s, the nearly 300 miles of bay edge had become a motley, unplanned assemblage of landfills, cement factories, salt ponds, water treatment plants, freeways, farmland, ports, marinas, rail lines, military bases, airports, industrial parks, and explosives factories.[68]

But according to a 1959 report commissioned by the US Army Corps of Engineers, there was still plenty more bay to be reclaimed for the rapidly growing metropolis.[69] This report, called *Future Development of the San Francisco Bay Area 1960–2020,* included a fold out map, Plate 21, titled "Areas Susceptible of Reclamation in San Francisco Bay." This iconic drawing of the bay depicted 325 square miles of additional potential reclaimed land as shaded, hatched areas extending from the current bay edges outward into the bay's shallows. A simple chart supplied on the map provides estimates of square miles of land gain potential for each of the sub-bays: San Francisco Bay proper, San Pablo Bay, and Suisun Bay. Had it been undertaken, this radical terraforming of the estuary would have reduced the bay by more than half. No calculation was provided of the enormous amount of fill material that would have been required to realize this future.

The report itself determined that the rate of bay reclamation had more than doubled, from an average of 1.6 square miles/year in 1850–1900 to 3.6 square miles/year in 1940–1957. At the time of publication, three quarters of all vegetated tidal marsh in the bay had already been reclaimed, but the tidal mud flats and sub-tidal shallows were largely intact. Prosaic language in the report described standardized techniques that would be used to reclaim these lands (diking and ditching, or full infilling), accompanied by cost calculations.

Plate 21's depiction of "Areas Susceptible of Reclamation" was a holistic representation of a future that could emerge through continued infilling of the bay, drawn by those that saw it as a potential solution to regional development pressures. Its tentative, peculiar title seems self-prophetic. Rather than enabling reclamation, though, the iconic depiction helped stop the unregulated filling of the bay by instigating a concerned public to fight to preserve what was left.[70] When citizens became aware of the bay's "susceptibility" to continued reclamation, many said no. The map helped galvanize the *Save the Bay* political movement, famously marshaled by bay resident-activists Sylvia McLaughlin, Kay Kerr, and Esther Gulick, who saw value in the bay as it was.[71] These women's activism gave voice to a larger cultural shift in aesthetic sensibilities toward the Baylands, a shift that rejected the past century's privatized, profit-driven transformations, and argued for preservation, use, and revitalization for the common good.

Responding to these cultural shifts, the McAteer-Petris Act was passed by the California State Legislature in 1965. The first two articles of this remarkable legislation are worth quoting in full:

> *66600. Declaration of Public Interest.*
> *The Legislature hereby finds and declares that the public interest in the San Francisco Bay is in its beneficial use for a variety of purposes; that the public has an interest in the bay as the most valuable single natural resource of an entire region, a resource that gives special character to the bay area; that the bay is a single body of water that can be used for many purposes, from conservation to planned development; and that the bay operates as a delicate physical mechanism in which changes that affect one part of the bay may also affect all other parts. It is therefore declared to be in the public interest to create a politically responsible, democratic process by which the San Francisco Bay and its shoreline can be analyzed, planned, and regulated as a unit.*
> *66601. Findings and Declarations as to Threat of Uncoordinated, Haphazard Filling.*
> *The Legislature further finds and declares that uncoordinated, haphazard filling in San Francisco Bay threatens the bay itself and is therefore inimical to the welfare of both present and future residents of the area surrounding the bay; that while some individual fill projects may be necessary and desirable for the needs of the entire bay region, and while some cities and counties may have prepared detailed master plans for their own bay lands, a governmental mechanism must exist for evaluating individual projects as to their effect on the entire bay; and that further piecemeal filling of the bay may place serious restrictions on navigation in the bay, may destroy the irreplaceable feeding and breeding grounds of fish and wildlife in the bay, may adversely affect the quality of bay waters and even the quality of air in the bay area, and would therefore be harmful to the needs of the present and future population of the bay region.*[72]

The McAteer-Petris Act reversed the state's century-long ceding of control of the Baylands to private ownership and development. And now, rather than wasteland, the bay was legally recognized as the "most valuable natural resource" in the region. The legislation's mandate to understand and choreograph action within the bay as a cohesive, relational whole was equally progressive. Decisions would now have to be vetted through a publicly accountable process and permitting system. To do this, the act created the Bay Conservation and Development Commission (BCDC), a state agency with jurisdiction over all fill of Baylands and uplands within 100 feet of the shoreline.

70 In defense of the report's authors, under the heading "Unfavorable Factors Affecting Reclamation of Tide and Submerged Lands," the report states the following: "it appears likely that only a portion of the tide and submerged lands susceptible of reclamation ever will be reclaimed. What portion eventually is reclaimed probably will depend largely upon future comprehensive and multipurpose plans for shoreline development" (91). In this language, they were affirmatively prophetic.

71 Their organization was initially called the Save San Francisco Bay Association and was later modified to Save the Bay, which it retains to this day. See "Save the Bay: A Brief History," Save the Bay.

72 "The McAteer-Petris Act," The Bay Conservation and Development Commission.

73 California State Coastal Conservancy, *The Baylands and Climate Change*.

74 Most significantly, Measure AA proposed a 20-year, $12 parcel tax to raise approximately $25 million annually, or $500 million over twenty years, to fund restoration projects in the Bay. It passed in 2016 with 70% approval across the region and went into effect in 2017.

75 Dusterhoff et al., *Sediment for Survival*.

This regulatory regime is still in place and has radically changed the development trajectory of the Baylands. The bay coastline today looks nothing like the techno-imaginary depicted in Plate 21. Instead, the bay has increased in overall size due to restoration. Extreme development pressure still exists, but there are new challenges. Rather than shrinking, the bay is expanding outwards. Accelerating sea-level rise is rendering the Bayland's airports, housing complexes, and other low-lying developments "susceptible" to a different type of reclamation, in which rising bay waters will take back the land with ever-increasing volumetric force. Over the next century, the increasing rates of reclamation proposed in *Future Development of the San Francisco Bay* will likely play out across those same terraformed lands—only in reverse. And given these novel conditions, faced not only in California's Bay-Delta, but by urbanized deltas, estuaries, and coastlines all over the earth, historical baselines and restoration targets are tricky to apply, are technically impossible, or may become entrenched in protocols that hinder innovation and adaptation to accelerated change.

In current Bay Area thinking, one of the best ways to respond to these climate change effects is by restoring the bay's wetlands and, where possible, providing them space to migrate landward. Wetlands provide an adaptive buffer to sea-level rise and storm effects, in addition to ecological and recreational benefits. Thus, restoring vast acreages of these habitats is pursued in the Bayland Ecosystem Habitat Goals, a set of established yet non-regulatory, science-based restoration goals for the region,[73] and is partially funded by local tax mechanisms widely approved of by residents.[74] But what is often not considered in setting these restoration goals is the sediment needed to restore wetlands at this scale. Massive amounts of imported earthen fill will be required to bring subsided land back up to wetland elevations, as well as increased upland river and creek sediment flows to sustain them.[75] All of this has to be done as quickly as possible, since the longer it takes to commence rebuilding the wetlands, the more sediment will be needed and the greater the risk that the marshes will be unable to establish and adapt to increasing rates of sea-level rise.

These challenges present BCDC with an inversion in conditions and paths of change. The McAteer-Petris Act has heavily restricted placement of fill in the bay for very solid public trust reasons. But now, saving the Baylands will require large emplacements of fill in the bay, at faster rates than ever before. Methods, means, and incentives are needed to get sediment selectively and safely into the bay to meet the same goals and values established by the act in the 1960s, but within today's very different climatic and landscape conditions.

The critical question is how fast policy and regulation can effectively change and adapt to these shifts, while remaining true to its charge of ensuring a "politically responsible, democratic process by which the San Francisco Bay and its shoreline can be analyzed, planned, and regulated as a unit." The nature of the Bay-Delta's emerging sediment shortfall is just beginning to be discussed, quantified, contested, and grappled with. Its implications extend across economic, ecological, cultural, and political domains. It will affect policy, dredging operations, and priorities for future investments and research. In many places, there may be a need to move beyond the past as precedent to more successfully prepare for a very different future. Sediment, as earthen infrastructure, takes on new and multiple valuations given the shortfall, restoration goals, and the bay's emerging expansion.[76] Incentivizing more projective research to model, spatialize, and lend more specificity to the qualities and parameters of these conditions will need to be a priority in future planning and design efforts.

Revisions in policy and regulation due to climate change, sea-level rise, landscape transformations, and shifts in public values are also needed at the federal level. Recall that the USACE's Federal Standard has long meant that if a beneficial reuse of dredged material is more expensive than the designated "least costly alternative" (which is usually open water disposal), then the burden of that extra cost must be picked up by a non-Federal partner, like the state in which the dredging is performed.[77] But if sediment is becoming an ever more valuable and limited material resource, and if wasting that resource when doing routine dredging increasingly puts more shoreline assets, infrastructures, and developed landscapes for which that dredging is performed at greater economic risk, is that really still sound practice? By what cost-benefit analysis?[78] By what value assessment? Is that value assessment public and participatory?

Similar adaptation challenges exist with the Clean Water Act. As an example, if a diked agricultural parcel (a reclaimed bayland marsh), subsided below sea level, were intentionally breached to reconnect it to the bay, it would become open water. Once thus hydrologically reconnected to the bay, even if a nearby cost-effective and clean source of dredged fill material became available, it couldn't be placed there to create the desired marsh landscape, because section 404 of the Clean Water Act regards sediment as a pollutant within water bodies. For this reason many subsiding lands slated for restoration are not breached, which means that they continue to subside while they wait for active fill placement and that they don't receive the suspended sediment flowing through the bay that would help bring them back up to marsh elevation. These policies need adaptation.

It may sound like heresy to question the Clean Water Act, or to suggest significant revisions to McAteer-Petris. But BCDC has been doing exactly that, responding to new conditions and values by collaborating with advocates like the Audubon Society and agencies like the federal Fish and Wildlife Service to change the law so that it is easier to fill the bay for the sake of creating marshes, mudflats, and uplands.[79] Much as landscapes are never "finished," neither is policy making, particularly in a time of emergent, radical landscape change. Whether or not environmental policies can adapt fast and effectively enough to get ahead of increasing rates of climate change (while not creating potential damaging loopholes for neoliberal economic exploitation) is a critical question.

76 We discuss this in depth in "Mud is Vital."

77 "The Role of the Federal Standard," EPA.

78 The 2020 Water Resources Development Act officially drew, for the first time, the economic benefits of the beneficial reuse of dredged sediment into the Corps' Federal Standard calculations. The full impacts of this shift are still emerging.
Water Resources Development Act of 2020, Section 125.

79 "Fill for Habitat Bay Plan Amendment," BCDC.

6. Adapting and Coevolving

Change brought about by technology has played a formative role in the evolution of the Bay-Delta and California. From indigenous peoples' techniques of land tending and stewardship, to hydraulic gold mining, the clamshell Tule Breakers, and thousands of miles of levees, to massively scaled water infrastructure, California's landscapes have been remade several times over. Going forward, technologies will need to address new landscape challenges, shifting conditions, and emerging public values.

80 Domurat et al., "Long-term Management Strategy."

81 Davis, Holmes, and Milligan, "Isthmus."

82 Graham, ed., *Disrupted Cities*.

83 See "Aesthetics and Publics" for more on this.

In the delta, land managers and scientists are experimenting with subsidence reversal, which uses wetland plants to make new soil, provide habitat for migratory birds, and store carbon. Flood control channels on the bay's tributaries are being redesigned to improve migratory fish passage and to direct sediment flows to where they are most needed to sustain marshes. Rather than being engineered to singular, isolated agendas, landscape technologies will need to be place-based and multifunctional, able to perform across diverse ecological, economic, and sociocultural criteria.

Innovation will require investment, incentivization, public involvement, openness to experimentation, and adaptive management. We will need to design, implement, assess, regulate, and improve upon how these technologies perform. But how can we push for this to happen? In this closing section of the chapter, we present some examples of adaptive technological innovation.

In the 1980s, San Francisco Bay was described as being in a state of "mudlock." Dredge disposal needs from shipping channel maintenance were exceeding the in-bay capacity to receive them. This shallowed the bay in disposal areas (which had once been deep water), creating public boating navigation hazards near Alcatraz Island. This generated public protest on both land and water, and this crisis shaped the emergence of regional sediment planning and beneficial reuse efforts among public agencies.[80] When existing structures fail in a sudden, public, and consequential way, there is often an opportunity to redesign policies quickly. This can be due to a "natural" disaster or climate event, a pivotal regime change in infrastructure, or when the capacity of a management framework is exceeded, leading to public outcry, as in "mudlock." Failures create incentives and constituencies for doing things differently.

Systemic change in standardized infrastructure can produce networked effects at global scales, as in the recent expansion of the Panama Canal, which initiated a shockwave of channel-deepening and port expansion across the Americas.[81] Infrastructural disasters, such as what New York City experienced after Hurricane Sandy, can provide incentives for rapid change in policy and design by exposing vulnerabilities and by inducing reassessment of what is publicly valued.[82] (This is not to ignore how horrible and inequitably experienced such events can be: disaster is never to be desired.) In contrast, the challenge with the emergent sediment shortfall in the bay (and in many deltas, estuaries, and coastlines around the world) is that it is a *slow* emergency that emerges incrementally over time, and thus one that is not aesthetically obvious or noticeable to many. Slow disasters require more intentional constituency building.[83]

It may be hard to accept that new technologies of humility will often emerge through failure and even disaster. But, in a world of ecological crisis, intense social disparity and poverty, on-going neoliberal pullbacks on public spending, and emergent climate instability, the lessons are likely to be tough. It will be crucial to learn from them, as there is no way out but through. It is going to be very challenging to design, inhabit, and co-evolve a different world.

Yet failure and disaster are not the only ways that new technologies can emerge. The increased volatility and mutability of landscapes, which were indeterminate even prior to a destabilized climate, is also going to require fundamental changes in policy premised on predictability. As researchers Adriaan Slob and Lasse Gerrits remark on the topic of "sedimentary systems and the whimsicality of policy processes":

As existing computational models are not able to predict results of measures in the sediment system to the degree of exactitude that is demanded by policy makers, sediment operations sometimes show unforeseen events and unpredicted outcomes. This non-linearity and unpredictability of the behaviour of sediment systems is hard to harmonize with the linearity of deterministic policies.[84]

Climate change will only increase this unpredictability, and beaches, estuaries, deltas, and river mouths—the sites of most sediment design—are particularly unpredictable. Slob and Gerrits point to the need for more adaptive policies that are premised on interaction and shortened informational feedback loops between the landscapes of concern and the people and institutional structures working with them. Their technological focus is on the capacity to learn, and how that learning is worked into protocols and decision-making processes, just like what is happening in the bay right now with changes in the McAteer-Petris Act. Monitoring and adaptively managing the actual outcomes of new policies will be critical in this learning.

A general guideline for safely testing out new technologies is to start small in application, observe what happens, refine the technology, and repeat as needed before scaling up. This type of experimental trialing is a *pilot*. Similarly, a *prototype* is an experimental test model made to test out a design. Small-scale experimentation allows for safe, quickened feedback loops for refinement of the technology, and to allow for failure (or learning) without high consequences. The same adaptive methodology should be utilized in making landscapes—we might call it landscape prototyping.[85]

In the Bay-Delta, we can find emerging examples of successful landscape prototyping. The California Department of Water Resources has prototyped wetland-based subsidence reversal techniques on deeply subsided polders in the central delta. Now, it is expanding the program based on experimental trials and testing to see if commercial wetland crops, such as rice, might be a viable way to build land elevation.[86] In San Pablo Bay, Sears Point is a tract of subsided agricultural land that is being restored to tidal marsh. Prior to breaching its levees, its flat topography, all below sea level, was resculpted by excavators into a gridded array of stacked earthen mounds.[87] When the levees were breached and the tract inundated, the mounds became tidal islands that could be colonized by wetland vegetation, as well as help accrue sediment and attenuate erosive wave action.

The Hamilton Wetlands are a former military airfield also on the bay edge, just west of Sears Point.[88] To restore these wetlands, dredged material from the channel deepening of the Port of Oakland was used to bring the subsided land surface nearly up to the mean high-water elevation of a tidal marsh, and the new surface was then hydrologically reconnected to the bay. The experiment was to let the action of the tides finish off the needed importation of suspended sediment and to observe the timeframe for that to happen. A collection of linear earthen berms was spread across the nascent, shallowly inundated landscape, their form and configuration based on computational models of how they might best assist with accruing sediment.[89] Both Hamilton and Sears Point are being closely observed and monitored to see how they evolve, including colonization

84 Slob and Gerrits, "The Dynamics of Sedimentary Systems."

85 In landscape architectural discourse "landscape prototyping" has generally referred to efforts to develop representational prototypes composed of images, digital models, or physical models. While those efforts are certainly also relevant to the design of new technological landscapes with sediment, we use the term here to refer to the full-scale making of new landscapes as experimental prototypes.
Robinson, *The Spoils of Dust.*
Harmon et al., "Tangible Topographic Modeling for Landscape Architects."

86 West Delta Program," California Department of Water Resources.

87 "Sears Point Wetland Restoration," Sonoma Land Trust.

88 We also discuss Hamilton Wetlands at length in "Mud is Vital."

89 "Hamilton Wetland Airfield Restoration," USACE San Francisco District.

90 "Announcement of the Selection," National Archives.
The Dredge Research Collaborative contributed a letter included in the California State Coastal Conservancy's submission, expressing support and the need for this research.

91 USACE ERDC, "Strategic Placement for Beneficial Use."

92 "History of the Montezuma Wetlands," Montezuma Wetlands LLC.

93 Ibid.

by different communities of vegetation, sediment accrual and erosion rates, and wildlife usage.

Landscape prototyping is also being recognized and pursued at the national level. For example, the Water Resources Development Act of 2016 required that the US Army Corps of Engineers competitively fund and test ten projects for the beneficial use of dredged material. In 2018, *Restoring San Francisco Bay's Natural Infrastructure with Dredged Sediment: Strategic Placement* was one of the 10 winners.[90] Strategic placement is the deposition of dredged material in a provisional location, with the expectation that forces like tides and currents will carry that sediment to desired locations.[91] In the Bay Area, strategic placement research has focused on the release of dredged material directly into the water column within shallow areas near the bay's edge. This technique involves determining where and when to release such sediment so that the tides and currents can effectively transport it up into the mudflats, where it will nourish both them and adjacent marshes. Learning how to do this well requires a high level of attunement and engagement with the bay's dynamics.

On the east side of Suisun Marsh near the transition from bay to delta, eight million cubic yards of dredged sediment have been emplaced to create 600 acres of new tidal wetland on subsided reclaimed farmland. This first cell (completed in 2020) is part of a larger operation called the Montezuma Wetlands, which aims to eventually restore 1,800 acres of tidal marsh. A key technology being used there is an on-site sediment offloader, which maximizes the efficiency of the placement process. The offloader is essentially a very large on-shore sediment vacuum. It sucks dredged material off a ship or barge and pumps it directly into the wetlands' phased cells. This offloader was a major capital investment and is the first and only offloader in the Bay Area. Once acquired, it made the beneficial reuse process much cheaper, and thus more economically viable.

Jim Levine is the entrepreneur and developer behind this effort. His company's goal is to restore these habitats while making money in the process.[92] Unlike most unconfined beneficial reuse projects, Montezuma's wetlands are able to receive both contaminated and uncontaminated material. Contaminated sediment is placed at the bottom of cells and covered by clean material, so the contaminants are contained and remain inert. Once contamination is safely capped and marsh elevations are established within a cell, the levees are breached to bring in the tides to create intertidal marsh.

The project required a ten-year permitting process overseen by paid experts and an interagency technical review team. Today, Levine's offloader is the most used "west of the Mississippi," as his website boasts.[93] And now regional public agencies are trying to find ways to emulate the success of the Montezuma Wetlands prototype, including potentially funding a publicly-operated offloader that could facilitate the beneficial reuse of sediment around the bay.

Montezuma Wetlands has been an innovation across policies, machines, and their enactment in landscape. It was a technological evolution in which a private entrepreneur saw opportunities for needed innovation that were being overlooked by others, including regulators. This example could be read as indicating the need for government and regulation to get out of the way of the market, and allow for private corporate investment. But we read it as demonstrating the need for greater public investment in technology and infrastructure, in order to realize the long-term benefits that can accrue through short-term capital expenditure.

Landscape prototypes can be precisely this kind of social and technological investment for our collective future. They can be places for societal learning, observation, work, play, wonder, enjoyment, and adaptation. In the Bay-Delta, prototypes are testing ways of doing things that are quite different from what was tried in those same places in the past. These design techniques attempt to rework or "unmake" past design and engineering of the land that does not perform as envisioned or does not match the set of values and concerns we have today.[94] There are so many landscapes in the United States that have been ambitiously engineered and transformed over the past two centuries—often unjustly—that don't fit with who we are now, nor what we have learned. There are so many relics, which have been shaped by ruinous spatial techniques that are out of synch with how we and the world are rapidly changing. These landscapes, ruins, and techniques present urgent opportunities for course correction, for exploring other ways of co-evolving with landscapes by design. These explorations must be approached with humility, embracing the fact that we cannot know all the implications of our actions. We must move forward with care and attention.

How these landscape prototypes might be designed is the focus of our final chapter. There, we will remain in San Francisco Bay, focusing on its creeks and the reworking of technologies of flood control and land reclamation. "Why Design?" illustrates how we might use transdisciplinary co-design practices to choreograph landscape change across a range of interacting scales. It positions health and ecology at the center of a new economy and culture of sustainment. It proposes earthworks and forms that are publicly inclusive, participatory, and aesthetically attuned. And it approaches technology as collective ways of learning and adapting.

94 Orff, "What is Design Now?"

Bibliography:

Barbara Adam and Chris Groves, *Future Matters: Action, Knowledge, Ethics* (Leiden: Brill, 2007).

Kat Anderson, *Tending the Wild* (Berkeley: University of California Press, 2005).

Hannah Arendt, *The Human Condition* (Chicago: University of Chicago Press, 2013).

Eric Ash, *The Draining of the Fens: Projectors, Popular Politics, and State Building in Early Modern England* (Baltimore, John Hopkins University Press, 2017).

"Fill for Habitat Bay Plan Amendment," The Bay Conservation and Development Commission, https://www.bcdc.ca.gov/BPAFHR/FillHabitat.html.

"The McAteer-Petris Act," The Bay Conservation and Development Commission, https://www.bcdc.ca.gov/plans/mcateer_petris.html.

Matthew Booker, *Down by the Bay: San Francisco's History Between the Tides* (Berkeley: University of California Press, 2020).

"About the Central Valley Project," Bureau of Reclamation, https://www.usbr.gov/mp/cvp/about-cvp.html.

"State Water Project," California Department of Water Resources, https://water.ca.gov/Programs/State-Water-Project.

California State Coastal Conservancy, *The Baylands and Climate Change. What We Can Do. Baylands Ecosystem Habitat Goals Science Update* (Oakland, CA: 2015), https://www.sfei.org/sites/default/files/biblio_files/Baylands_Complete_Report.pdf.

Vera S. Candiani, *Dreaming of Dry Land: Environmental Transformation in Colonial Mexico City* (Stanford University Press, 2014).

Coastal Protection and Restoration Authority of Louisiana, *Louisiana's Comprehensive Master Plan for a Sustainable Coast*, 2017, 94-107.

Andrew Cohen, "Nonindigenous Aquatic Species in a United States Estuary: A Case Study of the Biological Invasions of the San Francisco Bay and Delta," US Fish and Wildlife Service and the National Sea Grant College Program (1995), https://nas.er.usgs.gov/Publications/SFBay/sfinvade.html.

Matthew Coolidge and the Center for Land Use Interpretation, *Around the Bay: Man-Made Sites of Interest in the San Francisco Bay Region* (Blast Books, 2013).

Dilip Da Cunha, *The Invention of Rivers: Alexander's Eye and Ganga's Descent* (Philadelphia: University of Pennsylvania Press, 2019).

Brian Davis, Rob Holmes, and Brett Milligan, "Isthmus," *Places Journal* (2015), https://placesjournal.org/article/isthmus-panama-canal-expansion/.

"West Delta Program," California Department of Water Resources, https://water.ca.gov/Programs/Flood-Management/Delta-Conveyance-And-Flood-Protection/West-Delta-Program.

Steve Deverel and David Leighton, "Historic, Recent, and Future Subsidence, Sacramento-San Joaquin Delta, California, USA," *San Francisco Estuary and Watershed Science, 8*, no. 2 (2010).

George Domurat, David Dwinell, James Delorey, Lawrence Fade, and Scott Nicholson, "Long-term Management Strategy for the Placement of Dredge Material in the San Francisco Bay Region," in *30th PIANC-AIPCN Congress 2002*, (Sydney, NSW: Institution of Engineers, 2002), 428–37.

Scott Dusterhoff, Katie McKnight, Leticia Grenier, and Nate Kauffman, *Sediment for Survival: A Strategy for the Resilience of Bay Wetlands in the Lower San Francisco Estuary* (California, San Francisco Estuary Partnership, 2021), https://www.sfei.org/documents/sediment-for-survival.

Keller Easterling, *Extrastatecraft: The Power of Infrastructure Space* (London: Verso Books, 2014).

Jacques Ellul, *The Technological Society*, trans. John Wilkinson (New York: Vintage, 1964).

Mark Elvin, *The Retreat of the Elephants: An Environmental History of China (New Haven:* Yale University Press, 2004).

"The Role of the Federal Standard in the Beneficial Use of Dredged Material from U.S. Army Corps of Engineers New and Maintenance Navigation Projects," US Environmental Protection Agency, https://www.epa.gov/cwa-404/role-federal-standard-beneficial-use-dredged-material.

"Section 404 of the Clean Water Act: Permitting Discharges of Dredge or Fill Material," US Environmental Protection Agency, https://www.epa.gov/cwa-404.

"Fish Conservation and Culture Laboratory," Fish Conservation and Culture Laboratory, https://fccl.ucdavis.edu/.

Philip Garone, "Managing the Garden: Agriculture, Reclamation, and Restoration in the Sacramento-San Joaquin Delta," *Delta Narratives Project,* Delta Protection Commission (2015).

Amitav Ghosh, *The Great Derangement: Climate Change and the Unthinkable* (Chicago: University of Chicago, 2018).

Amitav Ghosh, *The Nutmeg's Curse* (Chicago: University of Chicago Press, 2021).

Liviu Giosan, James Syvitski, Stefan Constantinescu, and John Day, "Climate Change: Protect the World's Deltas," *Nature News* 516, no. 7529 (2014).

Stephen Graham, ed., *Disrupted Cities: When Infrastructure Fails* (New York, Routledge, 2010).

Don Hankins, "Ecocultural Equality in the Miwkoʔ Waaliʔ " *San Francisco Estuary and Watershed Science* 16, no. 3 (2018).

Brendan Harmon, Anna Petrasova, Vaclav Petras, Helena Mitasova, and Ross Meentemeyer, "Tangible Topographic Modeling for Landscape Architects," *International Journal of Architectural Computing,* 16, no. 1 (2018): 4–21.

James Hobbs, Peter Moyle, and Nann Fangue, "Is Extinction Inevitable for Delta Smelt and Longfin Smelt? An Opinion and Recommendations for Recovery," *San Francisco Estuary and Watershed Science* 15.2 (2017).

C.S. Holling and Lance Gunderson, *Panarchy: Understanding Transformations in Human and Natural Systems* (Washington, DC: Island Press, 2002).

Rob Holmes, "The Problem with Solutions," *Places Journal* (2020), https://placesjournal.org/article/the-problem-with-solutions/.

Sheila Jasanoff, "Technologies of Humility: Citizen Participation in Governing Science," *Minerva* 41, no. 3 (2003): 223–44.

Robin Kimmerer, *Braiding Sweetgrass: Indigenous Wisdom, Scientific Knowledge and The Teachings of Plants* (Minneapolis: Milkweed Editions, 2013).

Robin Kimmerer, "Restoration and Reciprocity: The Contributions of Traditional Ecological Knowledge," in *Human Dimensions of Ecological Restoration*, ed. Eric Higgs (Washington, DC: Island Press, 2011), 257–76.

Liz Koslov, "The Case for Retreat," *Public Culture* 28, no. 2 (2016): 359–87.

Bruno Latour, "Morality and Technology," trans. Couze Venn, *Theory, Culture & Society* 19, no. 5–6 (2002): 247–60.

Bruno Latour, *Science in Action: How to Follow Scientists and Engineers Through Society* (Cambridge: Harvard University Press, 1987).

Bruno Latour and Steve Woolgar, *Laboratory Life* (Princeton: Princeton University Press, 2013).

Jay Lund, Ellen Hanak, William Fleenor, Richard Howitt, Jeffrey Mount, and Peter Moyle, *Envisioning Futures for the Sacramento-San Joaquin Delta* (San Francisco: Public Policy Institute of California, 2007).

Brett Milligan, "Accelerated and Decelerated Landscapes," *Places Journal* (2022), https://placesjournal.org/article/accelerated-and-decelerated-landscapes/.

Brett Milligan, "Landscape Migration," *Places Journal* (2015), https://placesjournal.org/article/landscape-migration/.

Brett Milligan and Rob Holmes, "California Delta, USA," in *Polder Landscapes of the World*, eds. Steffen Nijhuis, Bart Schultz, and Michiel Pouderoijen (forthcoming).

Brett Milligan and Alejo Kraus-Polk, "Evolving the Evolving: Infrastructure, Place and Rewilding in the California Delta," *Urban Planning* 2, no. 4 (2017): 93–115.

"History of the Montezuma Wetlands," Montezuma Wetlands LLC, https://montezumawetlands.com/history/.

Jeff Mount and Robert Twiss, "Subsidence, Sea-level Rise, and Seismicity in the Sacramento-San Joaquin Delta," *San Francisco Estuary and Watershed Science*, 3, no. 1 (2005).

Peter Moyle, Karrigan Bork, John Durand, Tien-Chieh Hung and Andrew Rypel, "Futures for Delta Smelt," Center for Watershed Sciences white paper, University of California–Davis (Dec. 15, 2019).

"Announcement of the Selection of the Ten Pilot Projects Pursuant to Section 1122 of the Water Resources Development Act of 2016, Beneficial Use of Dredged Material," National Archives, https://www.federalregister.gov/documents/2018/12/28/2018-28306/announcement-of-the-selection-of-the-ten-pilot-projects-pursuant-to-section-1122-of-the-water.

Dick Netzer, "Future development of the San Francisco Bay Area 1960–2020. Economic Aspects of Comprehensive Survey of San Francisco Bay and Tributaries. Prepared and published for US Army Engineer District, San Francisco Corps of Engineers, by Office of Area Development, Business and Defense Service Administration, US Department of Commerce. Superintendent of Documents, US Government Printing Office, Washington 25, DC, 1959. xxi, 131 23 Maps. $11.," *National Civic Review* 50, no. 7 (1961).

Kate Orff, "What is Design Now? Unmaking the Landscape," *Architectural Design* 90, no. 1 (2020): 94–99.

Daniel Pauly, "Anecdotes and the Shifting Baseline Syndrome of Fisheries," *Trends in Ecology & Evolution* 10, no. 10 (1995): 430.

Orrin Pilkey and Linda Pilkey-Jarvis, *Useless Arithmetic: Why Environmental Scientists Can't Predict the Future* (New York: Columbia University Press, 2007).

Fabrice Renaud, James Syvitski, Zita Sebesvari, Saskia Werners, Hartwig Kremer, Claudia Kuenzer, Ramachandran Ramesh, A. D. Jeuken, and Jana Friedrich, "Tipping from the Holocene to the Anthropocene: How Threatened are Major World Deltas?" *Current Opinion in Environmental Sustainability* 5, no. 6 (2013): 644–54.

Alexander Robinson, *The Spoils of Dust: Reinventing the Lake that Made Los Angeles* (Applied Research & Design, 2018).

"Save the Bay: A Brief History," Save the Bay, https://savesfbay.org/who-we-are/a-brief-history.

James Scott, *Seeing Like a State* (New Haven: Yale University Press, 2008).

Adriaan Slob and Lasse Gerrits, "The Dynamics of Sedimentary Systems and the Whimsicality of Policy Processes," *Journal of Soils and Sediments* 7, no. 5 (2007): 277–84.

Masashi Soga and Kevin Gaston, "Shifting Baseline Syndrome: Causes, Consequences, and Implications," *Frontiers in Ecology and the Environment* 16, no. 4 (2018): 222–30.

"Sears Point Wetland Restoration," Sonoma Land Trust, https://sonomalandtrust.org/current-initiatives/sears-point-wetlands-restoration/.

Will Steffen, Wendy Broadgate, and Lisa Deutsch, "The Trajectory of the Anthropocene: The Great Acceleration," *The Anthropocene Review* 2.1 (2015): 81–98.

Thaddeus Sunseri, *Wielding the Ax: State Forestry and Social Conflict in Tanzania, 1820–2000* (Ohio University Press, 2014).

James Syvitski and Yoshiki Saito, "Morphodynamics of Deltas Under the Influence of Humans," *Global and Planetary Change* 57.3-4 (2007): 261–82.

James Syvitski, Charles J. Vörösmarty, Albert J. Kettner, Pamela Green, "Impact of Humans on the Flux of Terrestrial Sediment to the Global Coastal Ocean," *Science* 308, no. 5720 (2005): 376–80.

John Thompson, "Early Reclamation and Abandonment of the Central Sacramento-San Joaquin Delta," *Sacramento History Journal* 6 (2006): 41–72.

John Thompson, "The Settlement Geography of the Sacramento-San Joaquin Delta, California" (PhD diss., Stanford University, 1958).

John Thompson and Edward A. Dutra, *The Tule Breakers: The Story of the California Dredge* (Stockton, California: Stockton Corral of Westerners, University of the Pacific, 1983).

US Army Corps of Engineers, Engineer Research and Development Center, "Strategic Placement for Beneficial Use of Dredged Material," ERDC CHL SR-19-3, by Joseph Gailani et al. (Vicksburg, MS: Engineer Research and Development Center, 2019).

"Hamilton Wetland Airfield Restoration," US Army Corps of Engineers San Francisco District, https://www.spn.usace.army.mil/Missions/Projects-and-Programs/Projects-by-Category/Projects-for-Ecosystem-Restoration/Hamilton Airfield Wetland Restoration/.

Martijn van Staveren and Jan van Tatenhove, "Hydraulic Engineering in the Social-Ecological Delta: Understanding the Interplay Between Social, Ecological, and Technological Systems in the Dutch Delta by Means of 'Delta Trajectories,'" *Ecology and Society* 21, no. 1 (2016).

Water Resources Development Act of 2020, H.R. 7575, 116th Congress, Section 125 (2020).

Alison Whipple, Robin Grossinger, Daniel Rankin, Bronwen Stanford, and Ruth Askevold, *Sacramento-San Joaquin Delta Historical Ecology Investigation: Exploring Pattern and Process* (Richmond: San Francisco Estuary Institute-Aquatic Science Center, 2012), https://www.sfei.org/documents/sacramento-san-joaquin-delta-historical-ecology-investigation-exploring-pattern-and-proces.

Richard White, *The Organic Machine: The Remaking of the Columbia River* (New York: Hill and Wang, 1996).

Langdon Winner, "Do Artifacts Have Politics?" *Daedalus* (1980): 121–36.

Donald Worster, *Rivers of Empire: Water, Aridity, and the Growth of the American West* (Oxford University Press, 1992).

Joy Zedler and Michelle Stevens, "Western and Traditional Ecological Knowledge in Ecocultural Restoration," *San Francisco Estuary and Watershed Science* 16, no. 3 (2018).

San Francisco South Bay (US Geological Survey, National Agricultural Imagery Program)

Sediment Shortfall: the California Bay-Delta

The California Bay-Delta is a mélange of sedimentary events like no other. Pulses, droughts, dredging, diking, and reclamation intermingle to generate a complex landscape shaped so significantly by human action that conventional ideas of restoration no longer apply. The region struggles to adapt to a changing future, constrained by a legacy of upstream management decisions.

Bay-Delta Sedimentshed

Almost half of California's landmass drains to the San Francisco Bay-Delta, one of the largest estuaries on the Pacific coast of the Americas.[1]

Rain and snowmelt erode sediment from the slopes of the Sierra Nevada Mountains and wash it across the flatlands of California's Central Valley. Picked up by the meandering course of the Sacramento and San Joaquin rivers, sediment moves to the inland Delta, interacting with agricultural lands, water supply systems, and habitat restoration initiatives. As tides rise and fall, fine sediment flows to the San Francisco Bay, where it builds marshy edges and mudflats, deposits in navigation channels, or transports offshore to the Pacific Ocean.

UPLAND: Hydraulic mining in the late 1800s generated large amounts of sediment runoff from the Sierra Nevadas.

Sediment was transported downstream through valleys and rivers.

Downstream sediment transport today is limited as dam and water infrastructure projects trap sediment upstream.

DELTA: River flows provided water for agriculture but also threatened livelihoods with flooding.

In the 1850s, farmers dredged channels and used the material to build up levee edges. With time, the process was mechanized with clamshell dredges.

By the 1950s sediment inputs slowed to the Delta and Bay. Farm land subsides and levees are at risk of failure.

BAY: Historic industrial uses increased heavy metal and chemical sources of contamination.

Harbor deepening projects abound, generating contaminated and clean sediment.

Urbanized edges of the bay contribute to negative sedimentation impacts on bay ecosystems.

Pulse, Sink, Shortfall

The last two centuries have been marked by intense modifications to the Bay-Delta sedimentshed, yielding dramatic changes in sediment supply downstream.

Sediment flows spiked during the Gold Rush, when deforestation and hydraulic gold mining eroded entire hillsides and led to a nine-fold increase in the amount of sediment flowing downriver.[2] This sediment choked the Delta, reshaped the bathymetry of the bay, and sparked a cycle of levee building, land reclamation, and subsidence that continues today. While the Gold Rush sediment pulse continues to work its way through the regional system, mid-century water infrastructure projects have reversed the trend of sediment excess. Dams constructed for power and water supply trap sediment upstream, significantly reducing the volume of material making its way to the Bay-Delta. Within decades, an excess of sediment has reversed to a shortage, starving downstream wetlands and mudflats of this essential material.

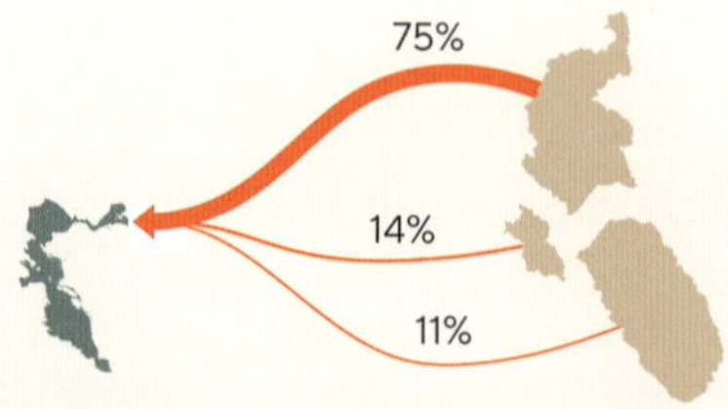

PRE-MINING
1.3 Million cubic meters annual sediment load

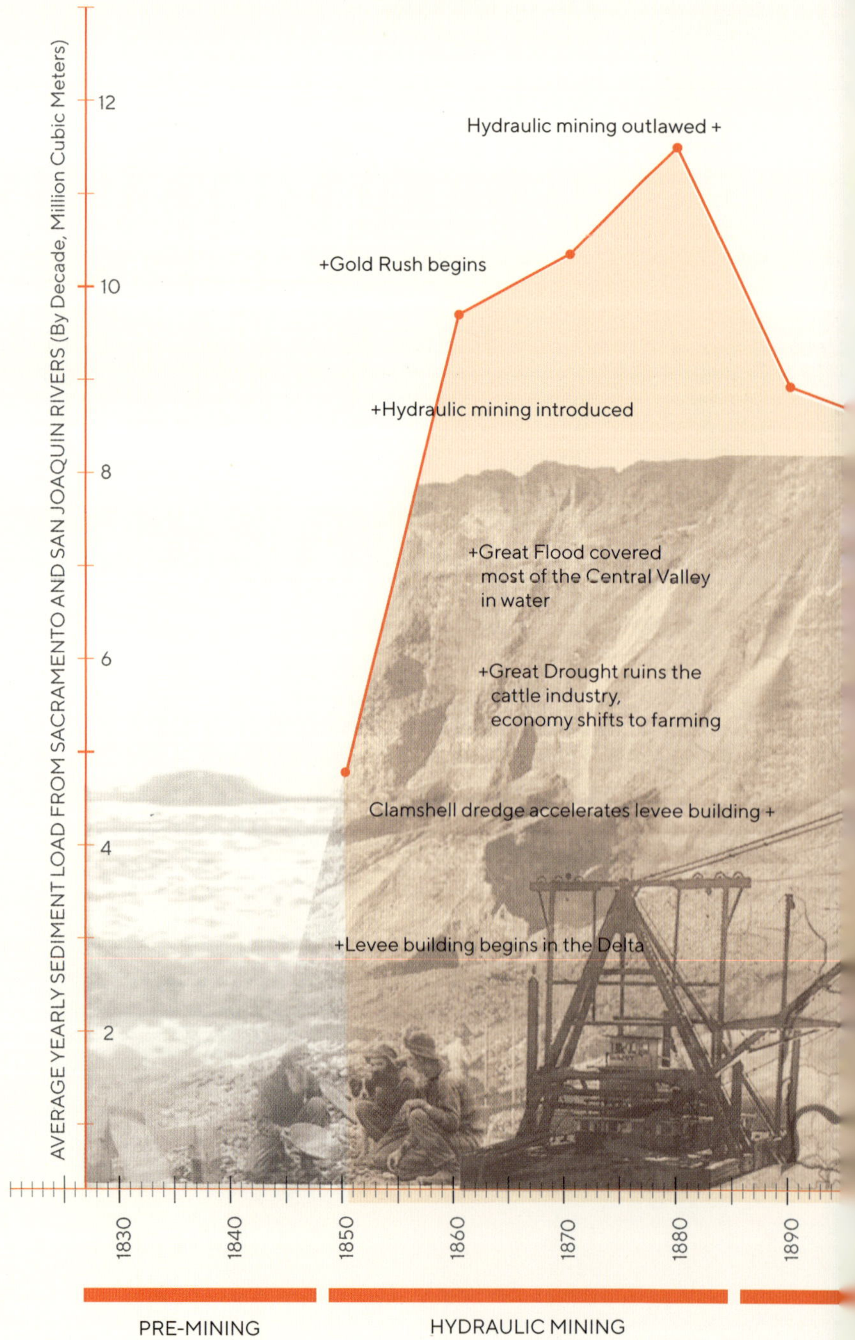

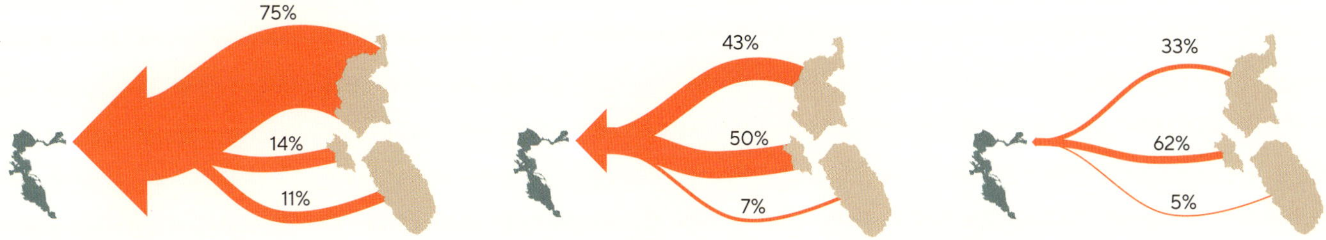

1849-1914 HYDRAULIC MINING
13.5 Million cubic meters annual sediment load

1914-1966 POST-MINING
6.6 Million cubic meters annual sediment load

PRESENT
2.7 Million cubic meters annual sediment load

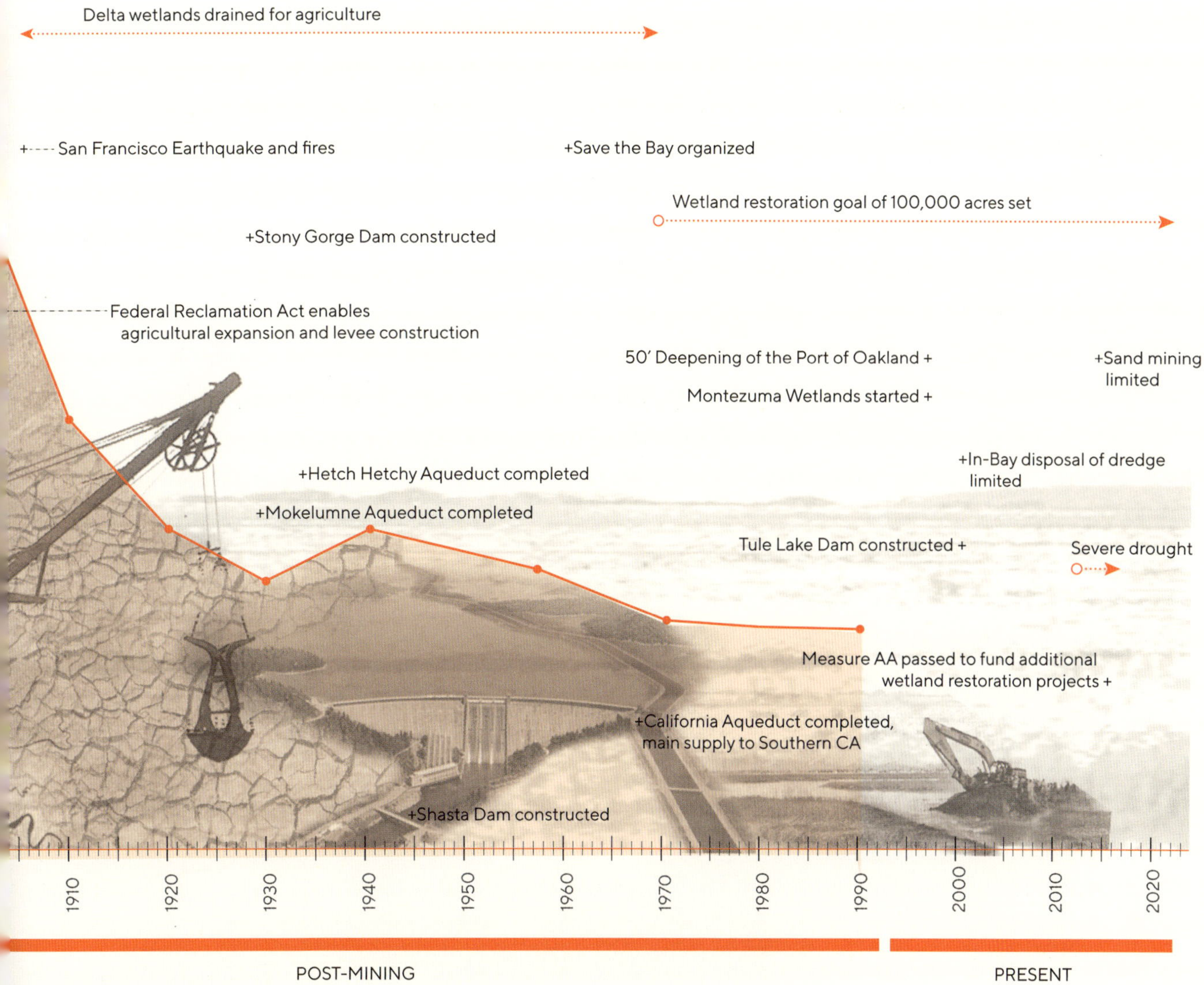

Upstream Sediment Traps

An upland network of more than one thousand dams collects, controls, and contains the state's surface water, decoupling the movement of water from the movement of sediment.[3] When water is released from the dams, sediment remains stuck in the uplands and unavailable to downstream ecosystems.

Hydrologic infrastructures like dams and flood control channels are essential to modern life in California. Water supply management, flood control, and power generation occurs at dam sites across the state. When sediment builds up in upland reservoirs, that not only impacts downstream ecosystems; it also reduces dam capacity and concentrates contaminants, such as mercury from historic gold mining operations. It is estimated that California's reservoirs have likely filled with 2.7 billion cubic yards of sediment over their lifespans, reducing the state's reservoir capacity by 4.5%.[4] Smaller reservoirs and watersheds most impacted by hydraulic gold mining are more susceptible to sediment buildup; over 200 reservoirs have likely had their capacity reduced by half.[5] The high costs of dredging, moving, and decontaminating sediment currently discourage the extraction and recycling of sediment from these reservoirs. California's dams have been quietly collecting mud for decades. The unintended consequence of this infrastructure comes into focus as reservoirs fill and water scarcity intensifies.

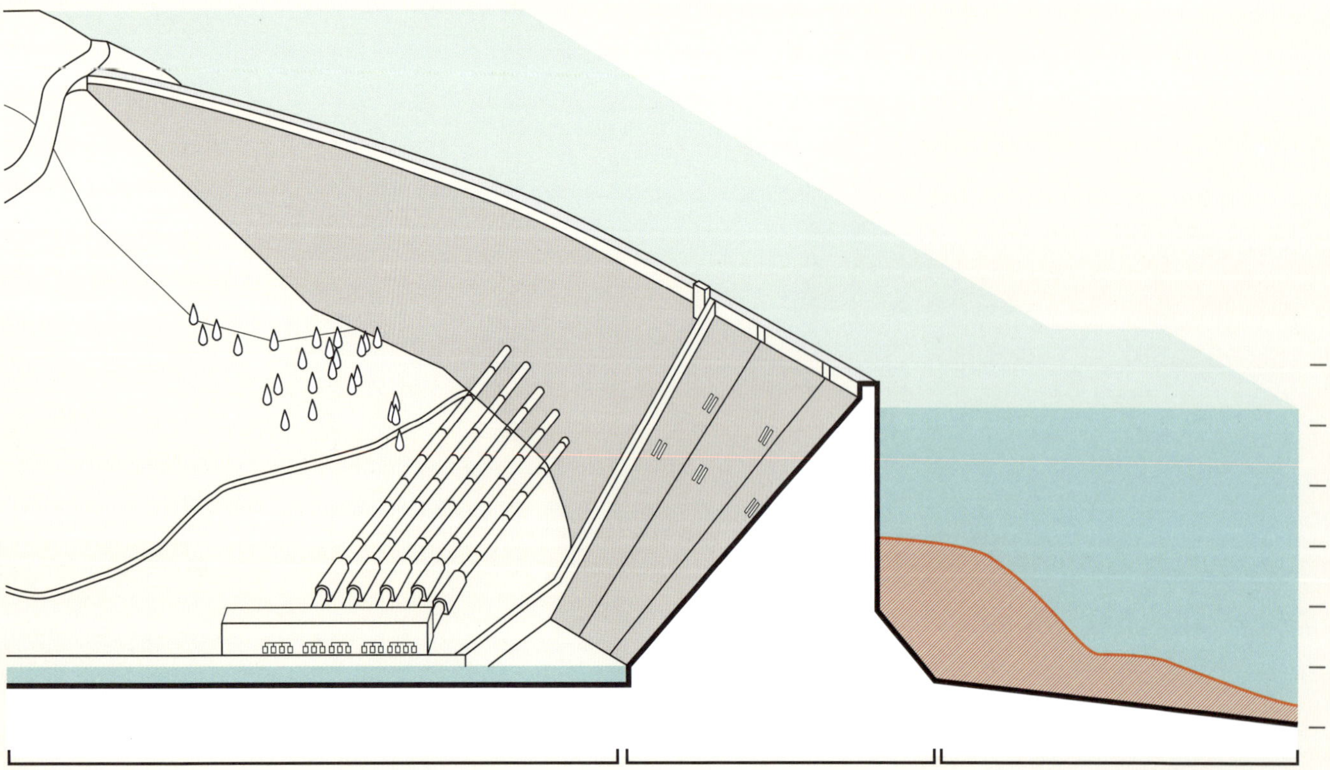

Sacramento River - Department of Water Resources

Shasta Dam - Montague Water Conservation District + U.S. Bureau of Reclamation

Shasta Lake and sediment buildup - Department of Water Resources

Shasta Dam during construction (Russell Lee, 1942)

Risks of Sediment Scarcity

The edges of the Bay-Delta have been constructed in response to a sediment-rich system. In the current climate of sediment shortfall and sea level rise, these economies and ecologies show signs of holistic collapse.

The reclaimed agricultural lands of the Delta are cut off from sediment flows and are subsiding at alarming rates, leading to conditions like the Netherlands, where polders are far below sea level. Outdated levees are susceptible to failure, and critical water supplies that rely on these levees as they cross the Delta are at risk of contamination.[6] Nearly all ecological conditions have radically changed in the Delta, including habitat, salinity, and hydrology.

This has led to the near extinction of species that once were common, like the delta smelt, and the rapid proliferation of non-native species. In the San Francisco Bay, a lack of suspended sediment is causing subsidence whereby marshes sink into the bay and convert to mudflats or open water.[7] Shoreline communities lose protective living

buffers and are at risk of flooding and sea level rise. While there is no consensus around sediment targets for the future, the present conditions hold great risks for the human and non-human inhabitants of the Bay-Delta.

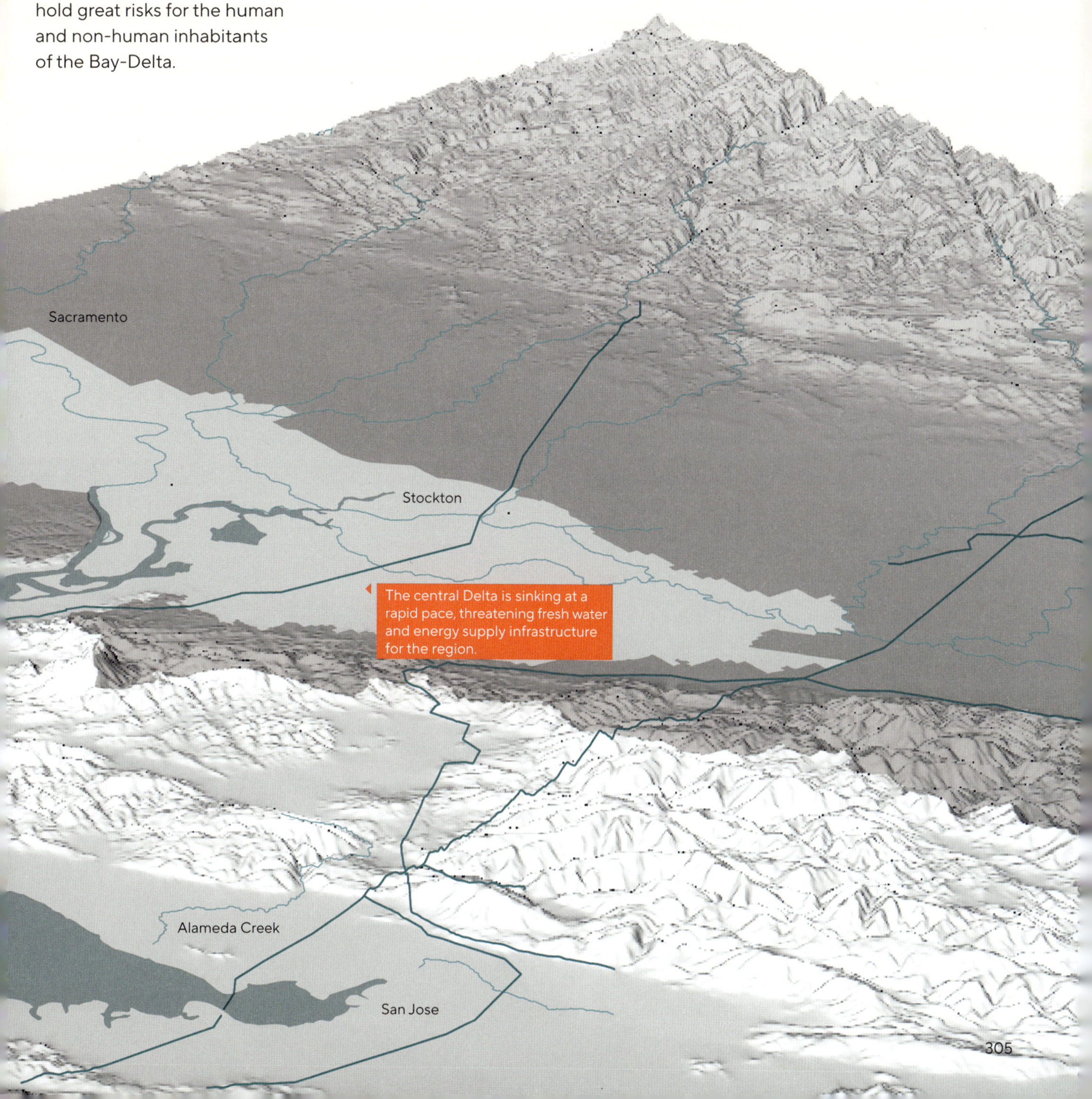

Delta Draining

The Delta was once a marshy inland delta, adapted to a seasonal cycle of freshwater inundation from mountain ranges in the rainy season and saltwater inundation from the Bay estuary in the dry summer months. Large-scale land reclamation transformed marshes to farmlands, irreparably changing the delta's morphology and function for future generations.

As homesteaders colonized the Delta flatlands in the late nineteenth century, the region was frequently inundated with the outwash of hydraulic gold mining, which delivered large quantities of mud and gravel with the wet winter rains. This slurry filled in the Delta's channels and covered flooded farmland and swampland in infertile gravel. As agricultural pressure grew, this material was harvested by Chinese laborers and used to fortify the edges of the farmlands, creating small levees to protect against frequent flooding.[8] Thus began the conversion of the Delta into one of the United States' first large-scale land reclamation projects, transforming wetlands (considered pestilent wastelands) to agricultural production.

HISTORICAL DELTA ECOLOGY (Before 1848): Low bank islands with regular tidal innundation

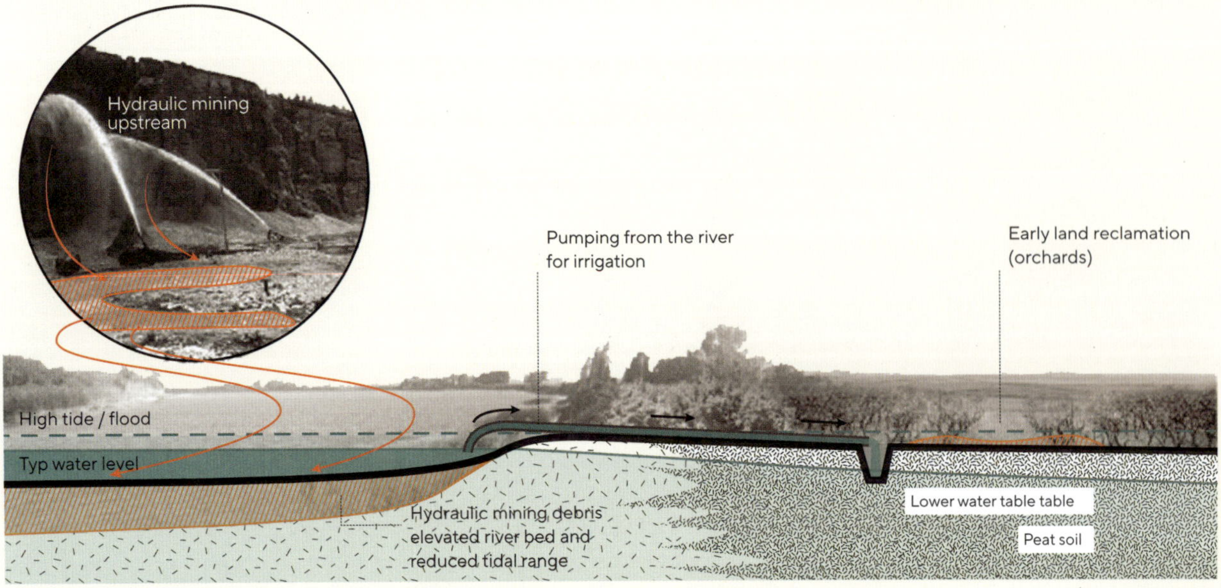

MINING AND AGRICULTURAL EXPANSION (1848-1894): Increase of sediment in waterways due to hydraulic mining

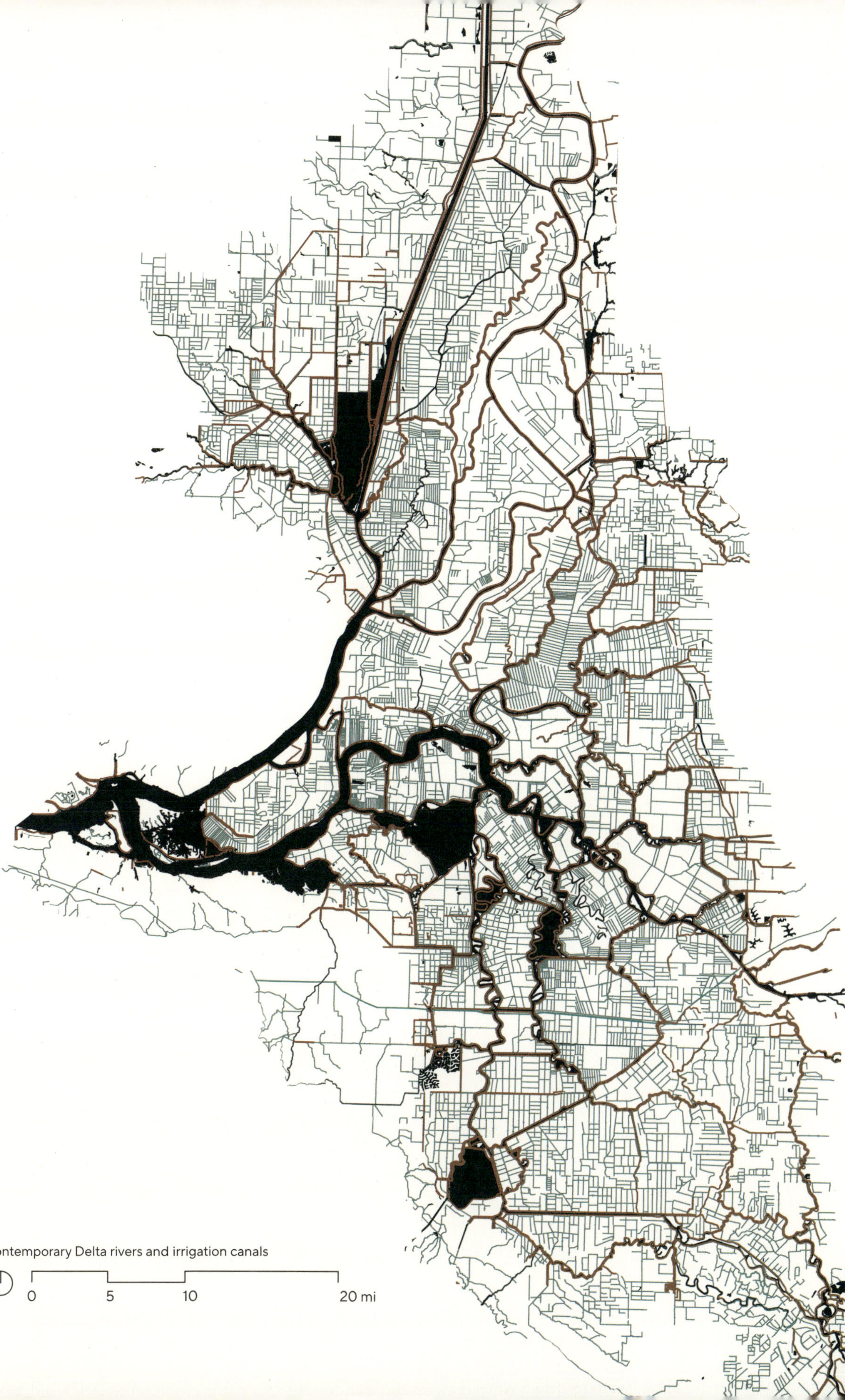

Contemporary Delta rivers and irrigation canals

0 5 10 20 mi

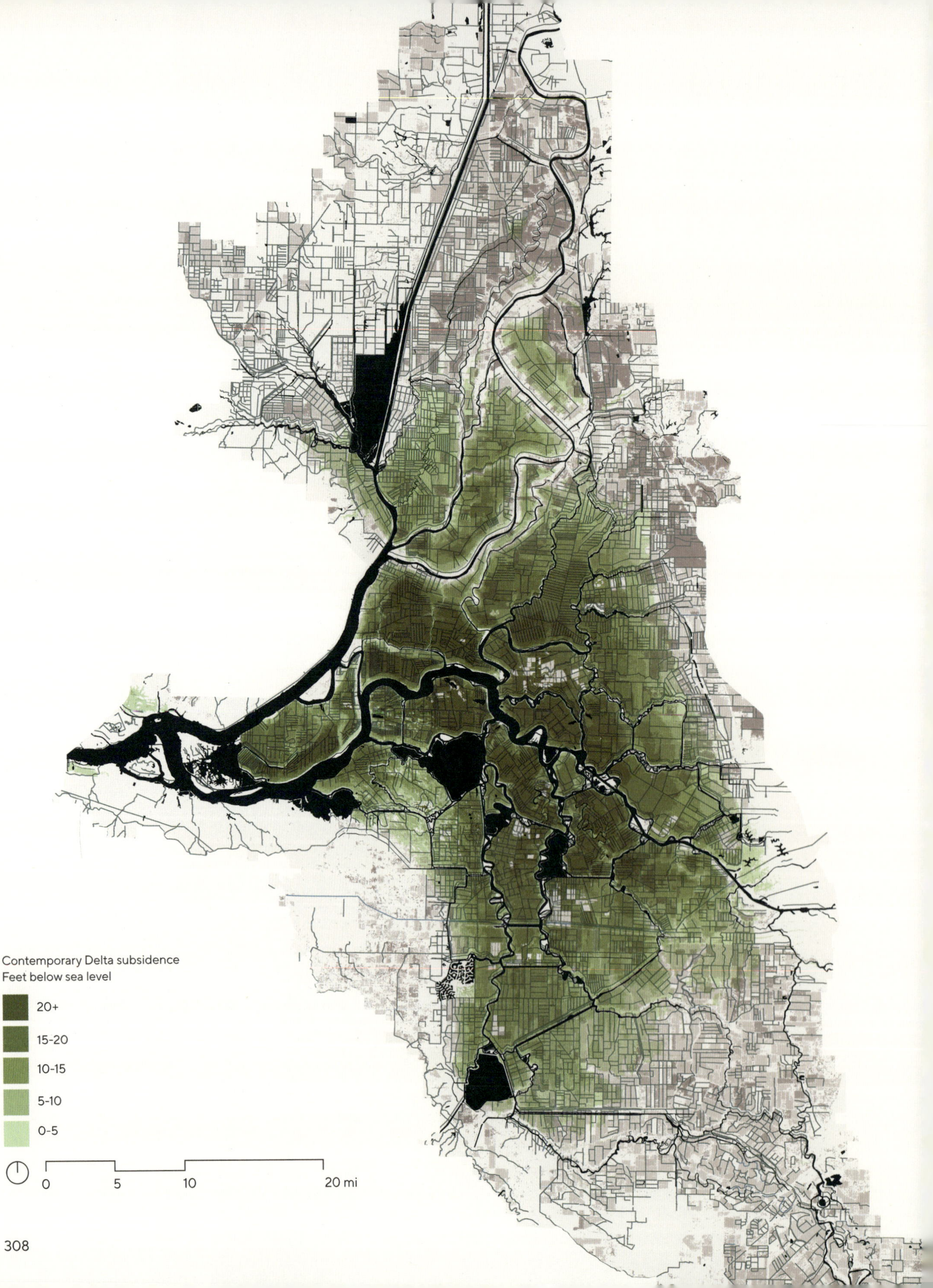
Contemporary Delta subsidence
Feet below sea level
20+
15-20
10-15
5-10
0-5
0
5
10
20 mi

Delta Sinking

The adoption of the clamshell dredge accelerated the process of land reclamation and, ultimately, the subsidence of the Delta landscape.

Designed to both deepen channels and build levees, the clamshell accelerated the conversion of the Delta landscape, resulting in a network of approximately 1,100 miles of levees.[9] A devastating byproduct of levee-building has been the large-scale subsidence of agricultural land. The pumping of water, the lack of sediment inputs, and the decomposition of the peat substrate has led Delta polders to sink 10 to 25 feet below sea level.[10] While levee construction has had serious ecological ramifications, including the near total loss of former tidal habitats, California has grown to depend upon these artificial structures. In the middle of the twentieth century, the Delta levees became a critical link in the state's water conveyance infrastructure, with pumps and aqueducts transporting drinking water from north to south. Over two-thirds of California's population gets at least some of its drinking water from the Delta, rendering the state highly reliant upon the fragile levee networks and the subsiding grounds behind them.[11]

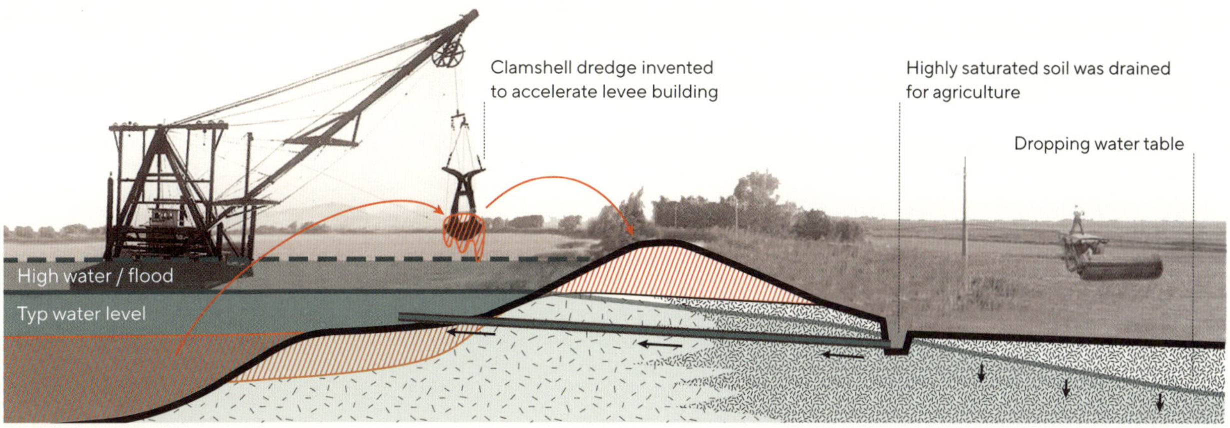

DREDGING AND LAND RECLAMATION (1850s-Present): Agricultural demands accelerate scale and pace of Delta modifications

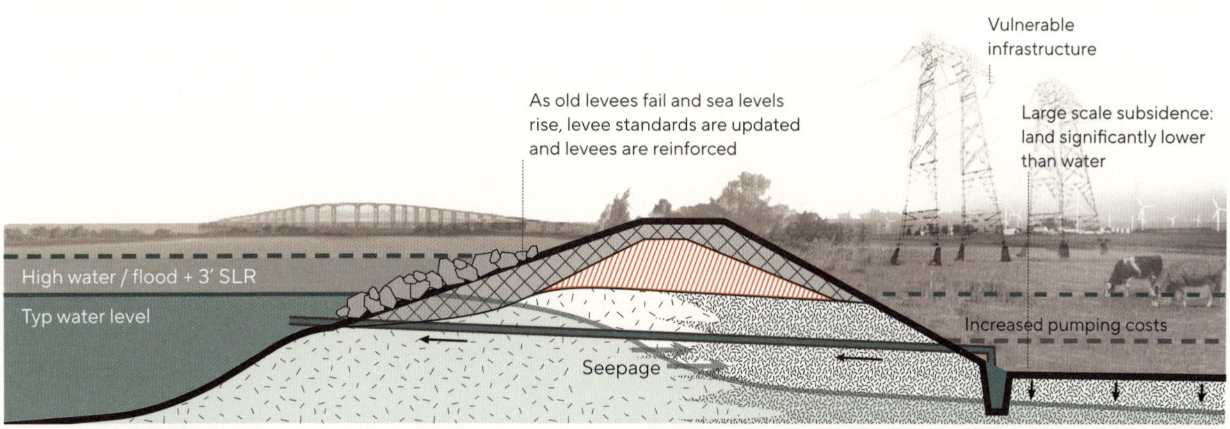

LAND SUBSIDENCE, RESTORATION, AND ADAPTATION (1900s-Present): Ongoing subsidence and severe loss of wetlands

Growing Land

Contemporary Delta land management practices accelerate subsidence, releasing large amounts of CO2 into the atmosphere. Submerged vegetation, in the form of re-established wetlands or rice farming, can slow peat decomposition and carbon dioxide release, allowing for decomposed plant material to build up, and thus slowing or reversing subsidence over long periods of time. Pilot projects are underway throughout the Delta to study "subsidence reversal" through the conversion of farmlands to shallow water wetlands and hybrid rice wetlands.[12] Cattails and tules are targeted as key species in this transformation.

Subsidence reversal at Whale Mouth Marsh (2016)

Building Consensus

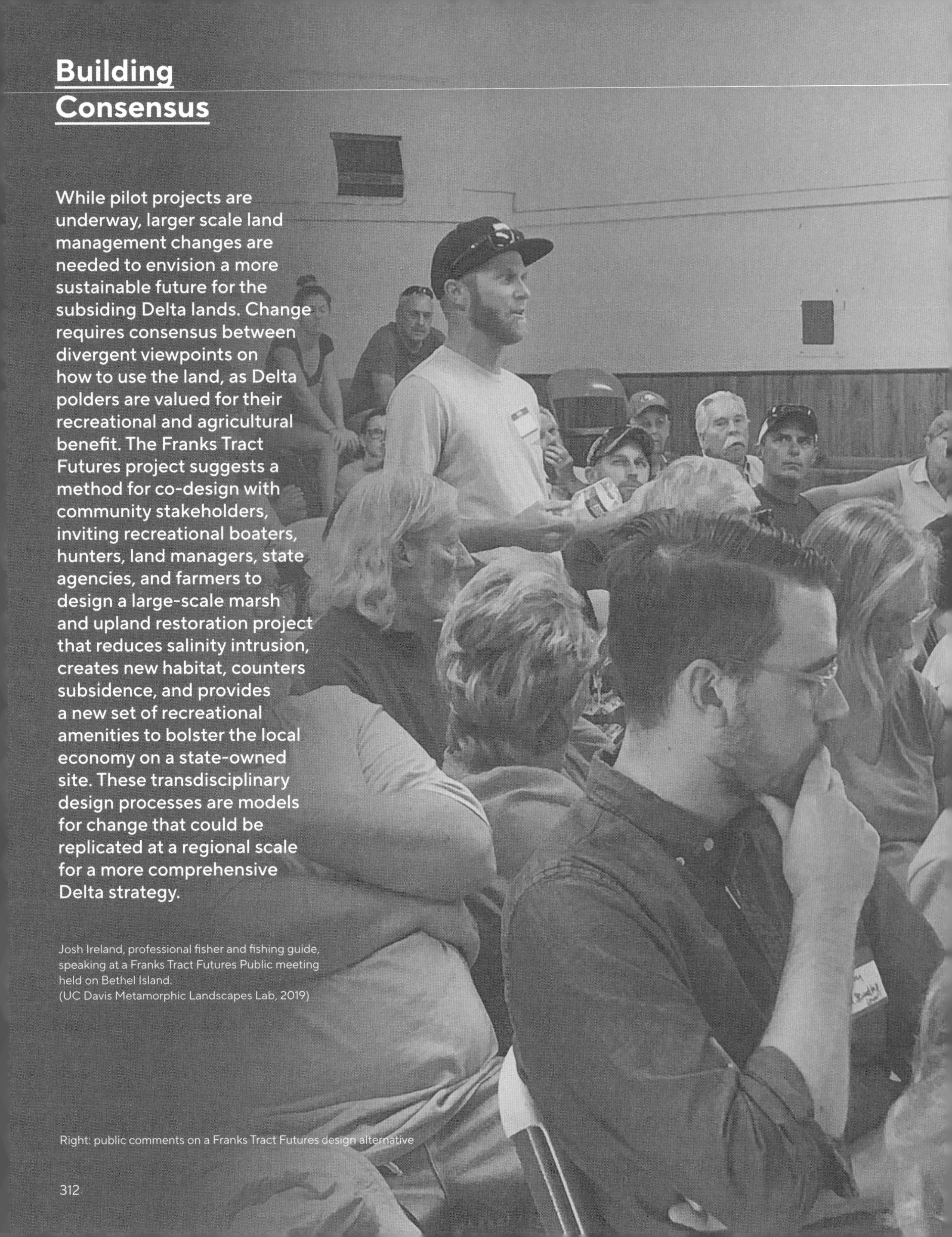

While pilot projects are underway, larger scale land management changes are needed to envision a more sustainable future for the subsiding Delta lands. Change requires consensus between divergent viewpoints on how to use the land, as Delta polders are valued for their recreational and agricultural benefit. The Franks Tract Futures project suggests a method for co-design with community stakeholders, inviting recreational boaters, hunters, land managers, state agencies, and farmers to design a large-scale marsh and upland restoration project that reduces salinity intrusion, creates new habitat, counters subsidence, and provides a new set of recreational amenities to bolster the local economy on a state-owned site. These transdisciplinary design processes are models for change that could be replicated at a regional scale for a more comprehensive Delta strategy.

Josh Ireland, professional fisher and fishing guide, speaking at a Franks Tract Futures Public meeting held on Bethel Island.
(UC Davis Metamorphic Landscapes Lab, 2019)

Right: public comments on a Franks Tract Futures design alternative

(UC Davis Metamorphic Landscapes Lab)

Building dunes with dredged sand (Antioch Dunes 2017)

Sediment as Life Support

While larger efforts are underway to rethink land management practices, sediment-dependent species are already at risk. The unique ecosystem of the Antioch Dunes, created by the sandy deposits of the San Joaquin River over 140,000 years ago, supports the last remaining populations of the endangered Antioch Dunes primrose, the endangered Contra Costa wallflower, and the endangered Lange's metalmark butterfly. Mining of sand, particularly for brick-making after the 1906 San Francisco earthquake, led to the rapid decline of this ecosystem and the species that depend upon the windswept dunes.[13] Since 2013, the Port of Stockton and the Fish and Wildlife Service have collaborated to build a new dune to support these species through the application of dredged material. When the Port of Stockton dredges their channels of clean sand for annual maintenance, they transport material to the Antioch Dunes.

Sediment in the Bay

The impacts of sediment scarcity trickle down to the San Francisco Bay, limiting the availability of sediment to nourish the Bay's marshes, beaches, and mudflats. As policymakers look to restore more of the region's lost ecosystems, sediment is emerging as a key ingredient of bayland regeneration.[14]

Like the Delta, the Bay's edges were diked and filled as the region expanded, leading to dramatic losses of marshes (79%) and mudflats (42%) between 1800 and 1998.[15] The loss of these estuarine ecosystems has led to declines in native species and game species, degraded water quality, and a loss of the wetland buffers that mitigate storm impacts. The Bayland Ecosystem Habitat Goals were developed by a collaborative agency and stakeholder process in 1999, and they continue to drive restoration policy with a goal of rehabilitating more than 100,000 acres of habitat.[16] The urgency has only intensified with climate change, with recent reports noting that marshes should be restored quickly to give them time to adapt, accumulate sediment, and keep pace with sea level rise.[17]

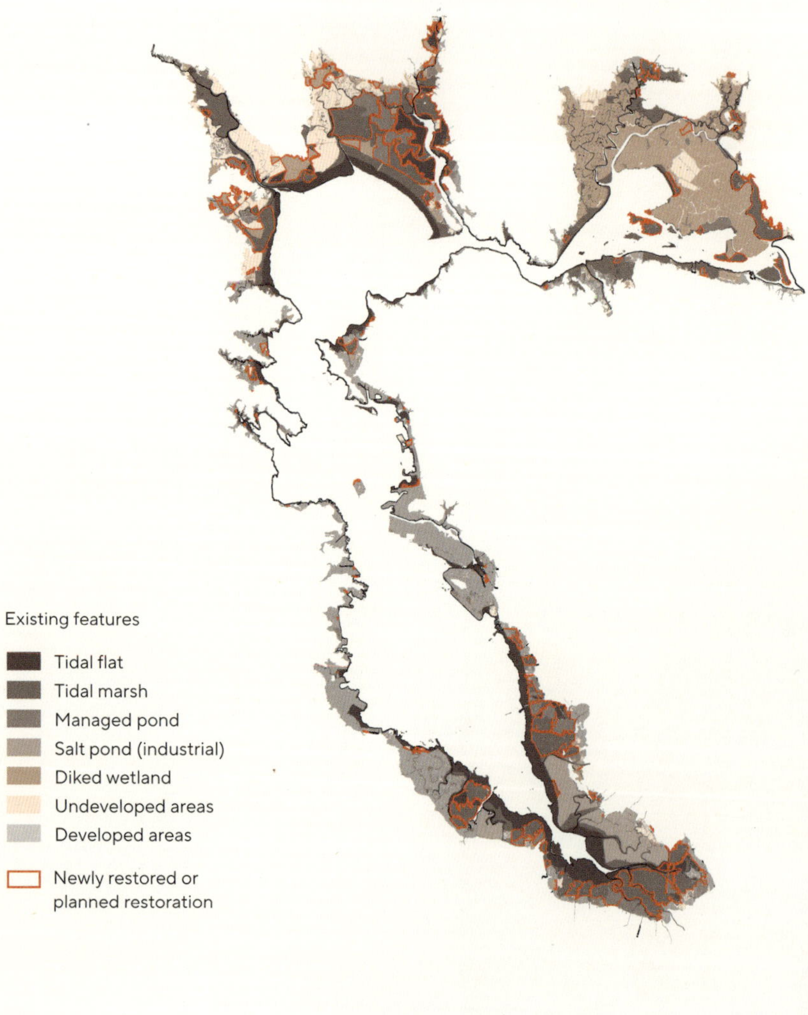

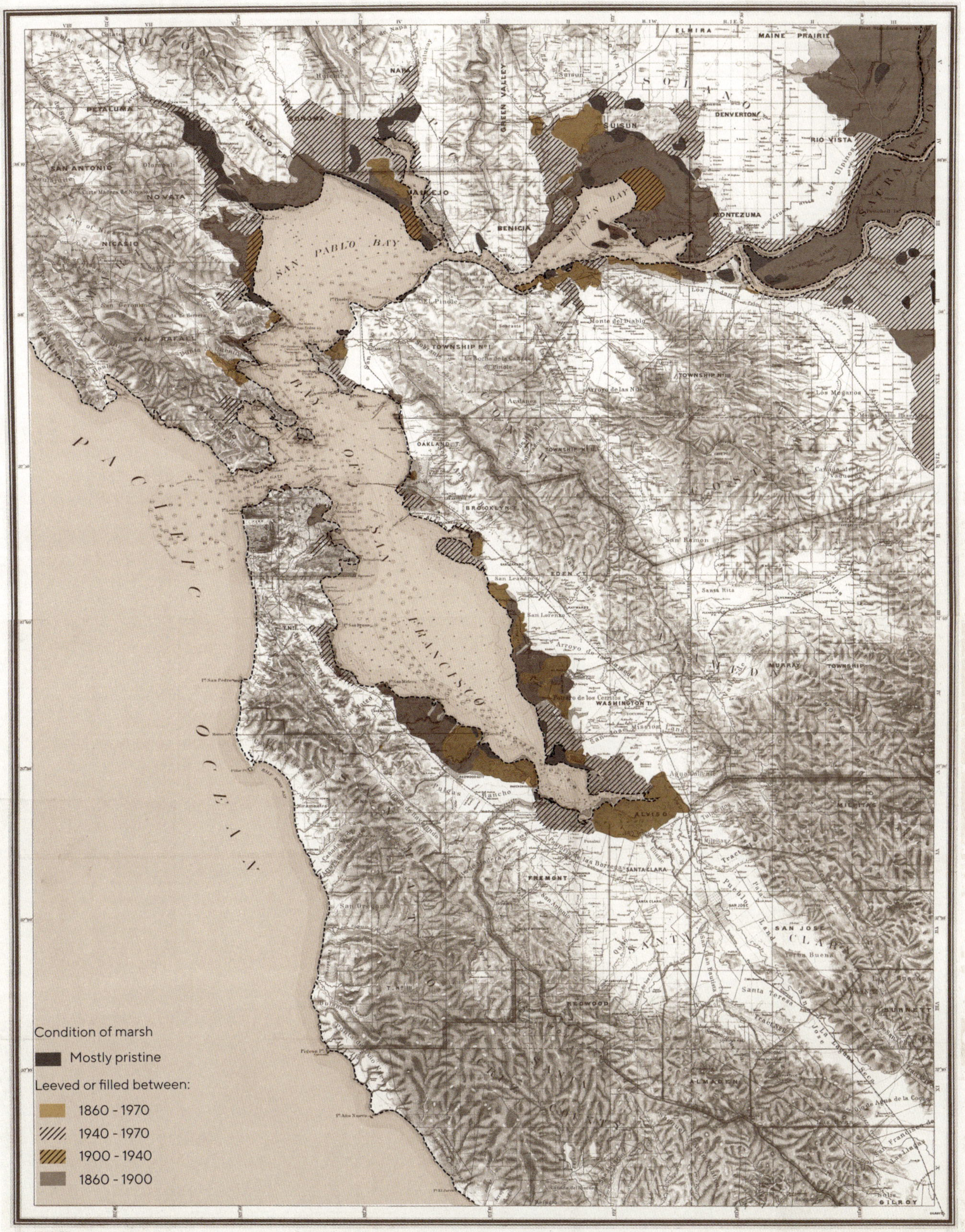

Condition of marsh
Mostly pristine
Leeved or filled between:
1860 - 1970
1940 - 1970
1900 - 1940
1860 - 1900
PETALUMA
SAN ANTONIO
NOVATO
NICASIO
SONOMA
NAPA
GREEN VALLEY
SUISUN
DENVERTON
RIO VISTA
MAINE PRAIRIE
ELMIRA
MONTEZUMA
BENICIA
SAN PABLO BAY
SUISUN BAY
SAN RAFAEL
OAKLAND
BROOKLYN
EDEN
WASHINGTON
MURRAY
TOWNSHIP
SANTA CLARA
FREMONT
SAN JOSE
ALVISO
REDWOOD
ALMADEN
BURNETT
GILROY
PACIFIC OCEAN
BAY OF SAN FRANCISCO

Marsh Creation

While dredged Bay sediment can be beneficially used to build mudflats, restore upland habitat, and nourish fine- and coarse-grain beaches, the highest volume of material goes toward wetland restoration projects, which need additional sediment to compensate for decades of land subsidence.

Hamilton Wetlands is one of the Bay's largest beneficial use sites, allowing for the placement of up to 24 million cubic yards of dredge for marsh restoration, including material excavated from the Port of Oakland's 50-foot harbor deepening project.[18] Previously farmland and a military airfield, the Hamilton Wetlands have subsided over time, and used the addition of dredge to amend the site to current marsh elevations in the Bay. While the project could have simply breached the levees and allowed the tidal land to slowly accrete sediment, the placement of dredged sediment likely accelerated the marsh-making process by decades. This project was realized through careful timing and planning. Advance notice of the harbor deepening project allowed for sediment matchmaking to occur, while economies of scale made the transport financially feasible.

Hamilton Wetlands (2016)

Moving Dredge

While sediment is being recast as a resource for bayland restoration, it is still a nuisance for port operators and boats. Every year, an average of 3 to 6 million cubic yards of sediments are dredged from the Bay to maintain safe navigation, and just over 40% of this material is beneficially used.[19 20]

Beneficial use is simplest and easiest to coordinate when dredging and restoration sites are in close proximity, so that material can be pumped or barged over short distances. Crucially, timeframes for dredging need to align with timeframes for receiving sediment, which are oriented around species' lifecycles and critical breeding and nesting seasons. Multiple restoration sites around the Bay need sediment for survival, but financial and logistical obstacles often prevent beneficial use, so dredged material is disposed offshore and exported from the system. As sea levels rise, it is especially critical to keep sediment in the system if the Baylands are to keep pace with a changing climate.

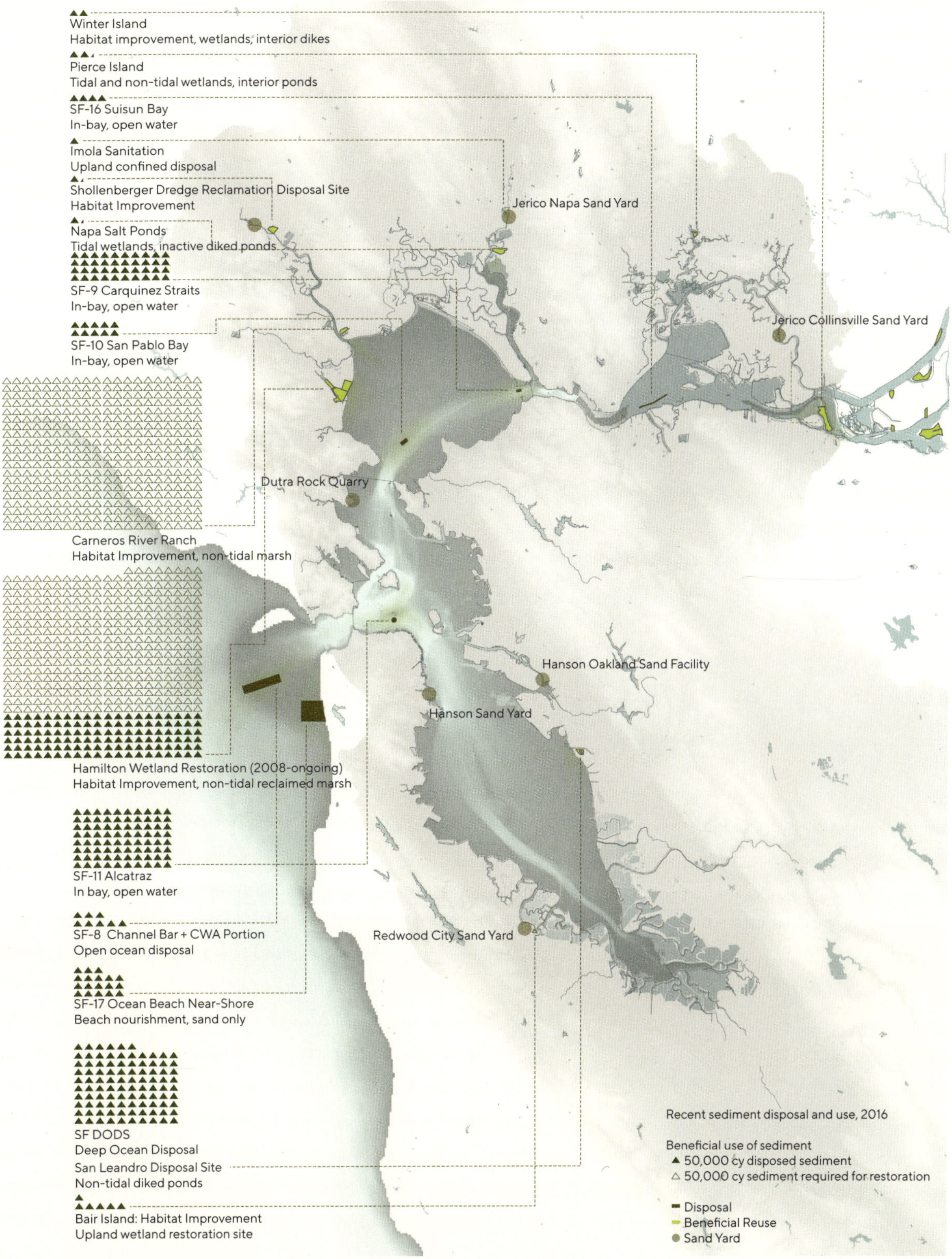
Winter Island
Habitat improvement, wetlands, interior dikes
Pierce Island
Tidal and non-tidal wetlands, interior ponds
SF-16 Suisun Bay
In-bay, open water
Imola Sanitation
Upland confined disposal
Shollenberger Dredge Reclamation Disposal Site
Habitat Improvement
Napa Salt Ponds
Tidal wetlands, inactive diked ponds
SF-9 Carquinez Straits
In-bay, open water
SF-10 San Pablo Bay
In-bay, open water
Carneros River Ranch
Habitat Improvement, non-tidal marsh
Hamilton Wetland Restoration (2008-ongoing)
Habitat Improvement, non-tidal reclaimed marsh
SF-11 Alcatraz
In bay, open water
SF-8 Channel Bar + CWA Portion
Open ocean disposal
SF-17 Ocean Beach Near-Shore
Beach nourishment, sand only
SF DODS
Deep Ocean Disposal
San Leandro Disposal Site
Non-tidal diked ponds
Bair Island: Habitat Improvement
Upland wetland restoration site
Jerico Napa Sand Yard
Jerico Collinsville Sand Yard
Dutra Rock Quarry
Hanson Oakland Sand Facility
Hanson Sand Yard
Redwood City Sand Yard
Recent sediment disposal and use, 2016
Beneficial use of sediment
▲ 50,000 cy disposed sediment
△ 50,000 cy sediment required for restoration
Disposal
Beneficial Reuse
Sand Yard

Developing a Sediment Economy

Increasingly, people, companies, and public agencies understand that dredged sediment is a valuable resource, not a waste product. So why is material still exported from the system?

Despite growing recognition that sediment is critical to protecting and nourishing shorelines, sediment is still not valuable enough to move twice. The material must be dredged and taken directly to its point of use. While regulations have begun to change, federal law still requires the US Army Corps of Engineers to use the most cost-effective disposal option that meets applicable federal environmental regulations or solicit a partner donation to cover additional costs. Projects like the Montezuma Wetlands attempt to develop a more efficient economy around the movement of dredge. Having an offloader on site maximizes the efficiency of the placement process, allowing multiple dredgers to unload dredged material and distribute it to a larger wetland restoration process over an expanded timeframe. Other programs, like Bay Area Sedi-Match, aim to improve sediment matchmaking between dredger and restoration entities, compiling an online database of sediment sources and supplies throughout the region.

Sediment off-loader at Montezuma Wetlands (2016)

Approximate costs of placing sediment in San Francisco Bay Area (2012)

*The diagram depicts costs generalized over a variety of conditions. Costs are highly specific to site and environmental conditions.

A Local Sediment Supply?

Beneficial use of dredge isn't enough. The volume of available sediment does not meet the needs of planned restoration projects, nor the need for suspended sediment with high rates of sea level rise. In the absence of sediment-rich river systems, interest has turned to local tributaries as potential sources of material for shoreline regeneration and adaptation.

In the early twentieth century, Delta waterways accounted for approximately 90% of the sediment entering the Bay.[21] Today, the large rivers no longer carry these sediment loads, and local tributaries—small waterways within the watershed of the Bay itself—account for half of the Bay's sediment supply.[22] But like their larger counterparts, these local tributaries have been modified over time; small dams and flood control channels trap, slow, and divert sediment. Recently, Bay tributaries have been gaining attention as critical sites of restoration, to reconnect sediment flows, encourage wildlife connectivity, and establish a gradient of connected ecosystem types from bayland to upland. Thoughtful management and redesign of Bay tributaries has potential to meet local sediment demand and recalibrate restoration goals to match available sediment supply.

Napa River
tributary and sedimentshed

San Lorenzo Creek
tributary and sedimentshed

Total annual sediment load

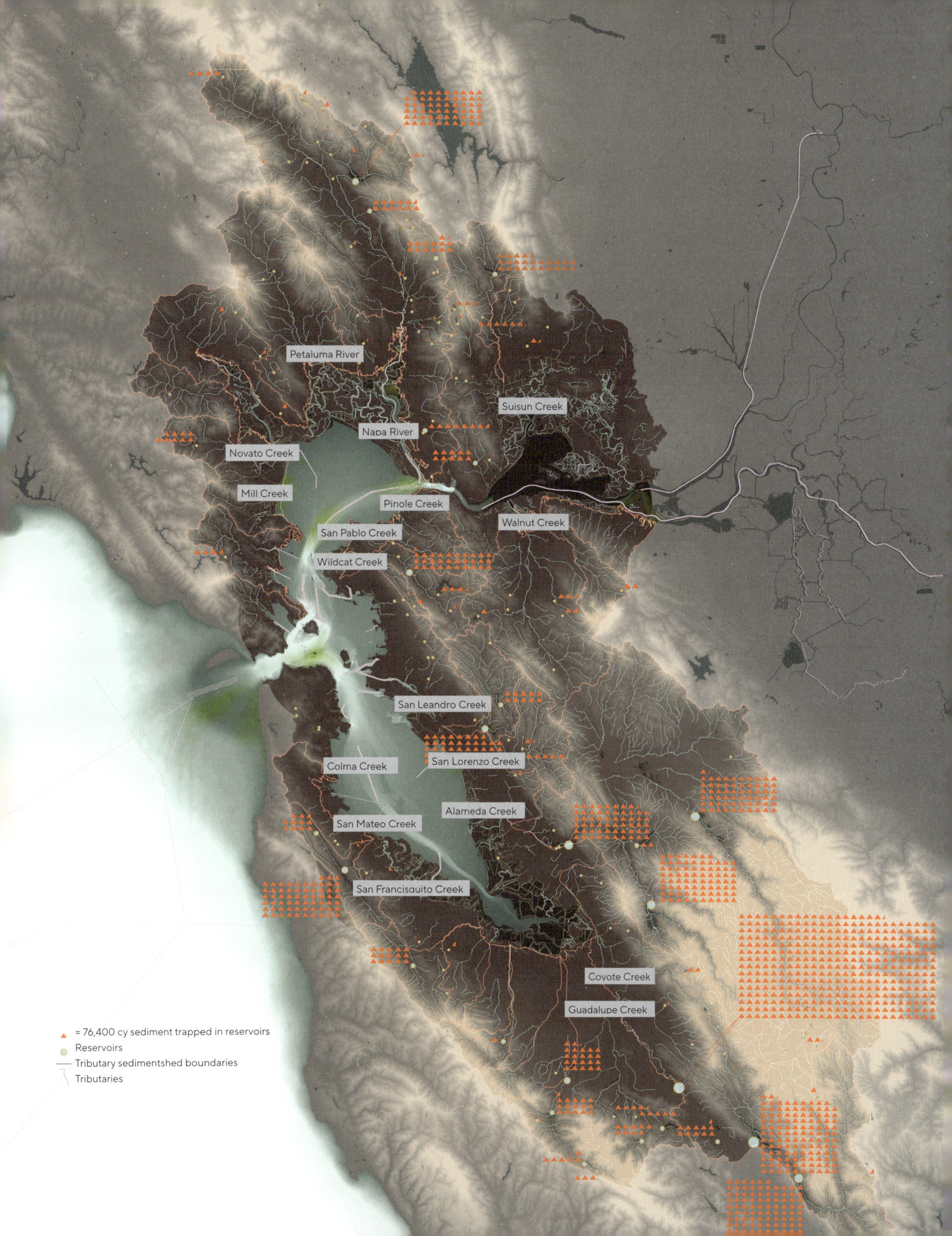

Petaluma River
Suisun Creek
Napa River
Novato Creek
Mill Creek
Pinole Creek
Walnut Creek
San Pablo Creek
Wildcat Creek
San Leandro Creek
San Lorenzo Creek
Colma Creek
Alameda Creek
San Mateo Creek
San Francisquito Creek
Coyote Creek
Guadalupe Creek
= 76,400 cy sediment trapped in reservoirs
Reservoirs
Tributary sedimentshed boundaries
Tributaries

Outfall of the Pinole Creek Flood Control Channel into the Bay (April, 2020)

Changing Bay

The region faces climate risks from all directions. The scarce resource of sediment will become even more critical as the Bay's edges are remade and the Delta adapts to future change.

Climate change is already impacting regional rainfall patterns and wildfire frequency, which erodes more sediment and puts stress on the outdated infrastructure of flood control channels and dams. While more sediment may move to the Bay over time, the volume needed to help the existing and proposed marshes and mudflats of the baylands keep pace with sea level rise far outpaces the anticipated supply. At the same time, the built infrastructures of the bayland edges—levees, elevated parcels, and stabilized shorelines—require sediment for construction and maintenance over time. As regional adaptation plans emerge, sediment will be a critical and contested resource to support human and non-human residents of the Bay-Delta. Regulatory adaptation that enables the testing of new sediment nourishment, rebuilding, and restoration strategies will be essential for the future of the baylands.

Endnotes:

1. "About the Watershed." EPA. Environmental Protection Agency. Accessed April 30, 2022. https://www.epa.gov/sfbay-delta/about-watershed.

2. Brett Milligan, Tim Maly, and Rob Holmes. "The Pulse, the Sink, and the Shortfall." *Ground Up 04: Out West*, 2015.

3. George Skelton. "Capitol Journal: California Should Stop Thinking about More Dams. the State Is Brimming with Them." *Los Angeles Times*, March 4, 2019. https://www.latimes.com/politics/la-pol-sac-skelton-water-storage-california-20190304-story.html.

4. Minear, J. T., and G. M. Kondolf. "Estimated Reservoir Sediment: (2009, Estimating reservoir sedimentation rates at large spatial and temporal scales: A case study of California), *Water Resources Research*, 45, W12502, doi:10.1029/2007WR006703. https://ca.water.usgs.gov/pubs/2009/MinearKondolf.pdf.

5. Ibid.

6. Port, Patricia Sanderson, and Samuel A. Hoover. "Anticipating California Levee Failure: The State of the Delta Levees and Government Preparation and Response Strategies for Protecting Natural Resources from Freshwater Oil Spills." *International Oil Spill Conference Proceedings* 2011, no. 1 (2011). https://doi.org/10.7901/2169-3358-2011-1-112.

7. Gies, Erica. "Fortresses of Mud: How to Protect the San Francisco Bay Area from Rising Seas." *Nature News*. Nature Publishing Group, October 9, 2018. https://www.nature.com/articles/d41586-018-06955-4.

8. Wong, William. "Chinese Pioneers and Delta Levees." SFGATE. San Francisco Chronicle, February 8, 2012. https://www.sfgate.com/news/article/Chinese-pioneers-and-Delta-levees-3130500.php.

9. US Geologic Survey and US Department of the Interior. Delta Subsidence in California: The Sinking Heart of the State. FS-005-00. April 2000. https://pubs.usgs.gov/fs/2000/fs00500/pdf/fs00500.pdf.

10. Ibid.

11. Ibid.

12. Robin Miller, Miranda S. Fram, Roger Fujii, and Gail Wheeler. "Subsidence Reversal in a Re-Established Wetland in the Sacramento-San Joaquin Delta, California, USA." *San Francisco Estuary and Watershed Science* 6, no. 3 (2008). https://doi.org/10.15447/sfews.2008v6iss3art1.

13. Haas-Wajdowicz, Julie. "Antioch Dunes - City of Antioch, California." City of Antioch. City of Antioch, California, February 14, 2019. https://www.antiochca.gov/environmental-resources/wildlife-preservation/antioch-dunes/.

14. Scott Dusterhoff, K. McKnight, Letitia Grenier, N. Kauffman. Sediment for Survival: A Strategy for the Resilience of Bay Wetlands in the Lower San Francisco Estuary. SFEI Contribution No. 1015. San Francisco Estuary Institute: Richmond, CA. 2021. https://www.sfei.org/documents/sediment-for-survival.

15. San Francisco Bay Area Wetlands Ecosystem Goals Project, California State Coastal Conservancy. The Baylands and Climate Change: What We Can Do. Baylands Ecosystem Habitat Goals Science Update 2015. 2015. Oakland, CA. 2015. https://www.sfei.org/sites/default/files/biblio_files/Baylands_Complete_Report.pdf.

16. Ibid.

17. Ibid.

18. California State Coastal Conservancy. Hamilton Wetlands Restoration Project Implementation: Bel Marin Keys Unit V Expansion. Project No 94-003. February 4, 2010. https://scc.ca.gov/webmaster/ftp/pdf/sccbb/2010/1002/20100204Board05_Hamilton_Wetlands_Restoration.pdf.

19. "Dredging and Sediment Management." EPA. Environmental Protection Agency. Accessed April 30, 2022. https://19january2017snapshot.epa.gov/www3/region9/water/dredging/ltms/index.html#:~:text=Every%20year%2C%20an%20average%20of,environmental%20issue%20for%20many%20years.

20. US Army Corps of Engineers, US Environmental Protection Agency, San Francisco Bay Conservation and Development Commission, and San Francisco Bay Regional Water Quality Control Board. Long-term Management Strategy for the Placement of Dredged Material in the San Francisco Bay Region: Management Plan 2001. July 2001. https://www.spn.usace.army.mil/Portals/68/docs/Dredging/LMTS/entire%20LMTF.pdf.

21. Patrick L Barnard., David H. Schoellhamer, Bruce E. Jaffe, and Lester J. McKee. "Sediment Transport in the San Francisco Bay Coastal System: An Overview." *Marine Geology* 345 (2013): 3–17. https://doi.org/10.1016/j.margeo.2013.04.005.

22. Ibid.

Image Credits:

Spread 1
US Geological Survey, National Agricultural Imagery Program (NAIP). "San Francisco South Bay." (Satellite Image) https://www.usgs.gov/centers/eros/science/usgs-eros-archive-aerial-photography-national-agriculture-imagery-program-naip

Spread 2
(left)
US Bureau of Transportation Statistics. U.S.Ports. Shapefile. 2016. https://koordinates.com/layer/22813-us-ports/
US Geological Survey, National Hydrography Dataset. Water Drainage Network and Watershed Boundary. Accessed 2016. https://www.usgs.gov/national-hydrography/national-hydrography-dataset
US Geological Survey. National Landcover Dataset. Shapefile. 2016. https://www.mrlc.gov/data?f%5B0%5D=category%3Aland%20cover&f%5B1%5D=year%3A2016
(right)
Composite imagery

Spread 3
(top)
Patrick L. Barnard, David H. Schoellhamer, Bruce E. Jaffe, and Lester J. McKee. "Sediment Transport in the San Francisco Bay Coastal System: An Overview." Marine Geology 345 (2013): 3–17. https://doi.org/10.1016/j.margeo.2013.04.005
H. R. Moftakhari, D. A. Jay, S. A. Talke, and D. H. Schoellhamer. "Estimation of Historic Flows and Sediment Loads to San Francisco Bay, 1849–2011." Journal of Hydrology 529 (2015): 1,247–61. https://doi.org/10.1016/j.jhydrol.2015.08.043.
A. Whipple; Grossinger, R. M.; Rankin, D.; Stanford, B.; Askevold, R. A. San Francisco Estuary Institute. Sacramento-San Joaquin Delta Historical Ecology Investigation: Exploring Pattern and Process. SFEI Contribution No. 672. SFEI: Richmond.2012.
https://www.sfei.org/documents/sacramento-san-joaquin-delta-historical-ecology-investigation-exploring-pattern-and-proces
(bottom)
Neil K. Ganju, Noah Knowles, and David H. Schoellhamer. "Temporal Downscaling of Decadal Sediment Load Estimates to a Daily Interval for Use in Hindcast Simulations." *Journal of Hydrology* 349, no. 3-4 (2008): 512–23. https://doi.org/10.1016/j.jhydrol.2007.11.026

Spread 4
Russell Lee, Farm Security Administration. "Shasta Dam Under Construction." (Photograph), 1942. http://loc.gov/pictures/resource/fsac.1a34236/

Spread 5
US Geological Survey, National Hydrography Dataset. Water Drainage Network and Watershed Boundary. Accessed 2016. https://www.usgs.gov/national-hydrography/national-hydrography-dataset

Spread 6
(right)
US Geological Survey, National Hydrography Dataset. Water Drainage Network and Watershed Boundary. Accessed 2016. https://www.usgs.gov/national-hydrography/national-hydrography-dataset

US Geological Survey, Surface-Water Network for the Central Valley Hydrologic Model (CVHM. Accessed 2016. https://water.usgs.gov/GIS/metadata/usgswrd/XML/pp1766_cvhm_sfr_network.xml#Metadata_Reference_Information

Spread 7
(left)
US Geological Survey, National Hydrography Dataset. Water Drainage Network and Watershed Boundary. Accessed 2016. https://www.usgs.gov/national-hydrography/national-hydrography-dataset
US Geological Survey, Surface-Water Network for the Central Valley Hydrologic Model (CVHM. Accessed 2016. https://water.usgs.gov/GIS/metadata/usgswrd/XML/pp1766_cvhm_sfr_network.xml#Metadata_Reference_Information

Subsidence information developed through analysis of:
California Department of Water Resources, URS Corporation, Furgo Earth Data. Elevation/SacDelta_LiDAR_DEM_2007 (ImageServer). 2008. https://gis.water.ca.gov/arcgisimg/rest/services/elevation/SacDelta_LiDAR_DEM_2007/ImageServer

Spread 8
Brett Milligan, Dredge Research Collaborative. "Subsidence Reversal at Whale Mouth Marsh." (Photograph), 2016.

Spread 9
(left)
UC Davis Metamorphic Landscapes Lab. "Josh Ireland, professional fisher and fishing guide, speaking at a Franks Tract Futures Public meeting held on Bethel Island." (Photograph), 2019.
(right)
Image courtesy of UC Davis Metamorphic Landscapes Lab

Spread 10
Brett Milligan. "Application of dredged sand from Stockton shipping channel maintenance to recreate dunes at the Antioch National Wildlife Refuge." (Photograph), 2017.

Spread 11
(left)
San Francisco Estuary Institute and Aquatic Science Center (SFEI ASC). Bay Area Aquatic Resource Inventory (BAARI) Version 2.1 GIS Data. Accessed 2022. http://www.sfei.org/data/baari-version-21-gis-data
(right)
B. F. Atwater, S. G. Conard, J. N. Dowden, C. W. Hedel, R. L. MacDonald, W. Savage. History, Landforms, and Vegetation of the Estuary's Tidal Marshes. In San Francisco Bay: The Urbanized Estuary, Investigations into the Natural

History of San Francisco Bay and Delta with Reference to the Influence of Man. (347-386). California Academy of Sciences, 1979.
https://www.waterboards.ca.gov/waterrights/water_issues/programs/bay_delta/deltaflow/docs/exhibits/dwr/dwr_exh3.pdf
Charles F. Hoffman, J.D. Whitney. Geological Survey of California. Composite: Map Of The Region Adjacent To The Bay Of San Francisco. (map). 1873. https://www.davidrumsey.com/luna/servlet/detail/RUMSEY~8~1~203546~3001695:Composite-Map-Of-The-Region-Adjace?sort=pub_list_no_initialsort%2Cpub_date%2Cpub_list_no%2Cseries_no&qvq=q:san%20francisco%20bay;sort:pub_list_no_initialsort%2Cpub_date%2Cpub_list_no%2Cseries_no;lc:RUMSEY~8~1&mi=64&trs=983

Spread 12
Brett Milligan, Dredge Research Collaborative. "Hamilton Wetlands." (Photograph), 2016.

Spread 13
Coast and Harbor Engineering, "4.2 Mineral Resources," in Sand Mining Resource Evaluation and Impact Analysis, San Francisco Bay and Delta Sand Mining Final Environmental Impact Report. Appendix G: Bathymetric and Hydrodynamic Study. Sept. 2012.
Elizabeth Christian. "Adoption of Reissued Waste Discharge Requirements and Clean Water Act Section 401 Certification for 2015-2019 Maintenance Dredging Program," Staff Summary Report 8, State of California Regional Water Quality Control Board, San Francisco Bay Region, May 13, 2015.
United States Environmental Protection Agency. Dredging and Sediment Management, Last updated Feb. 23, 2016, https://www3.epa.gov/region9/water/dredging/sfdods/table1.html
San Francisco Bay Area Wetlands Ecosystem Goals Project, California State Coastal Conservancy. Goals Project, The Baylands and Climate Change: What We Can Do. Baylands Ecosystem Habitat Goals Science Update 2015. Oakland, CA. 2015.
Dredged Material Management Office. San Francisco Bay Dredging and Disposal Database. http://www.dmmosfbay.org/.
US Army Corps of Engineers, US Environmental Protection Agency, San Francisco Bay Conservation and Development Commission, and San Francisco Bay Regional Water Quality Control Board, Long-term Management Strategy for the Placement of Dredged Material in the San Francisco Bay Region, 2001.

Spread 14
San Francisco Bay Conservation and Development Commission, San Francisco Bay Regional Water Quality Control Board, US Army Corps of Engineers, US Environmental Protection Agency. Long-term Management Strategy for the Placement of Dredged Material in the San Francisco Bay Region. Background Information for September 11th, 2012 meeting. Focus: Costs and Contracting. August 2012. https://www.spn.usace.army.mil/Portals/68/docs/Dredging/LMTS/Final%20Review/Appendix%20C.pdf
Brett Milligan, Dredge Research Collaborative. "Sediment Off-loader at Montezuma Wetlands." (Photograph), 2016.

Spread 15
Patrick L. Barnard, David H. Schoellhamer, Bruce E. Jaffe, and Lester J. McKee. "Sediment Transport in the San Francisco Bay Coastal System: An Overview." Marine Geology 345 (2013): 3–17. https://doi.org/10.1016/j.margeo.2013.04.005
Minear, J. T., and G. M. Kondolf. "Estimated Reservoir Sediment: (2009, Estimating reservoir sedimentation rates at large spatial and temporal scales: A case study of California"), *Water Resources Research* 45, W12502, doi:10.1029/2007WR006703. https://ca.water.usgs.gov/pubs/2009/MinearKondolf.pdf.

Spread 16
Brett Milligan. "Outfall of the Pinole Creek Flood Control Channel into the Bay." (Photograph), 2020.

Brenda Goeden: Regulating Sediment

Brenda Goeden became Sediment Program Manager for BCDC in 2008. Her focus at work, as described in her title, is all things sediment, particularly the beneficial reuse of dredged sediment in an effort to manage a healthy sediment system. Brenda has worked with navigation dredging through the Long Term Management Strategy program, and she has provided input to restoration projects around the Bay, including Hamilton, Bair Island, Cullinan Ranch, Montezuma, South Bay Salt Ponds, and Lower Novato Creek.

DRC: What is the Bay Conservation and Development Commission?

BG: BCDC is a state agency that plans for and regulates activities in and around San Francisco Bay.

While we are a state agency, our jurisdiction is only within the Bay and 100 feet of the shoreline. We are one of the oldest conservation agencies in the state, being formed in 1965 to stop haphazard filling and polluting of the Bay. Interestingly, we also have federal authority granted to us by NOAA through the Coastal Zone Management Act. This gives the ability to regulate federal projects under the Coastal Zone Management Program for the Bay. We were the first Coastal Zone Management agency in the nation in 1972. Our planning group develops regional policies and is working on sea level rise adaptation for the Bay; our regulatory group issues permits for activities in the Bay and along its shoreline. Our mandate is both conservation of resources and appropriate development.

DRC: What is your role with BCDC?

BG: I am the Sediment Program Manager, working with a small internal team, but also a larger external program, the Long Term Management Strategy for the Placement of Dredged Sediment in the Bay Region (LTMS), which focuses on maximizing beneficial reuse of dredged sediment. Our internal team works on all things mud and sand—including dredging, habitat restoration, flood protection, and aggregate mining.

We have both planning and regulatory functions.

DRC: What has BCDC's role been in moving toward beneficial reuse? Can regulation change fast enough to address sea level rise and expanding sediment need in the Bay?

BG: That's a fairly complex question. BCDC, along with our LTMS partners, the EPA, the Water Board, and the Corps, have been working with the navigation dredging community since the 1980's when "dredged lock" occurred, to increase the amount of sediment being beneficially reused. The whole program is about reducing in bay disposal and increasing beneficial reuse of the dredged sediment—basically treating it like the valuable natural resource that it is. Consider the LTMS Program a recycling program for sediment —and we've had some great successes— since the program began implementation in 2000 over 29 million cubic yards have been beneficially reused, primarily in landscape-scale wetland restoration projects. BCDC has been championing beneficially reusing sediment now for over 30 years, and we have become even more active on this front in the last several years due to some of the more recent challenges.

In the past 5 or so years, we've had a significant challenge with the Corps beneficially reusing its sediment—which makes up the lion's share of the sediment dredged annually from the Bay. This is primarily due to the Corps' most recent interpretation of a policy they call the "Federal Standard," which means the least costly dredged material disposal alternative or alternatives that both are consistent with sound engineering practices and meet the environmental standards established by the 404(b)(1) evaluation process or ocean dumping criteria. So the Corps views environmentally acceptable as legally being able to be done. Because they only focus on the navigation dredging, there is no consideration for additional benefits that can be gained by reusing the sediment. So, if it's cheaper to dispose of the sediment in the Bay, and it is, or the ocean, 55 miles out to sea, even by $1.00 per cubic yard, that's what they will do. This wasn't always the case, as the Federal Standard has been in place for a long time, but more recently the Corps has stepped away from the LTMS goals for its own projects in favor of cheaper disposal, despite the societal cost of doing so.

And yeah, societal cost—you ask about regulatory change and addressing sea level rise in San Francisco Bay—the EPA, the Water Board, and BCDC are aligned in maximizing beneficial reuse, and most specifically for tidal marsh restoration, because we recognize that tidal marshes are some of our best defenses for rising seas, and without sediment to raise elevations, the tidal marshes can't be restored fast enough to keep up with the predicted sea levels. So bottom line, the Corps is choosing cheaper disposal at the cost of loss of ability to adapt to sea level rise, which, by the way, it will end up addressing through their flood risk reduction mission—which undoubtedly will cost more and be more damaging over time. It makes little sense to have one policy for the whole country that does not consider the bigger picture or where federal dollars overall could be saved by taking a holistic approach to state or regional issues.

DRC: If you could revise the Corps' Federal Standard to help with beneficial reuse and climate adaptation, what would you change in that regulation?

BG: This is one of those areas where I wish I had a magic wand—I think the simple answer could be to change the evaluation language.

I think if we could, we'd change the phrase at the end from "meeting the environmental standards established by the 404(b)(1) evaluation process or ocean dumping criteria" to something like, "after an evaluation of the regional needs for sediment, selection, and use of the most beneficial placement option."

I think eliminating the language around disposal is important because, at least in the Bay Area, we are in agreement that we have a very significant sediment need, making sediment one of our most valuable resources. I think inclusion of regional needs evaluation is also important to help the Corps step away from the one size fits all comparison when the regional needs across the country are very different, though all coastal areas are now in the unhappy position of being threatened by rising seas. One issue with this suggested language is the evaluation—the Corps can take years to evaluate something that is fairly well established due to other bureaucratic requirements. So, one could try to work in the using the best available science rather than conducting additional evaluations. Of course the funding is the unspoken giant in the Federal Standard, as the Corps is trying to limit costs, which is understandable, but this type of change could at least eliminate some of the policy challenges. Funding is an appropriations issue that we will always have, but that's a different issue.

DRC: Should policy and regulation play a role in determining the best use of sediment as a public resource along the Bay edge?

BG: A big part of my job is working with regulations and policy—and yes, I do think both have a role. The LTMS program is interesting because while there are regulatory triggers built into it, the program overall has taken a voluntary approach—which works for many of our constituents, but, then again, the Corps has not been meeting those goals for its own projects in many years.

I think that policy has a significant role because in developing policy we look to the best available science, consider regional needs, work to balance resource and development needs, and address the concerns of the times. LTMS addressed some of the environmental impacts of dredging, but was created at a time when the system was sediment rich and didn't really address the loss of sediment to ocean disposal. It gave a nod to [the Bay Area's] commitment to 100,000 acres of tidal wetland restoration by seeking to beneficially reuse sediment. Now we are in a system that is in short supply of sediment, so we need to up our game, not only for the tidal wetlands restoration, but also because of rising seas. We know that sediment is needed for both, and disposing of it aquatically is contributing to the loss.

And yes, a public resource—much like water—it seems to me that there is also a right to sediment because like water it flows from one area to another and makes up the very fabric of habitat and areas that we enjoy recreationally. If we don't manage it properly, then we won't have the wetlands or beaches. I think if we change our thinking a bit more and think about not only what we gain from beneficially reusing sediment, but also what we lose by not using it, and the impacts of that, we could develop stronger policies. Regulations flow from policies, so there is a role for them as well. You're also speaking to priorities—that is again a regional discussion —and a conversation that is both important and ongoing.

DRC: From a regulatory perspective, how should we prioritize where sediment gets placed and for what reasons?

BG: So, that is an interesting question, and very complex though it seems simple. From a scientific perspective, we can identify what areas need the most sediment, and we might be able to evaluate which wetlands might have a better chance of surviving rising seas, and we can also think about ecological connections, like building a network of wetlands and shoreline habitats that create stepping stones for species to migrate with salinity changes. But the harder part about prioritizing is the human element—there are areas that are regional assets, like major transportation networks, and perhaps expensive properties, but there are also communities and its really hard to put a value on communities. Historically, vulnerable communities have gotten the short end of the resources, and as a result you might have a community that is already vulnerable become more vulnerable with rising seas—but still might not have the resources to advocate for a priority spot. I am not sure that we can or should create priorities by regulation. I think a better way to go is to work with communities and find ways to support them.

The other issue is that restoration projects happen where the land becomes available. So, you see habitat conservation and restoration where there is still land available for that use, and it tends to be agricultural land that is more remote from communities so far, but perhaps that will change.

DRC: Should we be more focused on managing sediment more locally, at sub-embayment and watershed scales, rather than the Bay as a whole?

BG: Yes and no. If you look at a map of the Bay and where dredging occurs, you will see that dredging occurs in the Central Bay and a bit of the northern portion of the Central Bay, as well as the main stem of the federal navigation channels. There is very little dredging in the South Bay or San Pablo Bay, a bit more in Carquinez due to the refineries there. There are also flood protection channels that could be considered a source of sediment as well. So, you can, and we at BCDC have, mapped out sediment management activities where local projects could collaborate with local sediment needs, but in many cases those restoration projects have been fairly small and may not be well enough funded to bring in sediment. The larger landscape-scale restoration projects that need lots of sediment create an economy of scale for the cost of the infrastructure needed to support offloading the sediment on to the site. These large projects create significant areas of habitat, as well as wave and flooding attenuation, but take several years to bring the sediment on to the site.

There are a mix of strategies that we could, as a region, employ to maximize this use of sediment as a resource. I don't think there is one solution.

Phoenix Love Armenta: Stories, Fun, and Climate Justice

Phoenix Love Armenta is an artist, activist, and environmental justice organizer living in Oakland, CA. They have worked on environmental, racial, and social justice campaigns around the world, including anti-nuclear organizing in Japan and India, homeless advocacy in San Francisco, and multiple civic engagement and political campaigns throughout the Bay Area.

DRC: You've been involved in climate change issues and sea level rise adaptation issues for a while now. Are there processes and methods that you've participated in that seem inspiring for the future?

PA: The Oakland 2030 Equitable Climate Action Plan. Out of any of the processes that I've been a part of, that has been one of the most expansive, trying to get the most number of voices included, and trying to incorporate the ideas and thoughts of so many different people and encapsulating it into one document. It had its rough patches too, but I think they invested more in community than is normal. They hired Marybelle Tobias from Environmental Justice Solutions and Colin Miller from the Oakland Climate Action Committee to be consultants and help design their community outreach strategy. And those two decided that what they were going to do was hire a group of residents from each of the different areas that they were targeting and have them do outreach in their community. The meetings were really community oriented, and they would have music sometimes, they would recognize the land, they would bring food out for the community so that folks would feel comfortable. They had childcare if we needed it. In terms of community outreach, that's the most thorough that I've seen.

DRC: If you could change how community engagement is happening in the Bay, what would you change so that it's more effective?

PA: I would just create a lot of jobs for community organizers.

Well-paid, well-funded, "all of the fixings" jobs, where people are paid to learn about their government and go into the community and teach people about how government works and get them engaged. I've been through so many different communities. I've talked to folks who were unhoused; some of the most political people that I've ever been involved with have been unhoused folks. Once they know that they can effect change and they know that the system is being so crappy to them, people will motivate and hustle.

The people who are the most disenfranchised have the most motivation to be involved. But there's not a lot of fun and easy ways to get involved. I would try to figure out how to make it easy for people to get involved. And fun.

DRC: Tell us more about fun.

PA: I'm a nerdy person for whom government is fun, and I don't even need all the bells and whistles... but [community outreach] should be more exciting. You're talking about how to build your city, and that's fundamentally what local government is about. We have this area around us and we have to decide what goes where, and how does this happen, and how are we addressing this problem. To me it's like a gigantic video game, like an in-person video game to be a part of that decision-making process. It's like, "well I think if you do x,y, and z the machine is going to change this way and it's going to be positive." And when you're part of these community engagement processes you can be part of moving that machine, but it has to be done in a way so that you feel like what you're doing has impact. I'm hoping that that happened with the Equitable Climate Action Plan.

There's a story to all of it. I've found time and again that if you can tell the story in a certain way, people get really interested and engaged in it. Young people. I have a group of fourth graders... they can talk about climate change so well. And I don't want to take any credit for that; whatever you tell them they just soak it up. And we've been talking a lot about climate change, so they know a lot about climate change. One of them was just recently shown talking about it on PBS News Hour. They did part of our class and he was able to interview about that kind of stuff. People are interested in their environment and what's happening around them. They just don't always get easy access to understanding how everything works.

DRC: How do you tell climate change stories to different community audiences to have them resonate?

PA: The funny thing is that I actually don't often talk about climate change. Because it is such a controversial thing that I decided a while ago that it doesn't matter if CO2 is actually causing it. That doesn't matter. What matters is that the air quality is poor and it's causing us to not be able to breathe. And water is rising, and it's going to be flooding our communities, and there are toxic sites out there. Our environment is polluted, and we can visually see that it's polluted. And even if we don't care about global warming, we care about the soot that comes out of the cars that causes the asthma for the young people in the community.

I have a workshop that I do that we've labeled "eco-mapping." It takes a series of government maps and tells a story about the pollution in the community, but also the racial disparity that comes along with it. We start with the redlining maps, and you can see it's all of the communities along the shoreline and Oakland, where the communities are that were redlined. They don't have it now, but it used to be that you could go and see how they chose to make it that way, because it would say something like "deficit," that "this has 'Orientals' and 'Negroes' in the community and that's why we've redlined it." It very specifically shows that it was racial discrimination that put people into these communities and these neighborhoods and kept them economically disenfranchised and not able to buy their homes. And then you take that and overlay it with the CalEnviroScreen, which shows environmental burden. And you can see that the communities that have the most environmental burden perfectly line up with those redlining maps. The communities that were allowed to buy their homes and work on their communities are, like, some of the least polluted communities in California. So you have that juxtaposition of the least polluted next to some of the most polluted. And it's hard not to see that this is a history of racial inequalities that have been played out through public policy, through business...

DRC: You advised on Resilient by Design, which tried to address some of these issues, and had a robust outreach effort. But what would be your overall assessment of that? What worked and what didn't?

PA: It was a really grandiose effort. A lot of investment, a lot of outreach. The start of the process was not [great]. People of color, people from disadvantaged communities weren't really there from the beginning, and so we were brought in further in the process at a time when a lot of decisions [had already been] made that we didn't necessarily agree with. And so there was a lot of tension. A lot of time was spent just trying to convince community members that something attached to the Rockefeller name was something they could trust and want to be a part of. There were some things that I thought were negative and some things that I thought were positive. I really felt like the investment in the People's Plan, Marin's People's Plan, was a great step in the process. The East Oakland Neighborhood Initiative —and I'm not saying that the other places didn't have similar success rates, but those are the ones that I'm most associated with — really worked to get community members from disadvantaged communities involved in the process. And I think to great success. I know EONI has brought in millions in planning dollars based on the plans that [RBD] created. And I've heard that the People's Plan is starting to get their projects funded as well.

A lot of that was the inspiration for the Oakland Shoreline Leadership Academy, which I was leading. It showed us how we could have community-led planning if we actually paid people to be a part of the process. I mean...it's a scale called "the principles of community engagement." Have you seen this before? It's a scale that starts from marginalization and takes you to community empowerment as outreach. And, for the most part, right now the Bay Area is stuck in tokenization. Where, you know, it's just like a lot of people who have power coming into the community, taking surveys, hiring people on to be consultants, but it's only one person from the community who doesn't necessarily represent the entire community. And these projects, like EONI or the People's Plan or the Oakland Shoreline Leadership Academy, are an attempt to get us to that community empowerment aspect of it. I don't know exactly what the outcome's going to be or if it's going to really have an impact. But I'm hoping.

We just had our final presentation for the Oakland Shoreline Leadership Academy. It felt really good to see so many people from the community who really were interested and engaged in this material. There always seems to be this feeling that people aren't going to really care about this, but one of the participants said that it was, like, hidden knowledge, you know, that they were getting. And they don't really have a way of knowing or understanding how to access this kind of knowledge. We're just hoping to be that funnel to get that information out into the community. And I guess it all started with Resilient by Design, so in that way I think it was a success.

DRC: One of the tensions we've been dealing with is the gap between the enormous, overwhelming scale of these systemic issues that we face today and the relatively small, halting efforts that have been made to address them. A marsh restoration project seems so small compared to the global climate crisis. In a 2018 essay for the California Adaptation Forum, you suggest that making a transition from resilience to regeneration can help us hold on to hope. Can you explain why you think that is necessary, and how that might happen?

PA: Yeah...we live in a world that's dominated by linear thinkers, you know. We have this extract-use-throwaway model. And the world is not created that way. Like, the way that most systems work in the world are circulative systems, regenerative systems. I think it's really important that people can visualize and see the pathway towards living in a world where we don't have to have every one of our decisions impact the Earth in a negative way.

Part of what I think I'd like to do with my career, what I'm trying to do, is to get people to envision that beautiful world so that we can start making it, so that we can get there.

I think that a lot more people would be interested in being in design if they understand what it really was, and the pathways to getting [to a regenerative world].

Development along Alameda Creek Flood Control Channel
(Image courtesy of Washington Township Museum of Local History, Fremont, CA, 1971)

THE PUBLIC SEDIMENT TEAM

SCAPE Landscape Architecture
Dredge Research Collaborative
Arcadis
UC Davis Dept. of Human Ecology and Design
TS Studio
Architectural Ecologies Lab
Cy Keener + Justine Holzman

WITH KEY STAKEHOLDERS

Alameda County Flood Control and Water Conservation District
South Bay Salt Pond Restoration Project
East Bay Regional Park District
State Coastal Conservancy
Alameda Creek Alliance
San Francisco Estuary Institute

Why Design?

Public Sediment for Alameda Creek

In 2017 and 2018, we—Brett, Gena, and Rob—joined a large team of colleagues, led by SCAPE Landscape Architecture, to develop a proposal for sediment design in the San Francisco Bay Area. Our team's proposal, Public Sediment for Alameda Creek, was one of nine generated for a design competition known as the Resilient by Design Bay Area Challenge. Other members of the Dredge Research Collaborative who participated in Public Sediment include Justine Holzman, who was both a core part of the design team and co-designer (with Cy Keener) of its sensing strategies, and Brian Davis and Sean Burkholder, who contributed through a physical modeling workshop. Here we report on this work and describe the design project. We aim to show how Public Sediment exemplifies the principles laid out in earlier chapters of this book. While we had many collaborators, this interpretation is, for better or worse, our own.

At the Mouth of Alameda Creek

The channelized creek continues all the way to San Francisco Bay, cutting between conspicuous polygons of open water, which are contained within a network of earthen levees.

These former salt production ponds, known as Eden Landing, are slowly sinking further and further below sea level, due to a lack of sediment supply. A federally approved plan for Eden Landing aims to reclaim the tidal marshes, which were themselves once reclaimed for industrial salt production. But it will be no small endeavor. For marshes to take hold again, the bottom of the ponds will need to be brought back up to the current tidal elevation. This will require over 6 million cubic yards of earth—much more sediment than is dredged from all of the bay navigation channels in most years.

Standing on the crest of the flood channel as it passes through the former slat ponds is an uncanny and disorienting experience. You are on a massive earthen infrastructure, which stretches to the horizon. The space between these levees is mostly filled in with sediment and covered by wetlands. Outside the levees lies a sea of industrial ruin. This portion of the channel no longer protects much of anything.

Although Alameda Creek is one of the bay's biggest and muddiest tributaries, it can't contribute any sediment to Eden Landing, because the levees get in the way. Instead, the channel dumps what sediment it can carry directly into the bay. Much of its watershed's upland sediment never even makes it that far, as it gets trapped along the length of the channel, where it must be removed by costly maintenance dredging.

What's more, the ecological costs of severing the connections between freshwater creek, floodplains, and salty Baylands were not factored into the channel design. This has caused severe habitat loss and endangered many aquatic species, like the steelhead trout that once swam Alameda Creek in great numbers. The creek's floodplains are blanketed in suburban development that relies on the false promise of forever flood control. This engineered vulnerability is a common inheritance for many creeks and communities along the bay. And as global warming increases the intensity of storm events, raises sea levels, and pushes tides further up the channels, the ability to control floods is increasingly compromised. Changing this trajectory will be possible only if the Bay Area learns to design with sediment.

Fluvial reach and upstream sediment accumulation in Alameda Creek flood control channel (Brett Milligan / Public Sediment Team)

Alameda Creek Flood Control Channel at the edge of San Francisco Bay (Brett Milligan / Public Sediment Team)

Former and Future Baylands

The Baylands today are very different from what they were before colonial settlement, and very different from what they will be in the future.

An extreme sediment shortfall threatens the littoral zone that rings the bay. If this landscape is lost, many species will be displaced, and shoreline communities will lose the protection of marshes that attenuate the impact of storms, flooding, and erosion. They will also lose a crucial opportunity to adapt to a changing climate. Unlike levees and other hard infrastructures, marshes have the capacity to accrete and grow, gaining elevation and mass over time, even in the face of sea-level rise. If the Baylands disappear, the region faces substantially exacerbated risk. Cities built on subsiding land could become dependent on ever-larger levees to protect them from ever-increasing risk of inundation.

To effectively maintain and rebuild the Baylands will require as much as 200 million cubic yards of sediment over the next 15 years.[1] This need vastly exceeds the amount of sediment flowing into the Bay-Delta, and that's before factoring in climate change and sea-level rise. Projecting sediment supply and demand out to the year 2100, scientists estimate that the supply for existing marshes and planned restoration will be less than half of the total need, and only a fraction of that potential supply will actually reach marshes. So the Baylands will drown unless action is taken soon. Such action will require new approaches to moving sediment and new levels of cooperation between citizens, regulatory agencies, and Bay Area governments.[2]

Yet sediment design remains largely invisible to the bay's publics, as the routine work of moving sediment is undertaken out of sight, by siloed agencies and private industry. The bay is dredged to maintain shipping channels, and its engineered tributaries are dredged to maintain flood capacity. Some of this material is beneficially reused for restoration projects within the watershed. But a lot of it is *discarded* in the Pacific Ocean, where it is lost indefinitely. Avoiding this waste is an urgent challenge.

Where and how the bay gets sediment has radically changed. Before colonization, as much as 90 percent of the supply came from the massive Sacramento-San Joaquin sedimentshed, through the delta. Today, less than half of the bay's sediment comes that way, due to extensive networks of dams and reservoirs that hold back sediment. The bay receives much less sediment overall, and the majority now flows from small local tributaries like Alameda Creek.[3] Most of these tributaries, though, have been channelized and lie within highly urbanized watersheds. Upland sediment supply also needs to be mobilized for the sake of the Baylands.

Conceptual mapping of Bayland transformation with sea level rise and low sediment supply (below)

- Diked Baylands: varying land use
- Diked Baylands: ponds and managed ponds
- Tidal Baylands
- Restoration project
- Mud flat
- Subtidal
- Impacted urban areas

Conceptual sections of Bayland transformation with sea level rise and low sediment supply

Baylands today

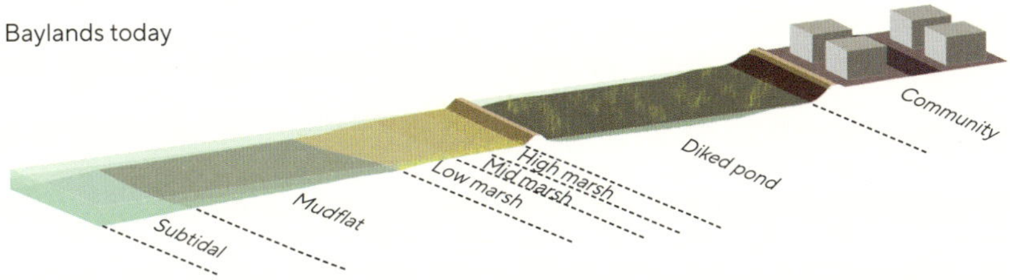

2100 with 3 ft of SLR and low sediment supply
Marsh converts to mudflat

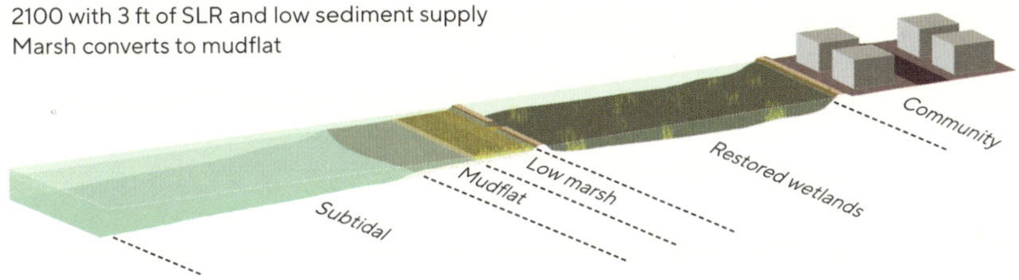

2100 with 7 ft of SLR and low sediment supply
Marsh drowning and urban inundaton

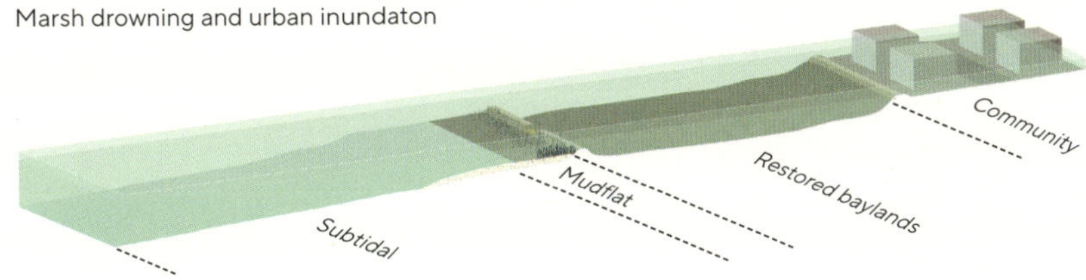

Bayland sediment supply and demand
2100, 3.5 ft SLR

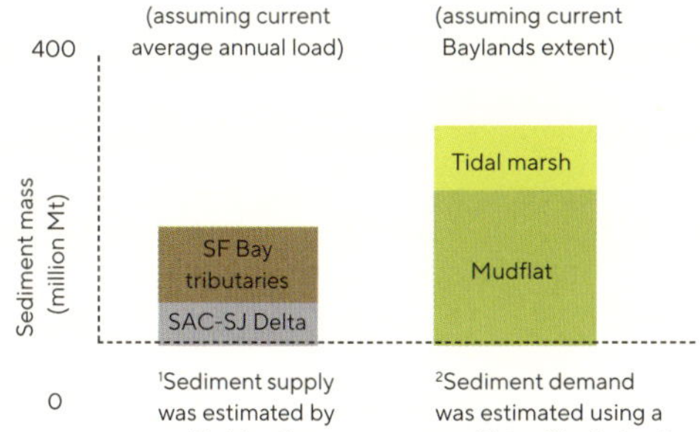

[1]Sediment supply was estimated by multiplying the current average annual sediment load values from McKee et al. by the number of years between 2017 and 2100.

[2]Sediment demand was estimated using a mudflat soil bulk density of 1.5 g sedintment/cm^3 soil (Brew and Williams 2010), and baywide mudflat and marsh area circa 2009 (BAARI v1).

Based on preliminary analysis by SFEI.
A more detailed analysis with different sediment and precipitation scenarios is available from the Healthy Watersheds Resilient Baylands project. (hwrb.sfei.org)

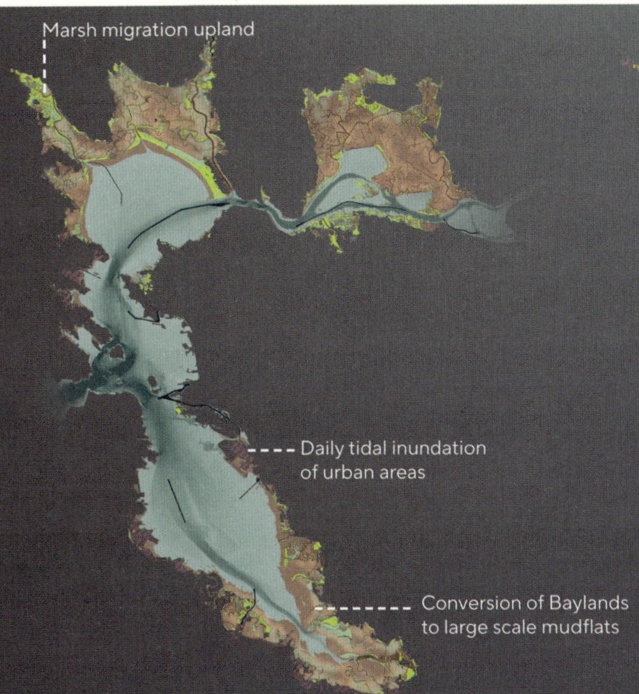

Bayland migration
3 ft of SLR and low sediment supply

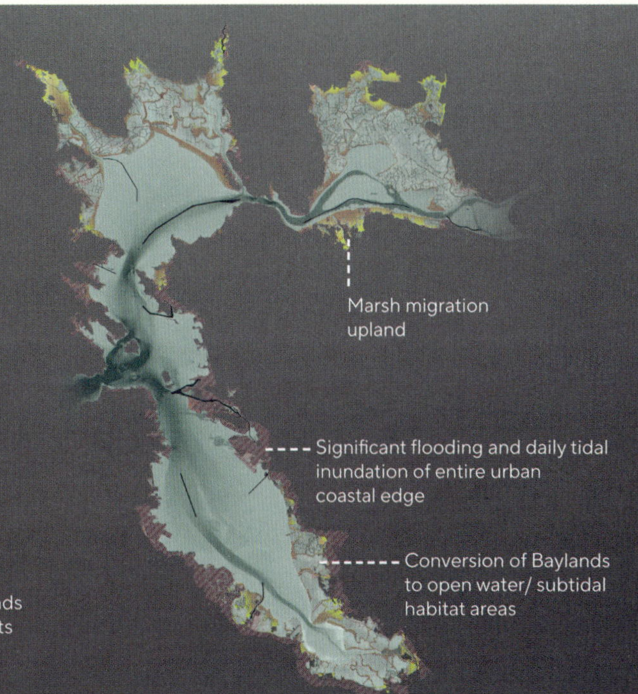

Bayland drowning
7 ft of SLR and low sediment supply

(SCAPE / Public Sediment team)

Public Sediment

Public Sediment is a design scheme that spans from sediment source to sediment sink.

Unlike designs for coastal adaptation that focus exclusively on the edge where land meets water, Public Sediment considers land, sediment, and water in a fluid, dynamic relationship across scales, from an individual creek to the Bay-Delta watershed, and across terrains, from the open waters of the bay to the uplands.

Now is a pivotal time for landscape design in vulnerable urbanized estuaries, deltas, and coastlines around the world. We are likely living in the final decades of a stable climate, and how we design for change now will dictate the living conditions that are possible in the future. If the static, extractive infrastructures of the past century are perpetuated into the next, the result will be technological lock-in, unsustainable path dependency, and inevitable failure. For example, the densely urbanized South Bay region of San Jose is on track to become the next New Orleans, encircled by levees and kept dry through a life-support system of water pumps. This is not real adaptation.[4]

Public Sediment embraces landscape change as the accelerating constant of our times, and with it the need to design living, equitable infrastructure. Design should not aspire to maintain a status quo of inequitable socio-technical, political, and institutional legacies. Rather, radical change is needed to create more inclusive worlds. Infrastructure can engender landscapes and ecosystems that sustain and protect human lives and communities in reciprocal fashion.

Public Sediment spans policies, machines, and landscapes at the interconnected scales of creek, sedimentshed, and region, across three corresponding scales of proposal: Unlock Alameda Creek, Rethink the Sedimentshed, and Plan and Pilot for a Future bay.

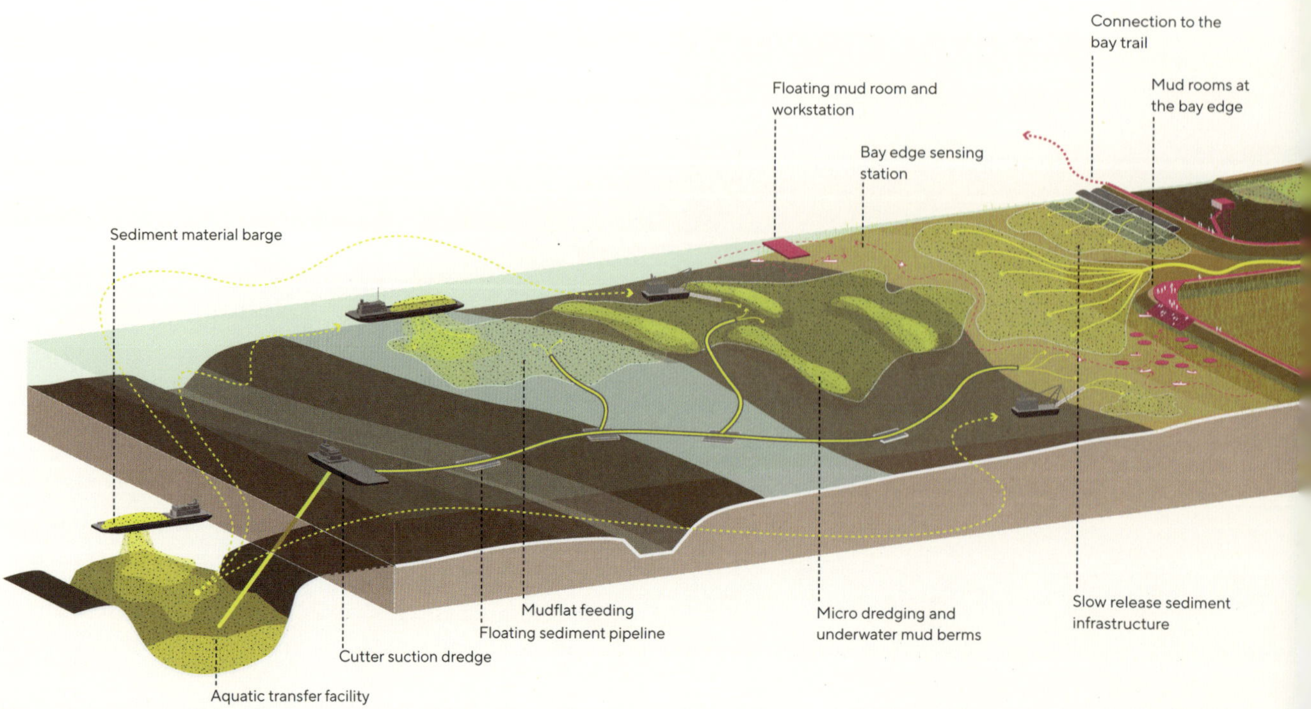

In keeping with the approaches we have described in earlier chapters of this book, Public Sediment choreographs landscape activities, working across multiple space-time scales and embracing indeterminacy. It applies a socio-ecological conception of economy by *internalizing* socio-ecological relationships and repairing damaged landscapes. It highlights the importance of aesthetic experience, public access, environmental equity, and intergenerational justice through co-design methods. And lastly, Public Sediment approaches technology as a means of active learning and landscape adaptation; this requires humility, observation, and attunement to the inevitable gaps that emerge between design intention and what becomes.

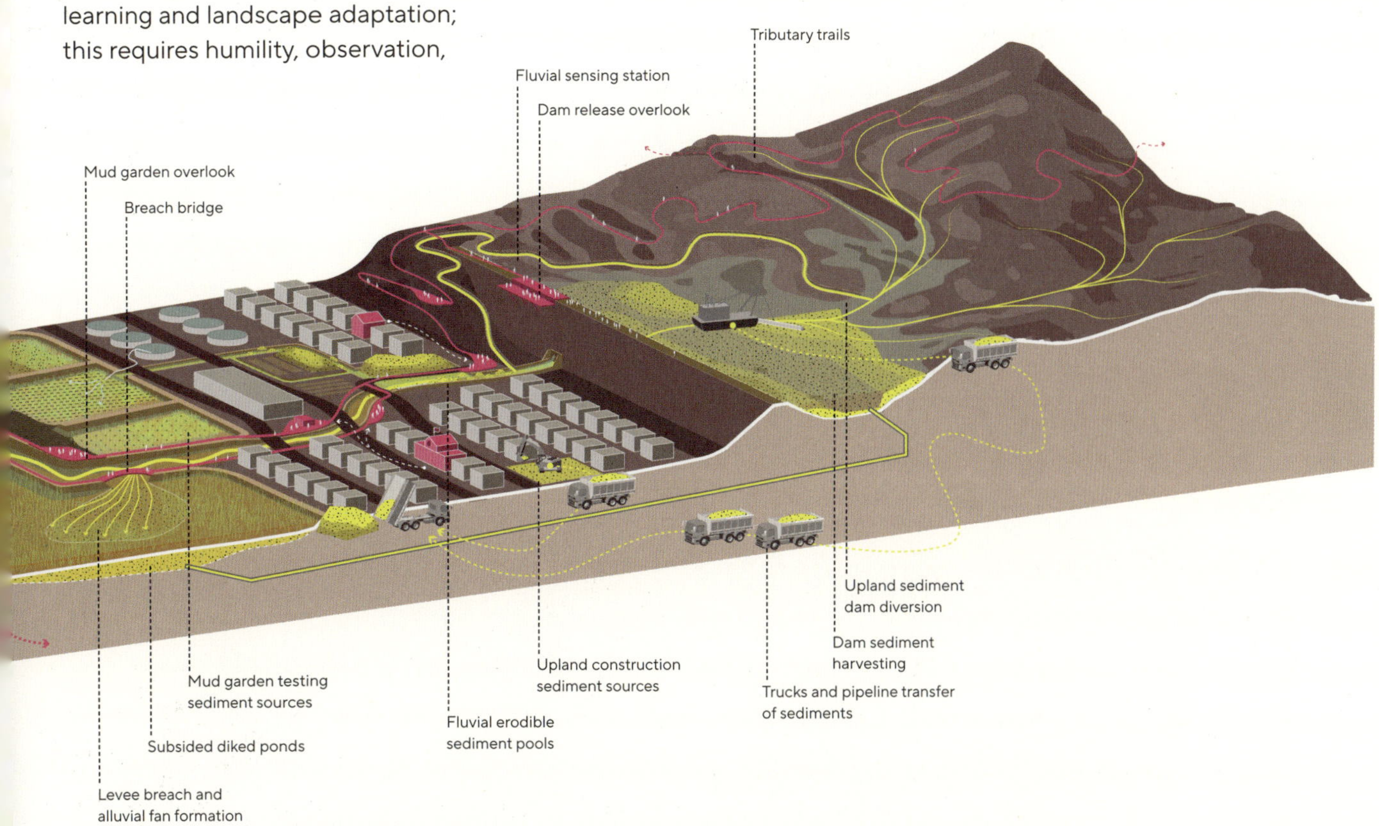

Conceptual transect of public sediment strategies (SCAPE / Public Sediment team)

The Bay's Largest Sedimentshed

Public Sediment focuses on Alameda Creek, the bay tributary with the largest watershed and the most untapped potential to deliver significant amounts of sediment to the bay. Today, the flood control channel of Alameda Creek traps sediment within its levees, is disconnected from adjacent Baylands, and is deprived of sediment by several large upstream dams. If the regenerative potential of Alameda Creek can be unlocked—for the sediment, the people, and the other species that share the creek—so can the potential of many other bay tributaries.

Alameda Creek sedimentshed (SCAPE / Public Sediment team)

Creek and Cone

Before the construction of the flood control channel, Alameda Creek migrated freely and wildly across a lowland floodplain, forming a broad alluvial fan called the Niles Cone.

During high winter and spring floods, the creek exited the confines of the East Bay Hills and deposited large quantities of sediment across the plain, from the western flank of the hills down to the bay. Over and over again, sediment built up along the creek's path, and its waters slowed, until the creek found a new, steeper path to the bay. As it switched courses, the creek began depositing sediments along a new path. Over time, this combination of sedimentation and course-switching created the cone. Heavier boulders and gravels deposited primarily at the top of the cone, as only the fastest floodwaters could carry them, while finer sediments were emplaced near the bay edge. This geomorphic action created the earthen foundation for the complexity and diversity of Alameda Creek—its varying depths, channel forms, and sinuosities, which ranged from meanders to thick braiding.

The creek, in turn, shaped lowlands and Baylands.[5] Across the cone, the creek's migration built grasslands cut by intermittent streams, wet meadows with their reeds and goldfields, and a dimpled topography of vernal pools and alkali playas. Some tributaries flowed year-round, while others were intermittent, flowing only in the wet season. Where the streams had room to spread out widely, they filled seasonal wetlands, which provided abundant forage for fish like the anadromous steelhead trout, which spawns and begins its life-cycle in freshwater before migrating to the ocean. The cone's waters wound between riparian forests, patchy stands of trees, and dense thickets; as they reached the Baylands, they snaked as sloughs, saturated tidal marshes, filled ponds, and covered mudflats, all ever-changing. In addition to steelhead, the Niles Cone was rich in amphibians like the yellow-legged frog, reptiles like the horned lizard, and birds like the yellow-billed cuckoo.

When Europeans arrived in the Bay Area, there were many Ohlone tribes living in these landscapes. The Ohlone sustained generations of village life by harvesting and hunting along the creek, and they stewarded the landscapes of the Niles Cone through the calibrated use of fire, building shell mounds along the creek, and harvesting salt from tidal salt ponds. In 1797, the tribes' lands were colonized and appropriated as the fourteenth Spanish Mission San José. Later, as California's immigrant population expanded during the nineteenth-century Gold Rush era, the cone's high soil fertility and close access to San Francisco markets made it a center of intensive ranching and agriculture. Sloughs, willow thickets, and vernal pools gave way to farms and grazing cattle.[6]

Still, rural development and town building on the Niles Cone was persistently jeopardized by floodwaters that overflowed the creek's banks and spilled across the plain. The flood of 1876 was so severe that the *Oakland Tribune* described the area as "one vast inland sea."[7] Catastrophic flooding didn't happen every year, but annual overflow was part of the regular pattern of Alameda Creek, and expanded development in the twentieth century inadvertently increased flooding.

In 1955, three days of heavy rain produced another huge flood that submerged homes and farms. This was the impetus to build a flood control channel that could contain the creek. Soon afterwards, the USACE conducted preliminary studies and surveys, and the Flood Control Act of 1962 authorized construction.[8] Designed for a standard project flood capacity of 52,000 cubic feet per second, the Alameda Creek Flood Control Channel would be twelve miles long, trapezoidal in section, and bound by levees lined with riprap on both banks. When it was completed in 1974, the creek was fixed in place.

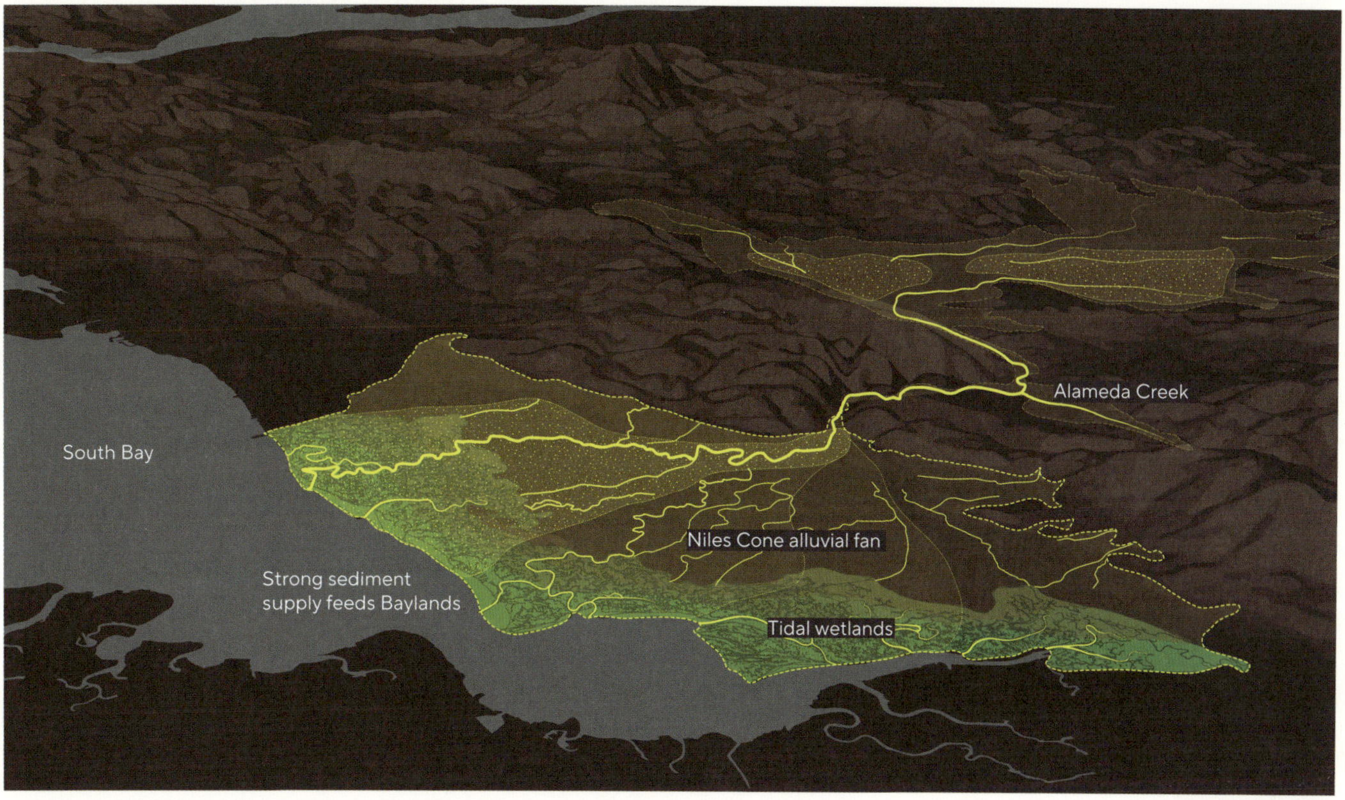

Alameda Creek watershed pre-urbanization (SCAPE / Public Sediment team)

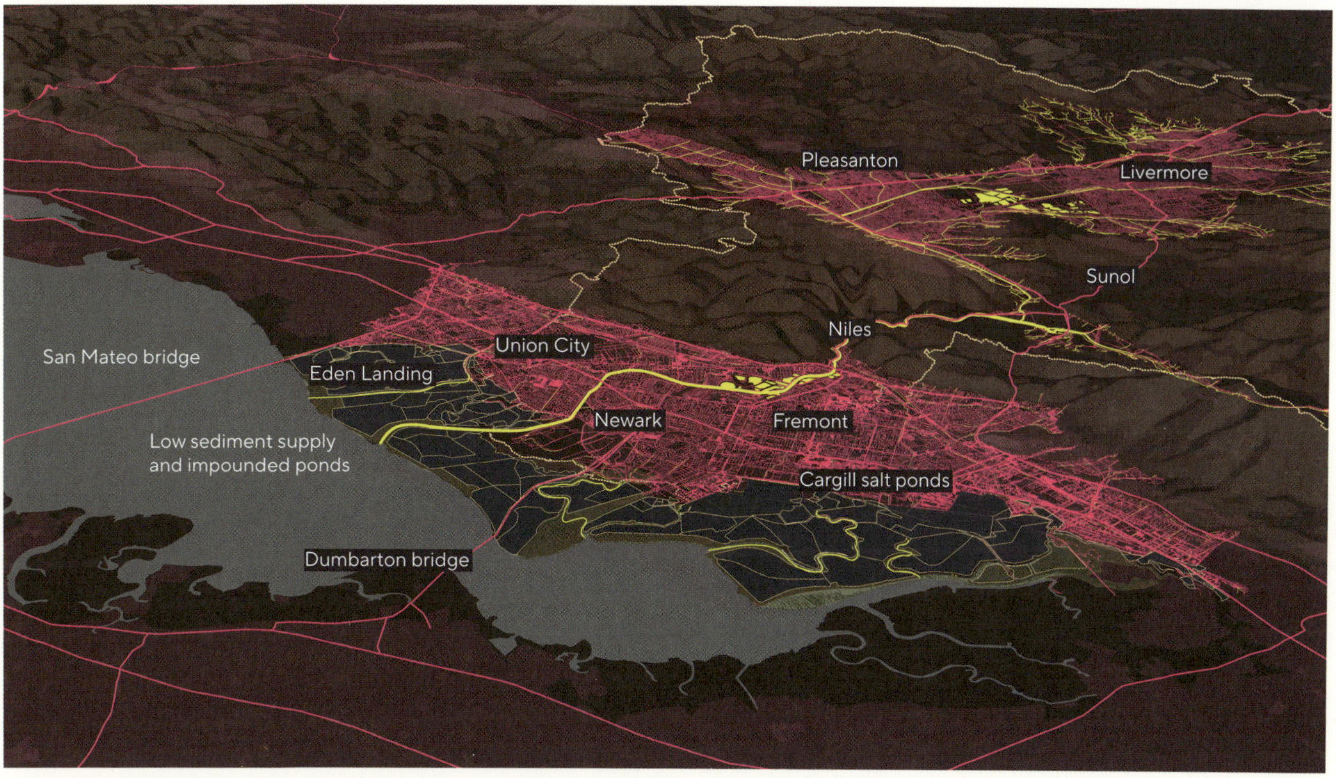

Alameda Creek watershed post-urbanization (SCAPE / Public Sediment team)

A Creek with One Purpose

The Alameda Flood Control Channel was designed to efficiently maximize flow capacity within a minimal footprint.

As with most flood engineering projects of the time, this was achieved by straightening, deepening, and widening the creek channel, while permanently confining it between levees to prevent flooding.[9] Flood control removed a great deal of land from the floodplain, facilitating the growth of a thick blanket of suburban development. With housing pushed right up to the edge of the flood channel, the infrastructure cannot be easily expanded to accommodate the stronger storms, more intense floods, and rising tides that accompany climate change. The channel is thus an example of technological lock-in, and it presents increasing risks over time.

The channel also severed the relationship between water conveyance and sediment transport. The USACE's flood modeling assumed a "clear water condition" that left out sediment. This is a remarkable exclusion, given that creeks and rivers do most of their sediment transport during floods. Unsurprisingly, the channel has been substantially impacted by sedimentation. Every year, more than half the sediment that enters the channel deposits within it. About a third of that is dredged out, in an effort to maintain flood conveyance capacity, but deposition continues, and so the channel does not meet the flood capacity it was designed for. It never really did.[10]

In large part, this is because the form of the channelized creek cannot effectively transport sediment downstream. In cross-section the channel is too shallow and even, and its length has been broken into a series of flattened steps by vertical barriers like sills and inflatable dams. Sediment that does make it all the way down the channel cannot reach the Baylands, because there is no connection to the diked reclamation ponds at Eden Landing. Single-mindedly optimizing the channel for flood control has in fact undermined flood protection and undone the once life-sustaining role of the creek as a transporter of sediment.

When Alameda Creek migrated freely across the Niles Cone, it was part of a rich tapestry of landforms and plant communities that supported animals and people.[11] Steelhead were at the center of this ecosystem, but the channelized creek is inhospitable to them. Since the channel has been smoothed and widened, the water is only a shallow scrim at low flow, rarely deep enough for fish passage. Maintenance crews regularly remove all the trees and shrubs that would provide cover from hunting birds. And at the edge of the Baylands, the transitional mosaic of ponds, sloughs, and vernal pools that helped juvenile trout mature into ocean-going steelhead is gone. People are no longer welcome here, either. A pair of recreational paths run atop the levees, but signs posted at intervals warn visitors that it is illegal to enter the channel. There are no trails that lead down inside, nor any places carved out that might welcome people to draw close to the creek, to touch its water, or step onto soft, recently deposited sand.

When the channel was designed, flood control was the only imperative. Other forms of value —social, sedimentary, ecological— were not considered. Now that exclusion is locked in. Unmaking the channel and regenerating Alameda Creek requires a different approach to design.

Sediment buildup in the creek bed (2018)

Limited public access (2018)

Urbanization and Alameda Creek (Brett Milligan / Public Sediment Team)

How To Design A Creek

Public Sediment's design process was based in fieldwork, synthesizing expertise, making sediment public, and finding forms that respond to key concerns.

Fieldwork: Through fieldwork we sought a direct understanding of the landscape.

The design team first set foot in the flood control channel in the late fall of 2017. There had not been much rain yet, and the creek was low; grasses, rushes, and forbs painted the channel in the yellows, browns, and ochres of dried seedheads and dead foliage. The space between the levee banks appeared featureless, aside from the clear distinction between dense vegetation and the reflective sheen of the tidal channel itself.

The next day it rained, and we returned to hike up the channel to the Dry Creek confluence, well above head-of-tide. Dried plants took on a different hue in the wetness, and greens were more prominent in the fully fluvial stretches of Alameda Creek. This was our first introduction to the creek as a fluctuating, dynamic entity. Even channelized, confined, and silted-in, the creek was lively, capable of surprising and delighting.

In the months that followed, the team came back often. For quick walks before morning meetings, and evening runs. With partners from the Alameda County Flood Control District, for a survey of plant communities with field biologists. To fly a drone, capturing images that could be turned into 3D models through photogrammetry. To walk and talk with residents who lived nearby. To sit by the channel and watch the sunset, slipping away through tall grass under moonlight. To install sediment sensors. The creek afforded so many varied and vibrant experiences; we were confident that a regenerated creek would offer more.

Synthesis: To synthesize expertise we had to be humble about our role as designers and eager to collaborate.

Many people, organizations, and public agencies have been working for decades to improve the design of Alameda Creek and other Bay Area channels. The first challenge we faced was comprehending the work that had already been done. This effort began months before the Public Sediment team decided to focus on Alameda Creek; in some sense, it began years before that, when the Dredge Research Collaborative hosted DredgeFest California. Much of it involved reading: journal articles, studies, reports. Our team also used mapping to understand spatial patterns within the data and relationships between disparate datasets.

Most fundamentally, this synthesis was social work: finding, meeting, and listening to partners at the Alameda County Flood Control District, the San Francisco Estuary Institute, the South Bay Salt Ponds Restoration Project, the San Francisco Bay Conservation and Development Commission, the Alameda Creek Alliance, East Bay Regional Parks, the Army Corps of Engineers, and more. Each partner brought unique expertise, informed by distinctive values, desires, and responsibilities.

Public: To make sediment public we had to become engaged landscape activists.

This meant fostering public support for the value of sediment and listening for how sediment design can respond to public needs and desires. Over the course of five months, the team engaged in fourteen public events. Sometimes we joined existing programs, like the City of Fremont's Earth Day celebration. Sometimes we organized our own. Cy Keener led a hands-on sensor workshop with the robotics club at nearby Kennedy High School. Co-design workshops were hosted at four local high schools as well as the Union City Teen Center. An intergenerational story share brought together youth and elders at the Ruggieri Senior Center. Our largest public event was the Creek Crawl, a walking tour of the upper reaches of the flood control channel. Events like these can build a foundation for the long-term stewardship of sediment as a public resource.

Each of these events was also an opportunity for the design team to listen. Common threads emerged. Most neighbors did not perceive the channel as "nature," nor did they regard it as accessible to them. Older residents, who remembered the creek before it was channelized, lamented the disappearance of the markers of a living landscape, like varied vegetation or abundant birds and fish. We often met people who lived within easy walking distance of the creek yet did not recognize its name or know how it once related to the larger living systems of the Baylands.

Forms: Then we shaped forms that responded to the challenges we had identified.

A path leads down from a levee to a creek, making near what was remote. A deeper channel concentrates water flow, mobilizing sediment toward the bay. Levees are set back at key points along a channel, providing space for floodwaters to expand and plants that require occasional inundation to take root. A room is carved out at the water's edge, signaling that this place welcomes visitors—and providing a chance, over the passage of seasons and years, to witness how sediment moves and what sediment does. Landscape form is not about static objects, but rather pattern and process, trajectory and transformation.[12]

Public Sediment's forms synthesize what the team learned from fieldwork, partners, and the public. Well before we began our work, the Alameda County Flood Control District was studying the design of a new "low-flow channel," and a modified version of this was included in our proposal. We also derived forms by drawing with a stream table, extrapolating from ecological restoration studies of Alameda Creek and other California streams, and studying pre-development landscapes, like the gravel beaches found on some south Bay shores. All of these forms were fed into an iterative process of refinement, testing, and discussion.

Vegetation Survey (SCAPE / Public Sediment Team, 2018)

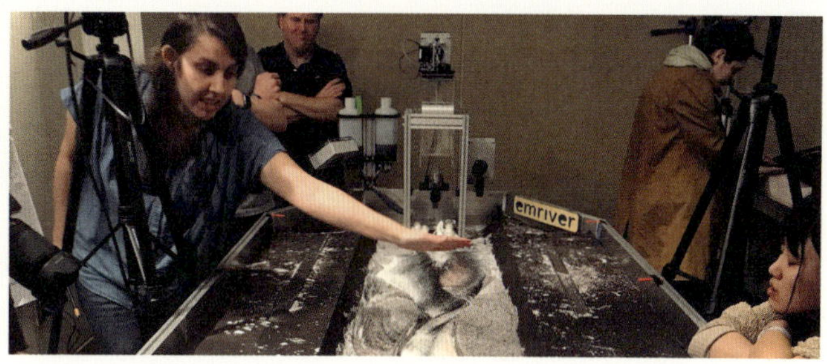

Creek physical modeling (SCAPE / Public Sediment Team, 2018)

Youth engagement on the creek (UC Davis / Public Sediment Team, 2018)

Stakeholder meeting (SCAPE / Public Sediment Team, 2018)

Unlock Alameda Creek

Public Sediment proposes remaking the Alameda Creek Flood Control Channel so that it mobilizes sediment to the bay, becomes a valued part of the public realm, and improves ecological function, particularly to support the migration of steelhead fish. A redesigned channel will directly link channel communities with sedimentary landscapes, feed the Baylands with desperately needed sediment, and contribute substantially to the ecological revitalization of the South Bay.

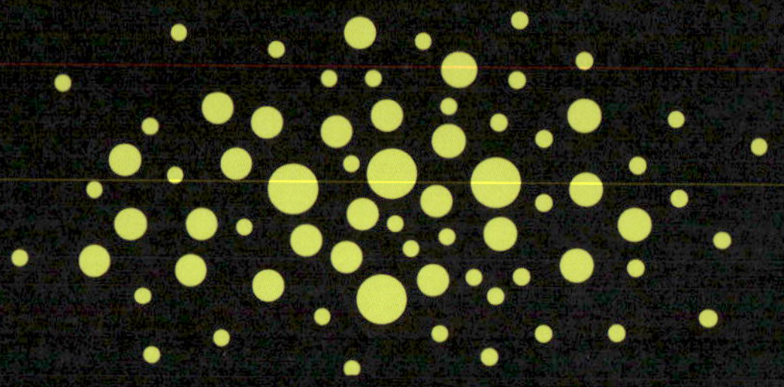

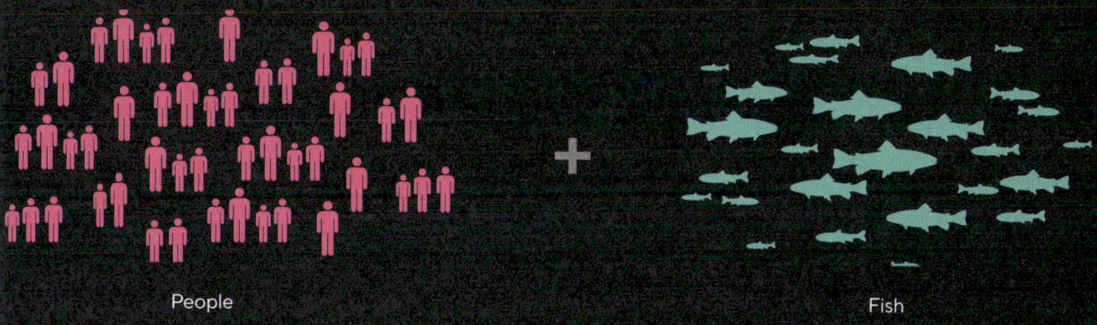

Public Sediment for Alameda Creek (SCAPE / Public Sediment Team)

The Active Channel

For Alameda Creek to again convey sediment to the Baylands, its channel must be remade.

This challenge is fundamentally a matter of landform. The current channel is too wide and too shallow. Without defined banks, floodwaters extend from toe of levee to toe of levee during the annual flood events that convey most sediment. As the floodwaters spread laterally, they lose velocity, encouraging sedimentation that traps sand and gravel.

Unobstructed and unmanaged, a stream like Alameda Creek would form its own banks through erosion and deposition, shaping a narrower, deeper channel that would carry more sediment to the bay. But within the engineered flood control channel, the creek cannot do this work. There are many obstructions along its length. Sills and inflatable dams break the channel with sudden vertical drops, greatly reducing the slope between breaks, which slows water and causes sediment to deposit. In several places, bridge piers within the channel force the creek to find its way between them. In the near term, all of these obstructions are locked in, as they serve the interconnected infrastructural purposes of groundwater salinity management (inflatable dams), flood control (sills), and regional transportation (bridge piers).

How vegetation is managed between the levees also prevents Alameda Creek from forming defined banks. Flood control management standards emphasize reducing friction within such channels, in order to rapidly convey floodwaters downstream. These standards permit only "lay down vegetation," or soft-stemmed plants like grasses and rushes that lie down during floods. Maintenance workers remove juvenile trees and shrubs that could stabilize banks before they grow large and impede floodwaters.

Because the creek cannot reshape itself, a new, sediment-conveying channel must be constructed for it. Building on an engineering design by the Alameda County Flood Control District, *Unlock Alameda Creek* regrades the channel along its full length, transforming the blunt, singular *flood control channel* into a multi-tiered, multi-purpose *active channel*. Between the broad shoulders of the two levees, a new, smaller *sediment channel*, capable of accommodating a 1.5-year flood event, will be excavated. This sediment channel will be deeper, narrower, and more steeply sloped than the existing channel. Just above it will be a floodplain, including slightly elevated banks akin to natural levees along the sediment channel, lower areas beyond the banks, and moderately elevated shelves. In most places, this floodplain will remain between the current levees because development is so dense along the creek. Where possible, though, *floodrooms* will be excavated alongside the existing channel, providing significant pockets of reclaimed floodplain.

Coupled with this earthwork, Public Sediment embraces plant diversity within the channel. Native riparian vegetation will be encouraged along the banks of the sediment channel, where the roots of sedges, rushes, and shrubs can help stabilize its narrower, deeper shape. In the floodplain, woodier shrubs and trees will be clustered in *planted pockets* atop the elevated shelves. These pockets will anchor the sediment channel at key points, shaping the evolution of its future course.

Remaking the channel is the infrastructural foundation for better sediment flows, and, in turn, improved ecological health and recreational opportunities. Earthwork and vegetation will diversify the channel, making it look and perform more like a living creek. The *active channel* will convey sediment to the Baylands; it will also be a better place to be, whether you are a resident of an adjacent city like Fremont or a steelhead trout.

Historic creek cross section

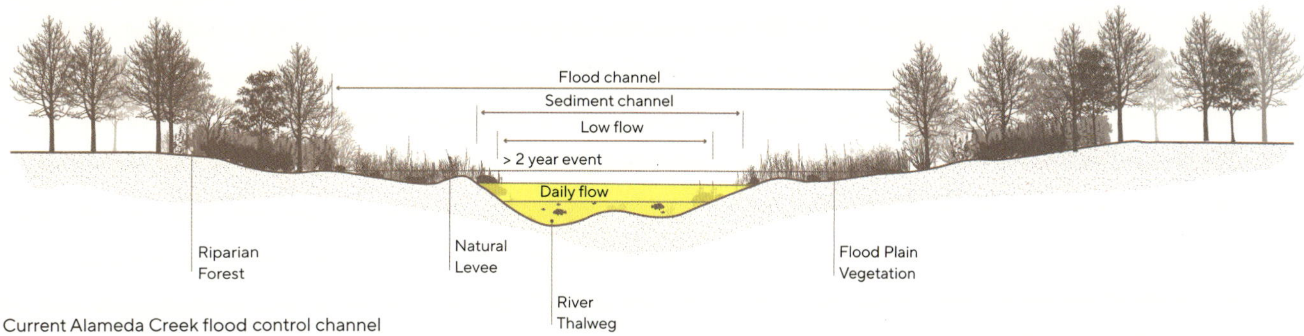

Current Alameda Creek flood control channel

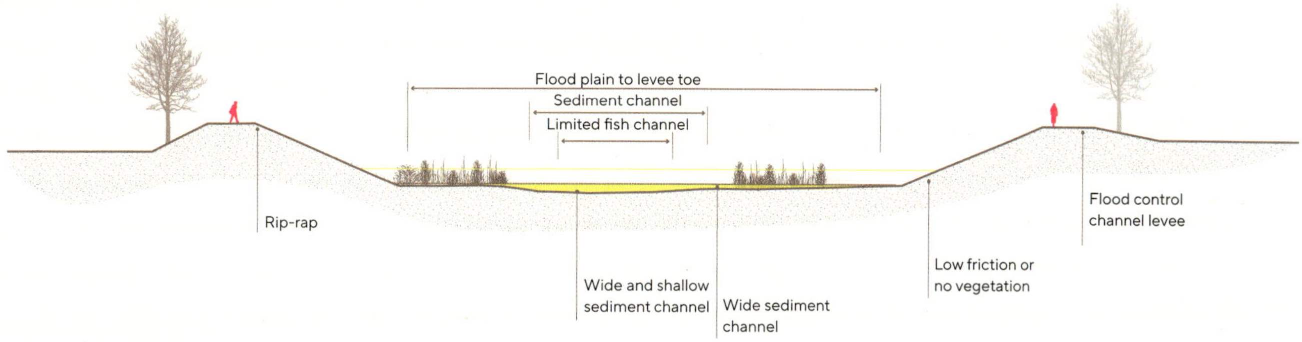

Proposed Alameda Creek active channel

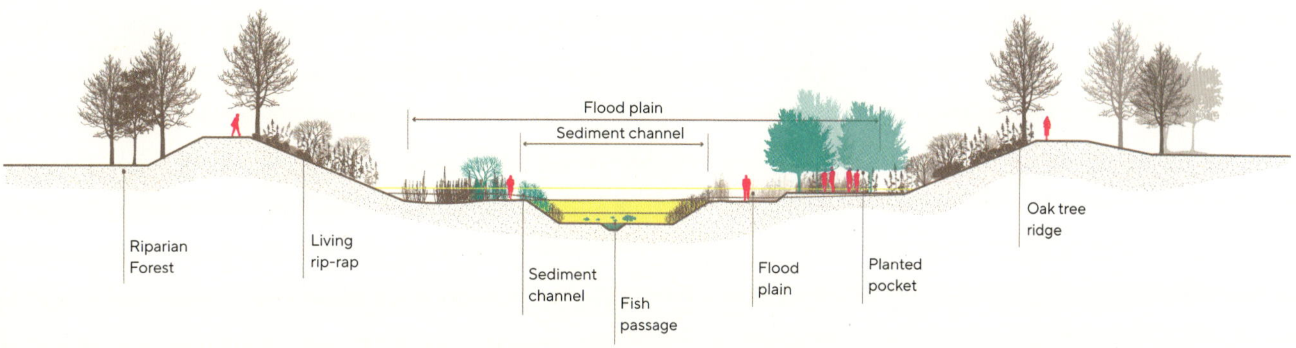

The active channel (SCAPE / Public Sediment Team)

A Channel for Fish (and Plants)

Reshaping the channel will also support the ecological regeneration of Alameda Creek.

Just as the active channel nests within the width of the broader flood control channel, an even deeper, even narrower channel will sit within the active channel: the *fish channel*. Where the active channel is designed to the scale of rainy-season floods, the fish channel is calibrated to the much smaller daily flows of the creek—and to the needs of steelhead trout.

Alameda Creek had one of the largest steelhead runs in the Bay Area. Now those fish must contend with obstacles like dams and sills; shallow water caused by the flat channel design; an absence of the braids, riffles, and pools that offered rest and forage; and a lack of the woody vegetation that provided cover. Only a small number of fish make the migration each year, and the once-robust run is functionally non-existent.

The excavation of a fish channel begins to address this. By defining a smaller course within the active channel, we can ensure that there will be a relatively deep waterway during both the in-migration and out-migration seasons. The same vegetation and planted pockets that anchor the active channel will also provide cover to fish. And the floodrooms excavated along the channel will offer forage and refuge opportunities during floods—while substantially diversifying the variety of plant species and micro-habitats available to other animals.

Designing to the needs of steelhead benefits other migratory fish that once inhabited the creek, such as lamprey and Chinook salmon. And a myriad of other animals and plants, from bobcats and osprey to willows and sycamores, will prosper from the ecological regeneration of Alameda Creek.

Migratory lamprey (SCAPE / Public Sediment Team)

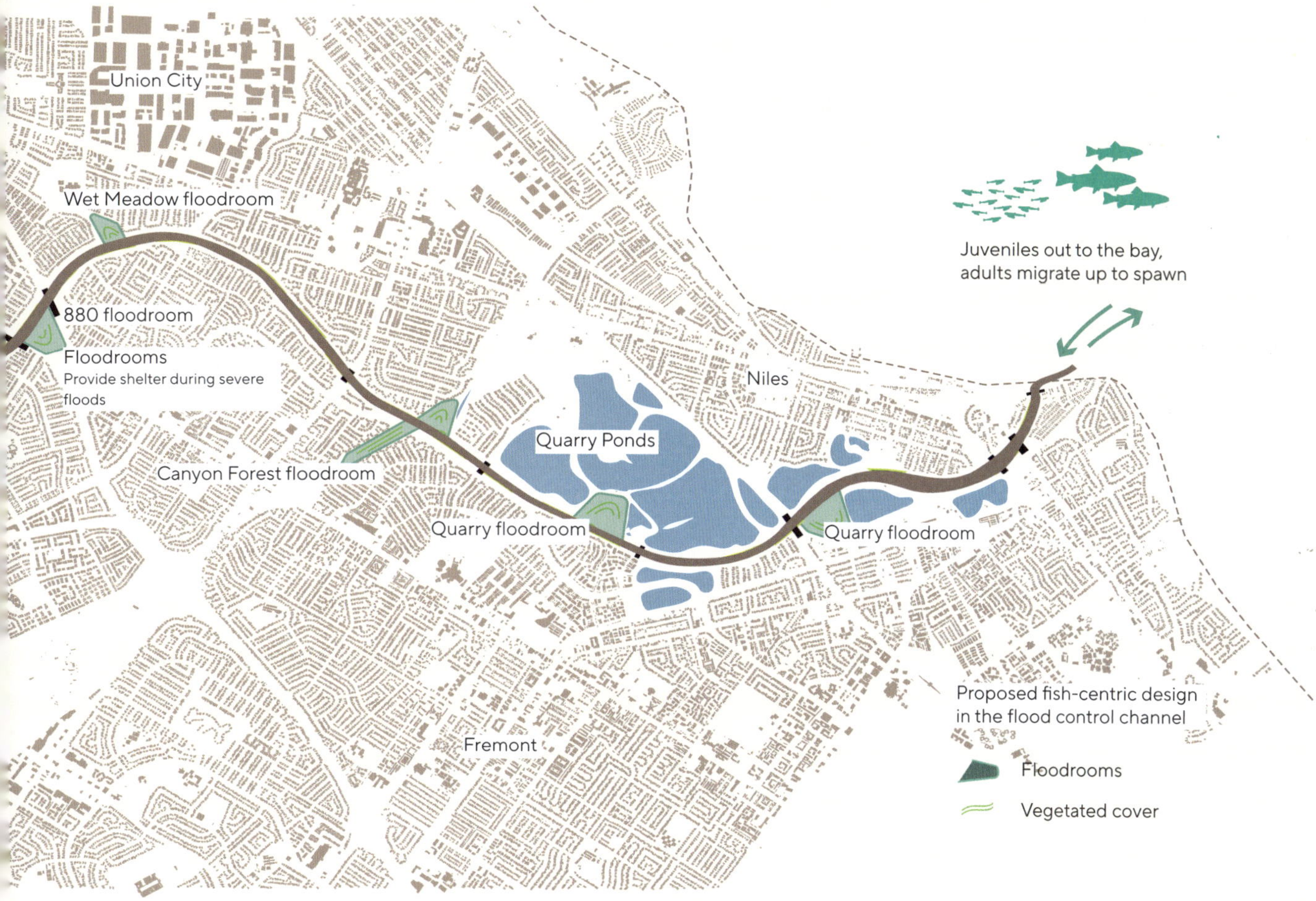

Juvenile steelhead (SCAPE / Public Sediment Team)

Design for People

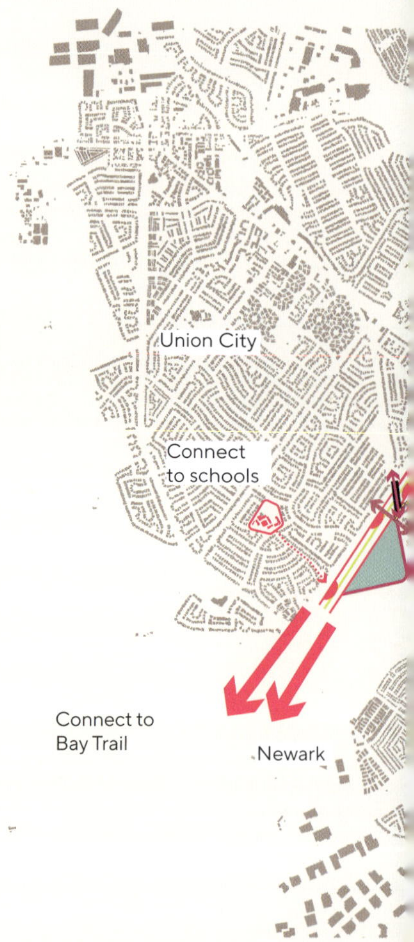

The choreography necessary to regenerate ecologically-healthy Baylands will only be successful—is only possible—if sediment is genuinely public.

Doing this requires direct public engagement, like the outreach work of Public Sediment, and it requires redesigning places like Alameda Creek so that people can experience sediment at work.

The suburbs surrounding the twelve-mile length of the Alameda Creek Flood Control Channel are highly diverse, including an Asian majority population and the largest concentration of Afghan residents in the Bay Area. Yet rather than stitch these communities together, as creeks often do, the channel keeps neighborhoods disconnected. Only half of the twelve roadway bridges that cross Alameda Creek provide for pedestrian or bicycle access, and the pedestrian trails along the crest of the levees never cross the channel or weave down into it.

Through public events, field interviews, and participatory mapping projects, our team found that many people who live nearby not only lack access to the channel but don't even know that the channel was ever free-flowing. The channel doesn't really look like a creek. Its high banks and flattened center suggest that water is a threat to be kept separate from people and moved away from homes and schools as quickly as possible. This is flood control infrastructure designed for infrequent events, and not for everyday well-being or pleasure.

Older residents recall a time when they could fish in the creek, when there were fish to catch. They remember there had been enough water for people to swim and play in. They have seen just how much the creek has changed since it became a flood control channel.

Many residents, old and young, want to see more fish, more water, and more vegetation. They want more ways to physically access the creek, rather than just look at it from a distance. They want more places to cross it and more public spaces to gather and recreate within. In other words, they want to recover its value as a commons.

Creek neighborhoods (Brett Milligan / Public Sediment Team)

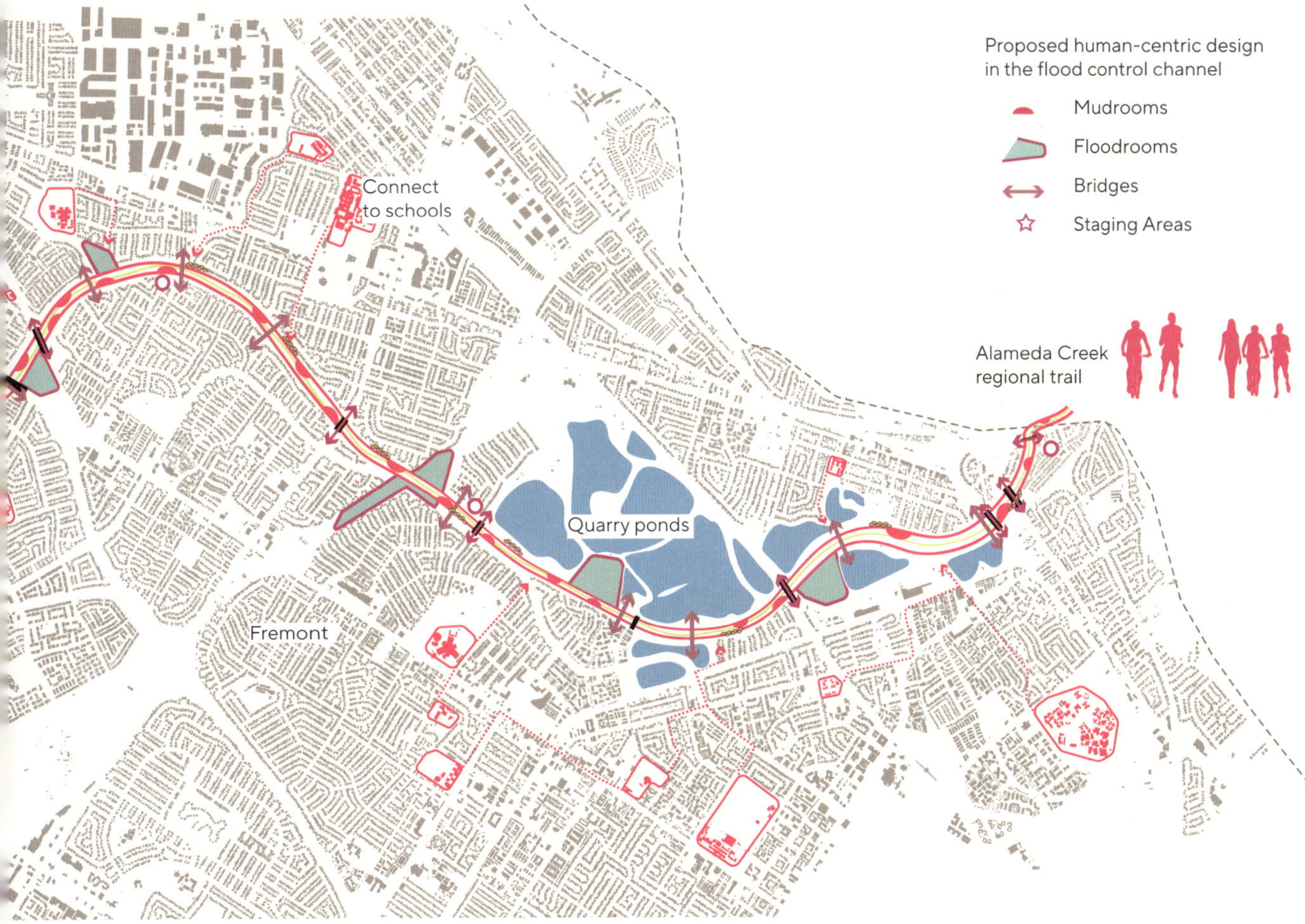

Alameda Creek Crawl (SCAPE / Public Sediment Team)

Accessing a Dynamic Creek

As the active channel is constructed and Alameda Creek begins to regenerate, a network of designed paths and nodes will open its increasingly diverse landscapes to residents.

Mudrooms are small public spaces along the channel and within the confines of its levees. Like the mudroom of a house, these are entry points and invitations to the creek. They will be strategically sited near schools, libraries, and parks. Some will be hardened extrusions at the top of the levees, accessible even in large floods, when the creek's water is laden with sediment. Others will be down within the channel, often in planted pockets. These mudrooms will provide an opportunity to get into the floodplain and be close to the active channel, to experience (see, touch, smell) the small dramas of a point bar forming, silt coating tree trunks after a flood, or even the mudroom itself being transformed by deposition and erosion. Local schools and advocacy organizations like the Alameda Creek Alliance will be able to use the mudrooms as venues for in-situ public sediment education or creek stewardship events. Collectively, the mudrooms will make the landscape-shaping work of the creek far more accessible.

Where the levees are set back to widen the creek's floodplain, *floodrooms* will be key nodes for public access. Given their siting off the main flow of the active channel, they will host substantially more diverse plant communities, including extensive marshes and riparian thickets of willow, poplar, sycamore, and alder. Like the lower mudrooms, their accessibility will be seasonal, determined by the Bay Area's rhythms of rain and flooding. During dry seasons, they will be the largest public landscapes along the channel. They will also make public the maintenance dredging that will be required occasionally along the creek; certain portions of the floodrooms will be designed to slow floodwaters and strategically collect sediment. When sediment is harvested from these reservoirs during dry months, floodroom paths and spaces will become overlooks permitting citizens to observe choreography in action.

Terrace trails will meander between levee tops, levee sides, and the channel bed, connecting floodrooms and mudrooms. As with those rooms, trail access will be tied to creek flows, making the pulsing rhythms of the creek evident over time. The trails' meanderings, meanwhile, will narrate the shaping of the landscape through sediment flows by weaving from delta to point bar to cut bank to flood terrace, and back again. As befits the shifting nature of the active channel, the trails can be provisional paths cut seasonally into the floodplain vegetation. At times and in places, the trails might be enveloped within dense stands of rushes and marsh grasses; in others, they would open onto a gravel bar or a long stretch where the creek runs calmly. Terrace trails will facilitate citizen science and grassroots environmental stewardship like species surveys or invasive plant removals, involving nearby communities in caring for the creek.[13]

Seasonal bridges will be located roughly every half-mile along the active channel, connecting communities and fostering equitable access to public resources on both sides of the creek. Given the low volume of water for most of the year, these bridges will have short spans and be relatively minor infrastructure—lighter and less-impactful than the piers of vehicular bridges. During infrequent floods, the seasonal bridges will fold up and retract to allow water to pass safely through the channel.

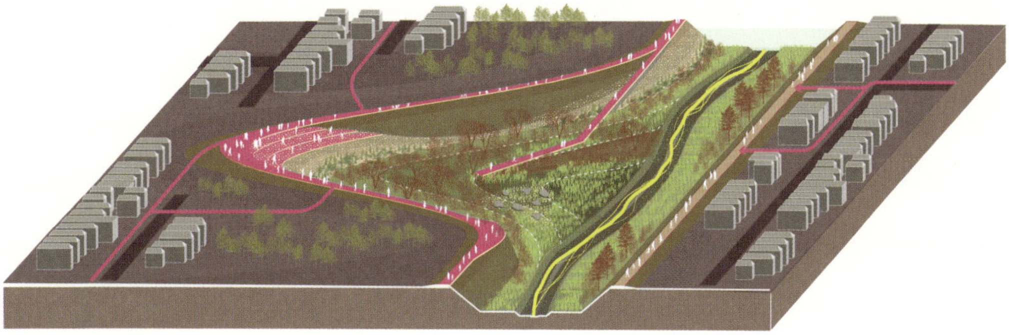

Floodrooms

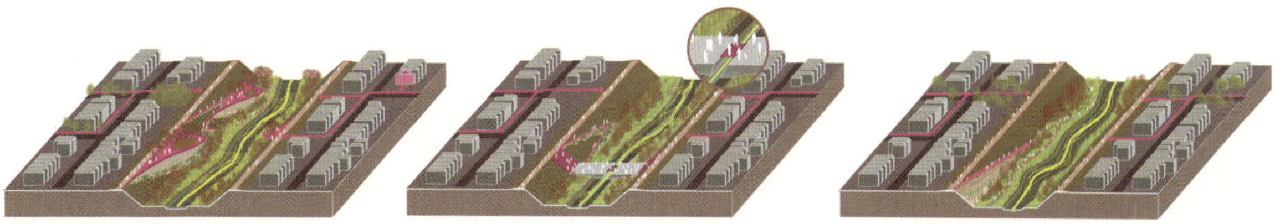

Mudrooms

Seasonal crossings

Terrace trail

Proposed public access in Alameda Creek (SCAPE / Public Sediment Team)

Walking along the flood terrace trail (SCAPE / Public Sediment Team)

The Dry Creek Confluence

Where Dry Creek meets Alameda Creek, we see how choreographing sediment can generate spaces that bring people and sediment together amid a regenerating ecosystem.

Dry Creek is a small tributary of Alameda Creek; it empties into the flood control channel about halfway down the fluvial portion. This confluence is set within dense residential housing. Dry Creek is itself channelized, and the flood control channel's levees wrap up into it. Suburban yards with their fences, swimming pools, and patios back right up to both creeks.

During storms, water and sediment, including heavy sands and cobbles, shoot out of Dry Creek toward Alameda Creek, at an angle roughly perpendicular to the larger channel. This causes a rapid deceleration as the two water flows merge, and a delta-like bar has formed to the southwest of Dry Creek's mouth. Substantial sediment is stuck here.

At this confluence, Public Sediment's active channel will be trained by a series of mudrooms, planted pockets, and rock baffles so that it accepts and carries to the bay as much sediment as possible. Upstream, the active channel will be encouraged to meander to the south levee, so that, as it approaches the mouth of Dry Creek, planted pockets and a mudroom extending into the channel from the south levee can redirect the flow of the active channel directly at the Dry Creek confluence, accelerating floodwaters and encouraging scour of the delta. The energy from Dry Creek will push water back toward the south bank again after the confluence, so a series of baffles will be constructed along the south bank just downstream to deflect the active channel's flow again to the north bank, restarting the process of meandering.

The large, multi-level mudroom that deflects Alameda Creek toward the confluence will step down into the channel from an existing public park, extending south from the levee into the suburban fabric of Fremont. This mudroom will not be armored by traditional riprap, but will instead be constructed as a "living levee," covered with a blanket of vegetated modular concrete units. From the mudroom, residents, school groups, and advocacy organizations will have direct access to one of the most dramatic moments along the entire length of the creek, and will be able to walk terrace trails upstream or downstream to further explore the active channel and its floodplains.

Riffle
Planted pocket
Living levee
Mudroom
Deposition
Siltation baffles
Mudroom
Terrace trail
Flood terrace
Fish Channel
Sediment Channel

Feeding the Baylands

Unlock Alameda Creek could mobilize as much as seven million metric tons of sediment by 2100. However, that sediment will only nourish the Baylands if the channel is reconnected to the subsided ponds of Eden Landing.

Currently, the flood control channel levees funnel sediment past Eden Landing and into the deeper waters of the bay. The same is also true for Old Alameda Creek, a waterway that runs past the north side of Eden Landing. Public Sediment will breach Eden Landing's levees on both the north and south, reconnecting subsided Baylands to channelized creek courses and, through the mouths of the channels, to the bay itself. Doing this will permit sediment—particularly marsh-nourishing fines—to flow freely again between creek, Baylands, and bay.

Even with a motivated, engaged, and informed public supportive of change, the implementation of projects like this requires a strategic approach to institutional, political, and technical protocols. Pilot projects have the potential to spur innovative landscape technologies and unlock calcified regulatory regimes, but they must be carefully gauged if they are to function effectively given what is practicable today, while still encouraging the evolution of current practices and opening the door to more radical climate adaptation in the near future.

Breaching the flood control channel was proposed in past planning efforts by public agencies working on the restoration plan for Eden Landing. But over the course of those efforts, breaching came to seem impossible due to disagreements between different agencies with siloed missions.

Breach of Alameda Creek (SCAPE / Public Sediment Team)

Although the subsided ponds have little ecological value in their current state, they and the levees along their western edge do provide a substantial buffer against flooding for neighborhoods to their east. To be feasible in the near term, any effort to reclaim the ponds as tidal marshes will need to provide adequate protection to those neighborhoods as well as nascent marshes. To be transformative in the long term, that protection will need to itself contribute to the goal of ecological regeneration.

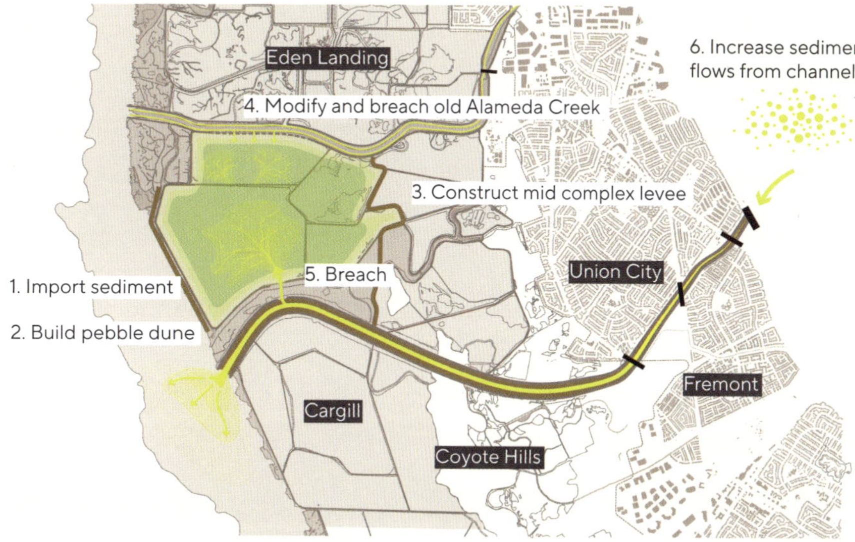

Proposed operations to breach Alameda Creek levee (SCAPE / Public Sediment Team)

The Pebble Dune

Short-term needs and long-term intentions can be balanced by piloting new landscape technologies.

At the western edge of the Eden Landing salt ponds, an old earthen levee armored with riprap protects inland homes from coastal wave erosion and inundation. It is a relic of salt pond infrastructure, now augmented and utilized for flood control.

The Pebble Dune transforms this ecologically barren wall into a living and adaptive infrastructure. It is modeled after the coarse-grained beaches, made of shells, pebbles, and gravels, that once lined long stretches of bay shoreline. Such beaches—often described simply as gravel beaches—have a unique tendency to gain elevation, stacking their crests vertically in response to storm waves. (Most other beaches erode during storms.) Coarse-grained sediment from the re-grading of the Alameda Creek Flood Control Channel will be deployed

Pebble Dune (SCAPE / Public Sediment Team)

here to produce an adaptive edge that rises as storms push the crests higher.

Many Bay species, such as nesting terns and harbor seals, used gravel beaches in the past. Exactly which species might use and benefit from them now is a question of great interest to local scientists.

To better understand this new technology, a small segment of Eden Landing's coastal edge is currently being used to test parameters for a gravel beach, including grain size, material, and slope configuration. This initial pilot-within-a-pilot, made possible by a 2018 grant from the National Coastal Resilience Fund, is proceeding through a partnership between the California State Coastal Conservancy, SCAPE Landscape Architecture, and the engineering firm Arcadis. The project will be monitored to see which species colonize and use the new shoreline, and the results will inform coastal adaptation techniques proposed for the larger Bay Area.[14] Like many pilot projects, the Pebble Dune spans design, planning, and policy evolution.

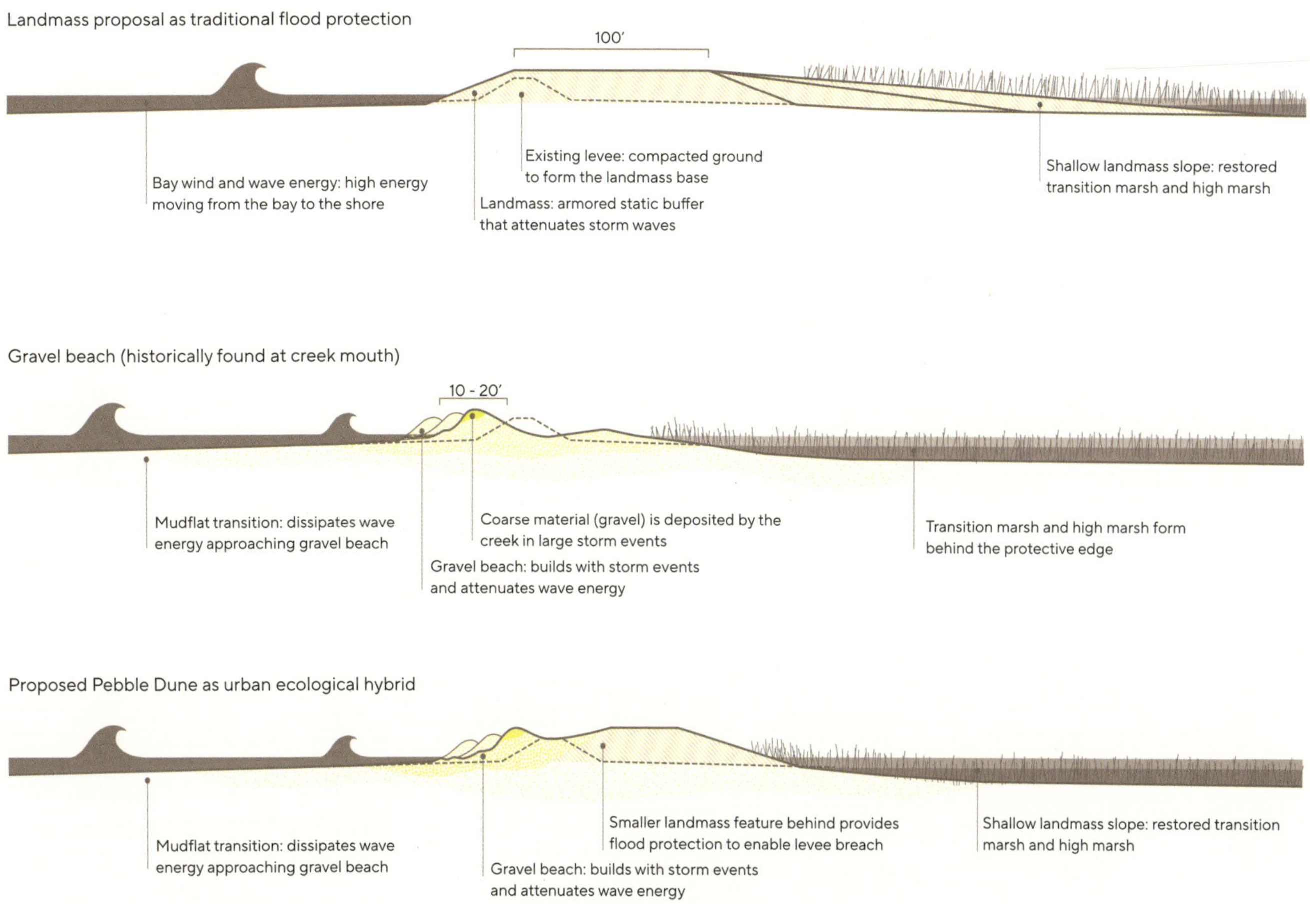

Alternatives considered for the Eden Landing perimeter levee (SCAPE / Public Sediment Team)

Monitoring and Adaptive Management

Every action needed to Unlock Alameda Creek—the active channel, the breaches, the pebble dune—will be a live experiment.

The altered landscapes will require observation and adjustment as they evolve. Some have already been studied with computational models in an effort to understand possible trajectories. But landscapes are dynamic material assemblies. Through their open, complex, and distributed agency, they defy predictions and linear causality. What is conceived in landscape design—no matter how rigorous and skilled the conception—is never the same as how the design manifests and performs as an actual landscape.[15]

This calls for a humble approach, a design fitted to the dynamic qualities of the landscape medium through direct field observation, monitoring, adaptive management, and small-scale experiments that can be tested and refined, before being broadly extended.

Before proceeding with construction, designers will need more comprehensive biophysical sensing, monitoring, and data gathering, so as to understand flood channel flows and sediment dynamics. One big unknown is how fine sediment moves and deposits within the tidal zone of the channel, and how this will be affected in the future as king tides extend farther up the channel due to sea-level rise. Breaches will only be effective if the tides actually pull sediment from the head of tide into Eden Landing. Such gaps in knowledge will be addressed in a manner consistent with existing bay-wide monitoring protocols (such as those for other channelized creeks) so that the information generated is comparable and transferable. Monitoring and sensing will also be done in a way that is aesthetically accessible to the public and that creates opportunities for citizen science. During Resilient By Design, a sensor workshop at Kennedy High School enabled the design team to link monitoring with its larger goal of making sediment public, and the team temporarily deployed a pair of sensors in the tidal portion of the creek, which is not currently monitored by scientists.

As Public Sediment is implemented, sensors installed for the pre-construction data collection will facilitate adaptive management. Tracking sediment flows and hydrological patterns will permit further adjustments to channel form or auxiliary infrastructures like the pebble dune, if and when they prove necessary. A living infrastructure like the Baylands cannot be expected to remain static into the distant future.

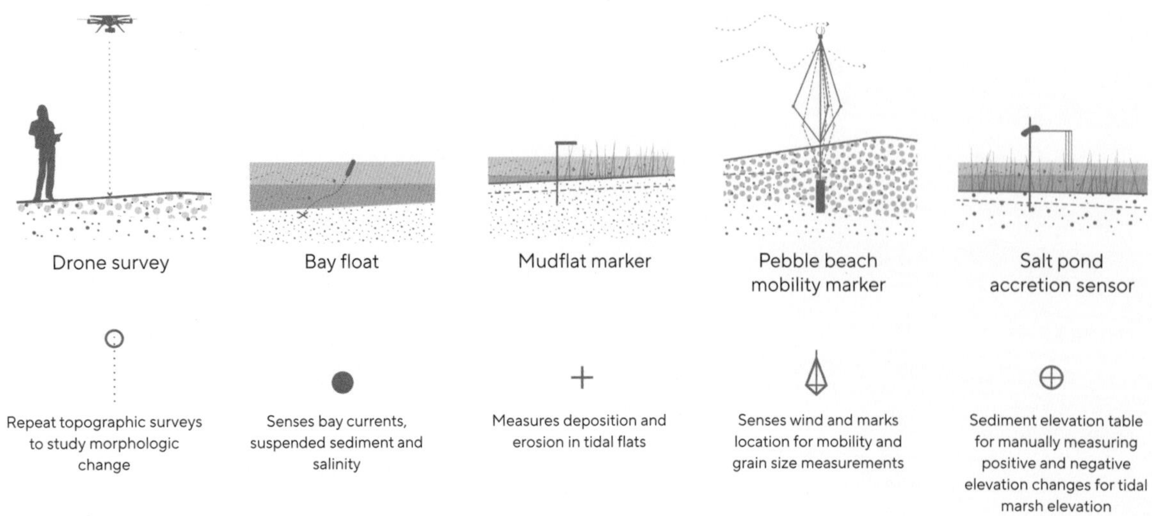

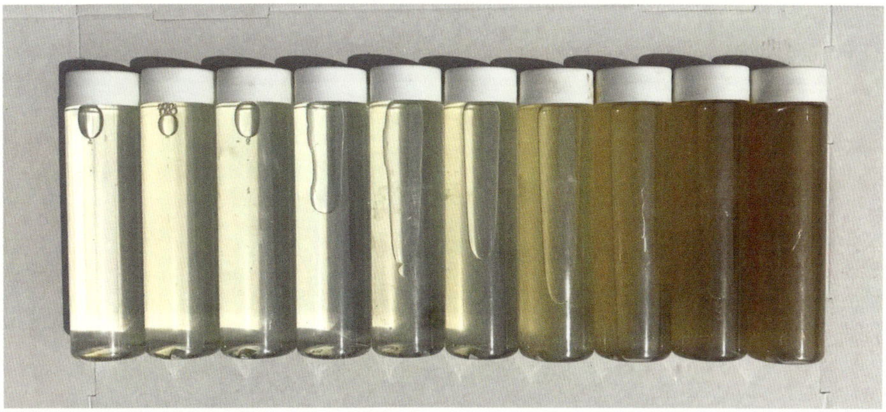

Sediment samples (Justine Holzman and Cy Keener / Public Sediment Team, 2018)

Sampling along the Creek (Justine Holzman and Cy Keener / Public Sediment Team, 2018)

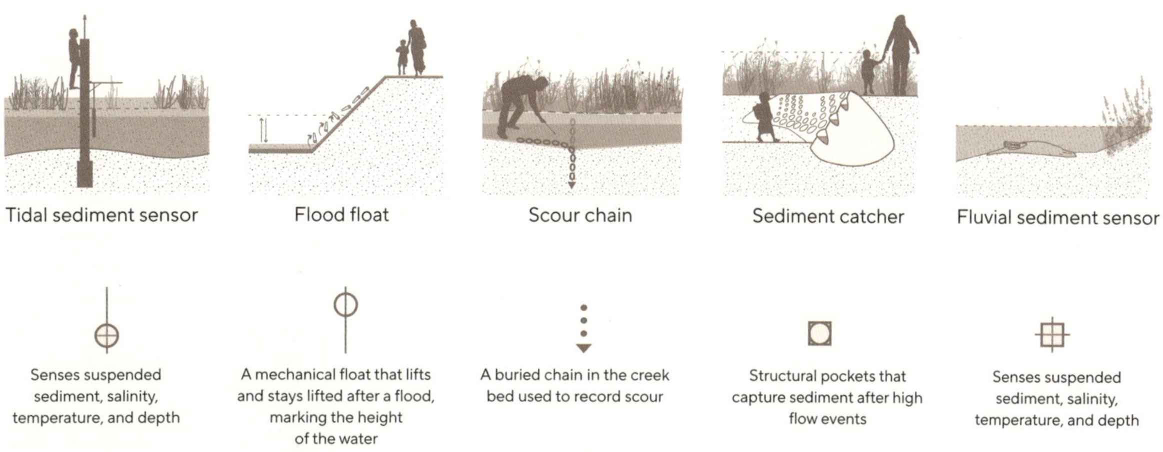

Proposed sediment monitoring methods (Justine Holzman and Cy Keener / Public Sediment Team, 2018)

Living Infrastructure

The bay urgently needs infrastructure that reconciles the well-being of human communities with the vitality of ecological systems.

Ecological health in a landscape like Alameda Creek is directly linked to morphological diversity. Steelhead and many other species thrive when a stream channel varies between deep, slow-moving pools; shallow, fast-moving riffles; and intermediate runs. Animals find shelter and food most readily when a stream is erratically disturbed by fallen tree trunks, intercepted by the root systems of adjacent trees, and pock-marked by boulders carried downstream in large floods. This complexity can be mimicked, to some extent, through construction: boulders can be placed, a varied topography can be excavated. But such an ecosystem performs best when formed and shaped over time by the action of a healthy creek itself.

The reshaping of the flood control channel as described in this chapter is not a singular action. It is meant to trigger a fundamental shift in the state of Alameda Creek. Currently, the creek is broken: too constrained to accomplish self-repair or rebuild the diversity it once fostered. The construction of the active channel will undo those bonds, elevating the creek into a new, more dynamic state, where it can once again evolve and change over time. Flooding, erosion, deposition, and plant growth will shape meanders, braids, pools, and riffles. Floodplain banks will accumulate in some places and be eroded in others. Planted pockets, constructed groins, and streambank vegetation will guide these transformations, particularly at points like the Dry Creek confluence and around bridge piers, where the dictates of sediment transport require particular paths.

Breaching the channel levees and constructing the pebble dune will similarly open the Baylands of Eden Landing to ongoing transformation. Tidal marshes will braid with runnels, depress into pools, and shallow into flats. The gravel beach will shift and pile up.

Unlock Alameda Creek is a "launching pattern" for evolution and diversification of landscape, not a static form to be maintained in perpetuity.[16] This is what it means to make living infrastructure. Not merely that biological life is grafted on to infrastructural systems, but rather that new forms of infrastructure are designed to support *life* in all its indeterminate, uncertain vibrancy.[17] Sediment is fundamental to this: it is the substrate for the dynamism of creek, marsh, and gravel beach.

Alameda Creek connected with the Baylands (SCAPE / Public Sediment Team)

One Piece of the Puzzle

Sediment mobilized by the active channel and diverted through the breach will help support the Baylands, but it is not nearly enough.

It is nowhere near the amount of sediment required to bring the Eden Landing salt ponds back to tidal elevation, so that they can again become saltwater marshes. Nor is it enough to sustain those future marshes in the face of accelerating rates of sea-level rise.[18] Under the most optimistic projections, it might be enough to meet half of the marshes' future needs, assuming that another (massive) source of sediment is used to refill the subsided ponds in the first place.

So channel design is only one piece of the puzzle for sediment supply. Obtaining more sediment requires that we *Rethink the Sedimentshed*, looking at the uplands that drain into flood channels like Alameda Creek, as well as investigating potential sources of sediment beyond the bay's tributaries.

Design must thus span many boundaries, both regulatory and physical. Public Sediment calls for the creation of a *Sedimentshed Vision Plan* for Alameda Creek, consisting of long-term scientific studies and a multi-agency planning effort. The aim of this work is to create a vision for the choreography of sediment across the creek's whole sedimentshed, balancing the lowland need for sediment with upland concerns. Comprehensive monitoring will build a picture of how sediment erodes and deposits within the Alameda Creek watershed, from the most remote hilltops all the way to the bay, and thus a picture of where sediment might be sourced and sustainably deployed. A consortium of public agencies, scientists, designers, and stakeholders—flood control districts, dam operators, landowners, local conservation districts, water managers, and policymakers—will explore how those scientific findings relate to concerns throughout the watershed, and how those concerns might be met or changed over time.

The vision plan will be time-based and projective, establishing a set of prioritized and phased actions to meet short- and long-term objectives. Strategies such as upland bank restoration and regenerative land stewardship practices supported by local conservation districts are already being implemented. But that work is not yet coordinated within a larger framework for sediment design and, ultimately, is not likely to deliver sediment in the necessary volumes. Greater volumes could be unlocked through large dam removal and dam retrofit projects, but those can be costly, technically challenging, and politically difficult, as Bay Area communities depend on dam reservoirs for their water supply.

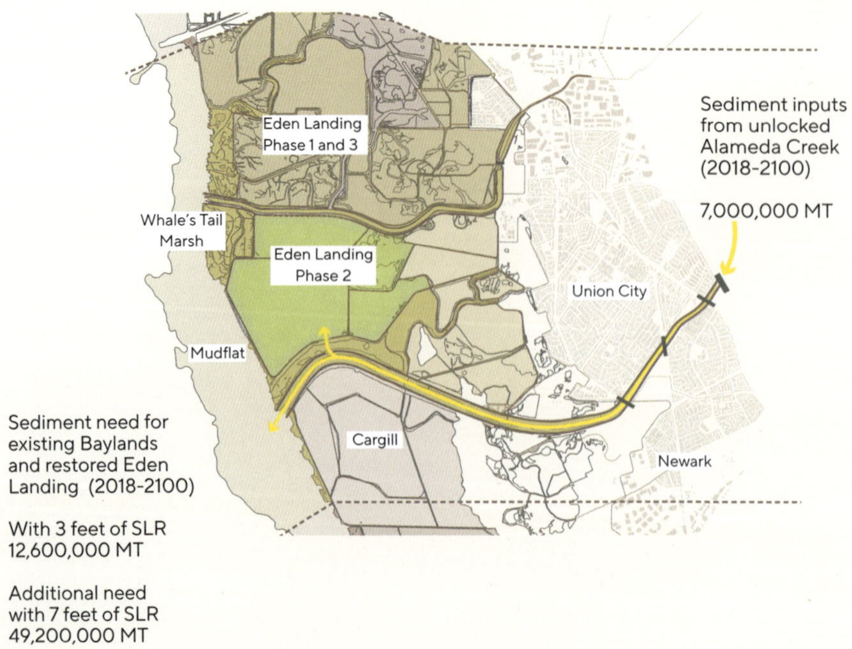

Future sediment budget scenarios (SCAPE / Public Sediment Team)

Creatively unlocking tributaries like Alameda Creek will be necessary, and better management of existing landscape technologies in their uplands will help, too. But choreographing a sustainable sediment supply for the Baylands will require other tools.[19]

Thin layer placement is a mechanical way to accelerate marsh accretion. It entails spraying inches-thin "lifts" of sediment on a subsiding or drowning marsh to help it adapt to sea-level rise. *Mudflat feeding* involves strategically placing dredged sediment within the lower portions of tidal mudflats, so that it can be transported by tidal action to adjacent wetlands. *Water column seeding* is similar, except that the sediment is placed directly in a moving column of water. Each of these shoreline sediment technologies has been applied in other regions but not yet in San Francisco Bay.

New technologies will also be needed for upland choreography. Public Sediment proposes *tributary feeding*: pushing stockpiled sediment near the outlet of a channelized creek during a flood event, so that sediments are carried out into the bay. On Alameda Creek, sediment could be sourced from maintenance dredging within the channel. Thus, dredging could serve a dual purpose, maintaining flood conveyance capacity within the channel (as it does now) and also rebuilding the Baylands.Another upland sediment source that could be beneficially used is *excess clean fill* from construction projects. This material is typically hauled to landfills, but it could be used locally as base material to fill subsided Baylands.

Public Sediment encourages testing all these methods at Eden Landing. A single landscape project can pilot many technologies and build adaptive capacity for a whole region.

Piloting sediment placement techniques (SCAPE / Public Sediment Team)

Plan and Pilot for a Future Bay

Design experimentation is needed not just in Eden Landing, but across most of San Francisco Bay.

San Francisco Bay is composed of several distinct bays, including San Pablo Bay in the north, and the shallow, muddy South Bay. Each of these sub-bays have their own geomophorphic, hydrological, tidal, urban, and cultural conditions, and their own markers of vulnerability. And within each sub-bay there are internal differences: land edges that are aggregating and eroding; variability in creek sizes, sediment loads, and the types and sizes of sediment that move and deposit in different areas; disparities in land use, population, and the degree of urbanization.[20]

New sediment design techniques must be tested and compared across a range of landscape conditions to learn how their performance varies. For example, mudflat seeding might work best in the South Bay, due to its unique tidal patterns. Or coarse-grained beaches might work on coarser mudflats, but sink under their own weight on siltier mudflats.

Plan and Pilot for a Future Bay calls for scaling up landscape prototypes around the bay so that they become living, vibrant classrooms of the present and future, where all residents can learn about and debate the Bay's future. Experiments have already been constructed in restoration projects like Hamilton Wetlands, Sears Point, and the South Bay Salt Ponds. This experimentation needs to be incentivized and ramped up quickly, while it can still have significant impact.

Many open questions swirl around the bay's efforts to salvage and regenerate the Baylands as sea-level rises. Foremost is how we adapt to climate change.

By what actions and processes is adaptation enabled, fostered, or dragged into existence? Adaptive capacity—a term that first gained traction in reports by the United Nations and the Intergovernmental Panel on Climate Change—should be understood to include not just economic and physical "resources," but also social learning, cooperation, and technical innovation.[21]

Public policy and regulation are important socio-technical levers. Much of the bay's environmental policy was created in the 1960s and 1970s, when regulators halted the indiscriminate diking and filling of the bay. Innovative for the time, this policy was motivated by a commendable desire to sustain a shared public resource for humans and others.

Today, however, the bay faces the opposite problem: rising water that will claim urbanized land with increasing pace and volumetric force. How should policy evolve given this reversal? In a public commons, indiscriminate filling is still undesirable, but there is also a shared desire to restore the bay's wetlands, with their ecological and recreational affordances and their ability to protect upland communities. To even begin to approach the stated goals, hundreds of millions of cubic yards of sediment are needed.

Can policies and regulations evolve fast enough to bend the curve on the bay's sediment shortage and attenuate the undesirable effects of climate change? Can they incentivize the beneficial reuse of dredged and upland sediment to restore bay marshes before sea levels rise so fast that these efforts become ineffective? Can sediment flows be choreographed at a sub-regional scale—like that of a single sub-bay—to more effectively work with the affordances and challenges particular to diverse Baylands and their sedimentsheds? Can funding and permitting of adaptive landscape projects be streamlined to enable urgent adaptation, rather than inhibiting it? The only way to answer these questions affirmatively is to begin the painstaking work of building pilot projects, learning from them, and scaling the earned knowledge up through changes in regulation and policy.

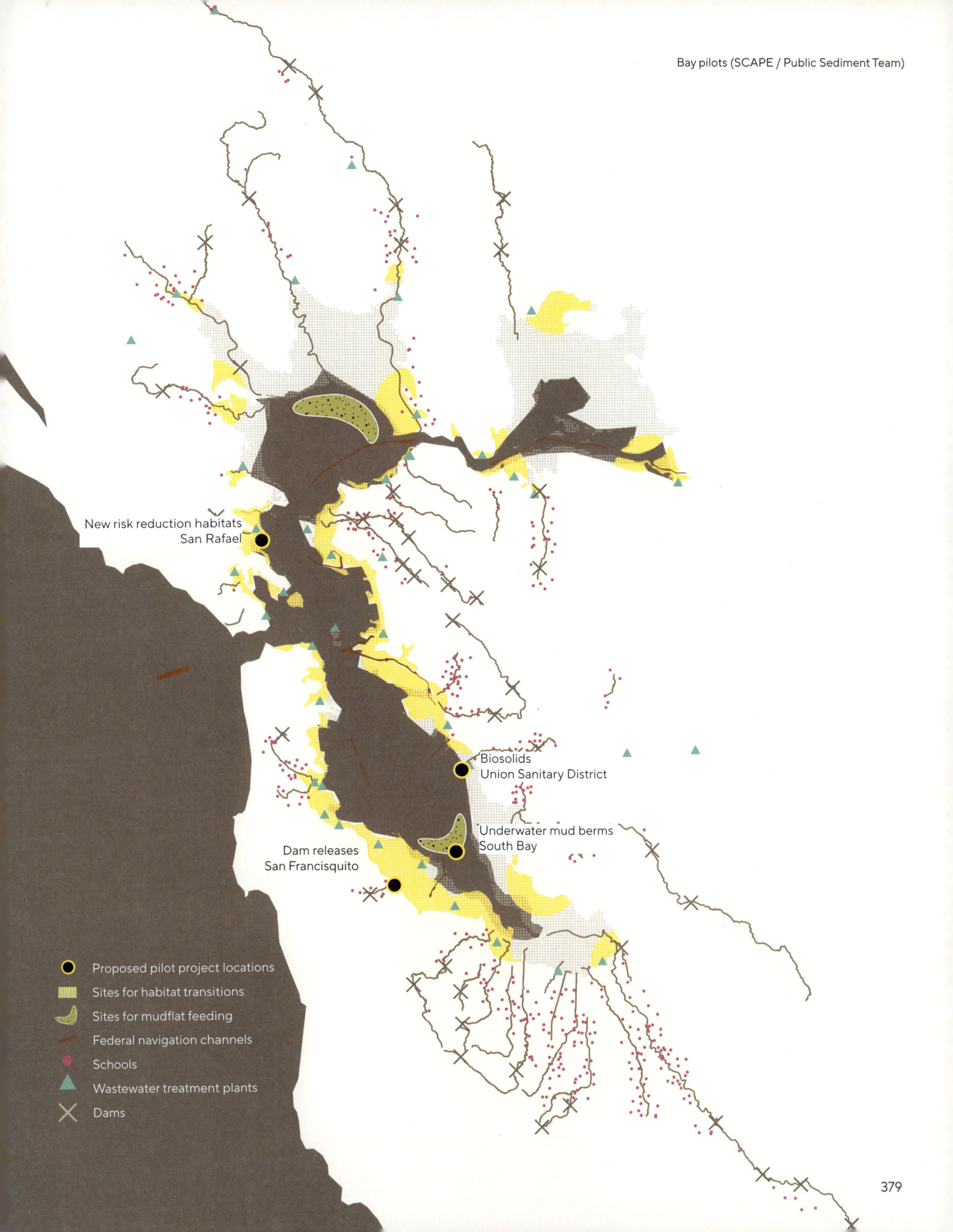
New risk reduction habitats
San Rafael
Biosolids
Union Sanitary District
Underwater mud berms
South Bay
Dam releases
San Francisquito
Proposed pilot project locations
Sites for habitat transitions
Sites for mudflat feeding
Federal navigation channels
Schools
Wastewater treatment plants
Dams

The Worlds we Could be Making

The bay has always been changing and it always will be.

From a geologic perspective, the current size and extent of the bay is ephemeral. Six thousand years ago, the bay was a terrestrial river valley. And in just the past two centuries, its coastlines and bathymetry have changed dramatically, as the bay was reclaimed, deepened, and extensively filled in along the edges. In the very near future, the bay will expand in tandem with sea-level rise.

The Public Sediment team conducted a projective geospatial analysis that looked at the interrelationship of bay sediment needs. We calculated the sediment that would be required to meet Bayland goals mandates, such as filling subsided polders, and to sustain and augment marshes in the face of increasing sea-level rise. All trends point to a huge and widening sediment supply shortage, which will become increasingly difficult to counter over time.[22] As sea-level rise accelerates, marshes are likely to drown, turning to mudflats and then open water. In a few places, tidal marshes may be able to migrate upland with the rising tides. But around most of the bay, migration space is limited by steep topography and urban development.

Thus, the bay cannot be "saved" in its current configuration. Plans that attempt to maintain the bay's current form and extent are examples of cultural resistance, not climate adaptation.

Hard decisions need to be made. If the entire bay edge cannot be saved, what should be protected and what should be let go? If retreat is inevitable, how should it be managed? How can such questions be decided equitably?[23] And since the effects of climate change are likely to accelerate and intensify over the long term, what time horizon should we consider in our adaptive designs? What looks promising according to twenty-year projections could be a complete failure in forty.

In the near term, projects like Public Sediment can bend the trajectories of sediment supply, so that the point when rising demand exceeds falling supply can be pushed further out. In doing so, we gain two things: time and landscapes.

The purpose of this work is not to solve the bay. It is to make time for society, institutions, and communities to adapt to radical change. We can make landscapes worth living in now, while opening paths to future worlds that will also be worth living in.

This asks us to fundamentally reconceive our understanding of design. We must abandon the idea of stabilizing landscapes in a permanent present and embrace the reality that every landscape is always in-between.

Here we must reconcile the need for radical change with a posture of humility. We can unmake infrastructures, like flood control channels, that are disposed toward separation and unsustainable development. We can make landscapes that write new dispositions into our infrastructure—dispositions toward landscape dynamism, regenerated ecosystems, and the holding of our coasts as truly public commons. We can do all of this without pretending that we can or should control what will become of the landscapes that we make.

Releasing the expectations of stability and control does not mean giving up on making landscapes that are worth living in. Designing with sediment is a crucial tool for doing this.

Sediment is a common inheritance for all living things; it links us all with the earth. Marshes, beaches, mudflats, riverbanks, floodplains, deltas, and estuaries—many of the most dynamic and life-abundant places on our planet—are all sustained by sediment. These have always been places people sought to live, and they always will be.[24] But we cannot live in them much longer the way we are living now, because we are destroying them and imperiling ourselves in the process.

Regenerating sedimentary landscapes means learning to follow the requests of sediment—listening to its propensities, accommodating and growing its dynamism, freeing it to build living landscapes. Marshes, mudflats, and point bars must replace levees, seawalls, and channelization. Not because marshes will keep rising seas out (although they do offer real protection), but because landscapes shaped by sediment can foster life without the false promise that things will never change.

Designers, by and large, have not engaged with sediment, and especially not with silt and sand in slurried motion. We landscape architects—the designers who claim landscape as our medium—have ceded sedimentary concerns to scientists, engineers, and planners.

It is time for that to change. Design has unique tools for synthesizing the knowledge of disparate partners; for experimenting, evaluating, and revising amid deep uncertainty; for making sediment public. Design means asking what should be made before asking how to make it.

This is crucial now. Landscape architects, we must get involved. But these questions are too important to be left only to professional designers. All of these other fields—including the engineers, scientists, and planners already knee-deep in sediment design—must take on this charge too, in collaboration. A world that has been technocratic and exclusive, that has served the needs of wealthy corporations and privileged interests over the well-being of humanity, and of all the living things on earth, must be opened up. It must become accountable to the worlds that we could be making together.

Public Sediment (SCAPE / Public Sediment Team)

Endnotes:

1. H. Perry, A. Lyndon, P. Soumoy, and B. Goeden, San Francisco Bay Conservation and Development Commission, "San Francisco Bay Sediment: Challenges and Opportunities" (poster presentation, 12th Biennial State of the San Francisco Estuary Conference, Oakland, 2015).

2. Scott Dusterhoff, Katie McKnight, Letitia Grenier, and Nate Kauffman, Sediment for Survival: A Strategy for the Resilience of Bay Wetlands in the Lower San Francisco Estuary, SFEI Contribution No. 1015 (Richmond, CA: San Francisco Estuary Institute, 2021).

3. Lester J. McKee, Mik Lewicki, David H. Schoellhamer, and Neil K. Ganju, "Comparison of Sediment Supply to San Francisco Bay from Watersheds Draining the Bay Area and the Central Valley of California," *Marine Geology* 345 (2013): 47–62.

4. J. A. G. Cooper and Jeremy Pile, "The Adaptation-resistance Spectrum: A Classification of Contemporary Adaptation Approaches to Climate-related Coastal Change," *Ocean & Coastal Management* 94 (2014): 90–98.

5. For a more detailed account of the geomorphic history of Alameda Creek and the Niles Cone, see Bronwen Stanford et al., Alameda Creek Watershed Historical Ecology Study, SFEI Contribution No. 679 (Richmond, CA: San Francisco Estuary Institute, 2013).

6. Stanford et al., Alameda Creek, 156-157.

7. Quoted in Stanford et al., Alameda Creek, 203.

8. Pun Lok Raymond Wong, "Federal Flood Control Channels in San Francisco Bay Region—A Baseline Study to Inform Management Options for Aging Infrastructure" (PhD diss., UC Berkeley, 2014).

9. Brett Milligan, "Accelerated and Decelerated Landscapes: On the Techniques, Knowledges, and Ethics of Bending Time," *Places Journal* (2022).

10. Wong, "Federal Flood Control Channels."

11. Stanford et al., Alameda Creek, 330.

12. See the discussion of form as trajectory in Catherine Dee, "Form, Utility, and the Aesthetics of Thrift in Design Education," *Landscape Journal* 29, no. 1 (2010), and the discussion of landscape as "a plurality of forms" in Brett Milligan, "Landscape Migration," *Places Journal* (2015).

13. Marianne E. Krasny and Keith G. Tidball, *Civic Ecology: Adaptation and Transformation from the Ground Up* (MIT Press, 2015).

14. See discussion of gravel beaches as an adaptation measure in the San Francisco Bay Shoreline Adaptation Atlas (Julie Beagle et al., San Francisco Estuary Institute and SPUR, SFEI Publication #915, April 2019), 72–75.

15. On this gap, see Brian Davis, "Landscapes and Instruments," *Landscape Journal* 32, no. 2 (2013): 293–308.

16. George Descombes uses the term "launching pattern" to describe the lozenge-shaped islands that the Superpositions team designed for their project "Renaturation of the River Aire." The term refers to the manner in which the islands both set initial conditions for a new course for the river and materially participated in the evolution of that new course.
George Descombes, "Designing a River Garden," in *Aire: The River and its Double* (Zurich: Park Books, 2018).

17. In "Choreographing Sediment," we noted that landscape architecture has been transformed in recent decades by a focus on the dynamic qualities of landscape and the virtue of landscape processes as instruments of landscape architecture. As Julian Raxworthy has noted, this "process discourse" has unfortunately prioritized formal mimesis of the effects of dynamic transformation over the actual agency of landscapes in making and re-making themselves. See, for instance, how Hargreaves Associates' Guadalupe River Park has been largely represented in landscape architectural media via photographs of the constructed wave-field topography, which looks like water in motion, rather than understood in terms of the calibration of the park's riverine elements to the hydrology of Bay Area flooding. For further discussion of Guadalupe River Park, see Brian Davis's essay in this volume.

18. Dusterhoff et al., Sediment for Survival.

19. Note the similarity here between what is needed in the Bay Area and the approach of the Louisiana Coastal Master Plan. In both cases, no single technology can address the full array of challenges and timings.

20. Beagle et al., San Francisco Bay Shoreline Adaptation Atlas.

21. Colette Mortreux and Jon Barnett, "Adaptive Capacity: Exploring the Research Frontier," *Wiley Interdisciplinary Reviews: Climate Change* 8, no. 4 (2017): e467.
Charlotte K. Whitney et al., "Adaptive Capacity: From Assessment to Action in Coastal Social-Ecological Systems," *Ecology and Society* 22, no. 2 (2017).

22. Since the time of the Public Sediment effort, the San Francisco Estuary Institute has confirmed the team's overall findings across a range of future climate change scenarios. See Dusterhoff et al., Sediment for Survival.

23. Questions like this are not limited to the Bay Area.
Kian Goh, *Form and Flow: The Spatial Politics of Urban Resilience and Climate Justice* (The MIT Press, 2021).
Carolyn Kousky, Billy Fleming, and Alan M. Berger, eds., *A Blueprint for Coastal Adaptation: Uniting Design, Economics, and Policy* (Island Press, 2021).

Lizzie Yarina, "Your Sea Wall Won't Save You," *Places Journal* (2018), https://placesjournal.org/article/your-sea-wall-wont-save-you/.

Lizzie Yarina, Miho Mazereeuw, and Larisa Ovalles, "A Retreat Critique: Deliberations on Design and Ethics in the Flood Zone," *Journal of Landscape Architecture* 14, no. 3 (2019): 8–23.

24. See for instance John Gillis, *The Human Shore* (University of Chicago Press, 2012).

Image Credits:

Spread 1;
Alameda Creek, US Army Corps of Engineers District, San Francisco, 100 McAllister St., San Francisco. Photograph. Courtesy of Washington Township Museum of Local History, Fremont, CA. 1971.

Spread 2:
Top:
Brett Milligan / Public Sediment Team. Upstream sediment accumulation in Alameda Creek Flood Control Channel. Photograph. Fremont, CA, November 13, 2017.
Bottom:
Brett Milligan / Public Sediment Team. Alameda Creek Flood Control Channel within the Baylands. Photograph. Union City, CA, November 13, 2017.

Spread 3:
Top Left:
SCAPE / Public Sediment Team. Conceptual Diagram of Habitat Transformation with Sea Level Rise. 2017.
Top Right:
SCAPE / Public Sediment Team. Conceptual Diagram of Sediment Supply and Demand for the San Francisco Baylands. Drawing. 2017.
Dusterhoff, Scott. High Demand and Low Supply by 2100 with 3.5 Feet of Sea Level Rise. San Francisco Estuary Institute. Personal communication, provided. October 11, 2017.

Note: This diagram was based on preliminary analysis by the San Francisco Estuary Institute in 2017. Since the time of project development, a more detailed analysis was conducted as a part of the Heathy Watersheds Resilient Baylands project (hwrb.sfei.org), below:
Scott Dusterhoff, K. McKnight, L. Grenier, and N. Kauffman. Sediment for Survival: A Strategy for the Resilience of Bay Wetlands in the Lower San Francisco Estuary. A SFEI Resilient Landscape Program. A product of the Healthy Watersheds, Resilient Baylands project, funded by the San Francisco Bay Water Quality Improvement Fund, EPA Region IX. Publication #1015, San Francisco Estuary Institute, Richmond, CA, 2021. https://www.sfei.org/sites/default/files/biblio_files/Sediment%20for%20Survival_v1.3_med_res.pdf

Bottom:
SCAPE / Public Sediment Team. Bayland Transformation and Impacted Urban Areas with Projected Sea Level Rise and Current Sediment Supply. Drawing. 2017.

Note on Methodology:
Sediment needs were derived from a spatial elevation analysis of the most recent Adapting to Rising Tides LiDAR data. Existing elevations were subtracted from target marsh plain elevations to obtain sediment needs. Those needs were multiplied by the Bayland area obtained from SFEI BARRI data to produce a sediment need volume. Using bulk density conversions, sediment volumes for marsh and mudflat habitat types were converted into mass to obtain numbers that reflect a need that considers soil composition. Need was compared with supply data provided by SFEI and the resulting landscape change based upon land elevation was depicted in the graphic.

San Francisco Estuary Institute. Bay Area Aquatic Resources Inventory (BARRI). Shapefile. 2015. https://www.sfei.org/data/baari-version-21-gis-data#sthash.7MbjqEXl.dpbs
San Francisco Estuary Institute. California Aquatic Resources Inventory (CARI). Shapefile. 2016. https://www.sfei.org/data/california-aquatic-resource-inventory-cari-version-03-gis-data#sthash.jJLS6lHY.dpbs
Adapting to Rising Tides. Bay Area Sea Level Rise and Shoreline Analysis Maps. Shapefile. 2017.
NOAA National Geophysical Center. San Francisco Bay, California 1/3 arc-second NAVD 88 Coastal Digital Elevation Model. NOAA National Centers for Environmental Information. Shapefile. 2010. https://www.ncei.noaa.gov/access/metadata/landing-page/bin/iso?id=gov.noaa.ngdc.mgg.dem:741
US Geological Survey Topobathymetric Elevation Model of San Francisco Bay Area, California. Shapefile. Accessed 2017. https://www.usgs.gov/special-topics/coastal-national-elevation-database-%28coned%29-applications-project/science-4#overview

Spread 4:
SCAPE / Public Sediment Team. Making Sediment Public in Infrastructural Sediment Systems. Drawing. New York, NY, 2017.

Spread 5:
SCAPE / Public Sediment Team. Alameda Creek Sedimentshed. Drawing. 2017.
San Francisco Estuary Institute. Bay Area Aquatic Resources Inventory (BARRI). Shapefile. 2015. https://www.sfei.org/data/baari-version-21-gis-data#sthash.7MbjqEXl.dpbs
San Francisco Estuary Institute. California Aquatic Resources Inventory (CARI). Shapefile. 2016. https://www.sfei.org/data/california-aquatic-resource-inventory-cari-version-03-gis-data#sthash.jJLS6lHY.dpbs
San Francisco Estuary Institute. Draft Bay Watersheds. Shapefile. 2017.
NOAA National Geophysical Center. San Francisco Bay, California 1/3 arc-second NAVD 88 Coastal Digital Elevation Model. NOAA National Centers for Environmental Information. Shapefile. 2010. https://www.ncei.noaa.gov/access/metadata/landing-page/bin/iso?id=gov.noaa.ngdc.mgg.dem:741
US Geological Survey Topobathymetric Elevation Model of San Francisco Bay Area, California. Shapefile. Accessed 2017. https://www.usgs.gov/special-topics/coastal-national-elevation-database-%28coned%29-applications-project/science-4#overview

Spread 6:
Top Right:
SCAPE / Public Sediment Team. Historic Alameda Creek bringing sediment to Baylands in the Niles Cone Alluvial Fan. Drawing. 2017.
San Francisco Estuary Institute. Alameda Creek Historical Ecology GIS Data. Shapefile. 2013. https://ssl.sfei.org/content/alameda-creek-historical-ecology-gis-data#sthash.hSqOBw5D.dpbs
San Francisco Estuary Institute. California Aquatic Resources Inventory (CARI). Shapefile. 2016. https://www.sfei.org/data/california-aquatic-resource-inventory-cari-version-03-gis-data#sthash.jJLS6lHY.dpbs

US Geological Survey Topobathymetric Elevation Model of San Francisco Bay Area, California. Shapefile. 2019. https://www.usgs.gov/special-topics/coastal-national-elevation-database-%28coned%29-applications-project/science-4#overview
Bottom Right:
SCAPE / Public Sediment Team. Urbanization of Alameda Creek and the Niles Cone Alluvial Fan. Drawing. 2017.
Alameda County Flood Control District. Streets. Shapefile. 2017.
US Government Open Data. TIGER/Line Shapefile, 2019, state, California, Primary and Secondary Roads. Shapefile. 2017. https://catalog.data.gov/dataset/tiger-line-shapefile-2019-state-california-primary-and-secondary-roads-state-based-shapefile
San Francisco Estuary Institute. Bay Area Aquatic Resources Inventory (BARRI). Shapefile. 2015. https://www.sfei.org/data/baari-version-21-gis-data#sthash.7MbjqEXI.dpbs
San Francisco Estuary Institute. California Aquatic Resources Inventory (CARI). Shapefile. 2016. https://www.sfei.org/data/california-aquatic-resource-inventory-cari-version-03-gis-data#sthash.jJLS6lHY.dpbs
US Geological Survey Topobathymetric Elevation Model of San Francisco Bay Area, California. Shapefile. 2019. https://www.usgs.gov/special-topics/coastal-national-elevation-database-%28coned%29-applications-project/science-4#overview

Spread 7:
Left:
SCAPE / Public Sediment Team. Sediment Buildup in the Creek Bed. Photograph. Fremont, CA, 2017.
SCAPE / Public Sediment Team. Restricted Public Access to the Alameda Creek Flood Control Channel. Photograph. Fremont, CA, 2017.
Right:
Brett Milligan / Public Sediment Team. Urbanization and Alameda Creek. Photograph. Hayward, CA, 2017.

Spread 8:
Top Left:
SCAPE / Public Sediment Team. Plant Identification Field Work with San Francisco Estuary Institute and Alameda County Flood Control District. Photograph. Hayward, CA, 2018.
Bottom Left:
UC Davis / Public Sediment Team. High school students studying a physical model of Alameda Creek while on site. Photograph. Hayward, CA, 2018.
Top Right:
SCAPE / Public Sediment Team. Iterative stream table modeling at Cornell University. Photograph. Ithaca, NY, 2017.
Bottom Right:
Dredge Research Collaborative/ Public Sediment Team. Stakeholder Meeting to advance the Alameda Creek Conceptual Design. Fremont, CA, 2018.

Spread 9:
SCAPE / Public Sediment Team. Unlocking sediment movement, public access, and fish passage along Alameda Creek. Drawing. New York, NY, 2017.

Spread 10:
SCAPE / Public Sediment Team. Historic, Current, and Proposed Cross Sections of Alameda Creek. Drawing. New York, NY, 2017.
Alameda County Flood Control and Water Conservation District. Alameda Creek Restoration Project (Phase 1). Drawing set for construction. Fremont, CA, December 2016.

Spread 11:
Top:
SCAPE / Public Sediment Team. Expanding habitat and passage for fish along Alameda Creek. New York, NY, 2017.
Alameda County Flood Control District. Buildings. Shapefile. 2017.
San Francisco Estuary Institute. Bay Area Aquatic Resources Inventory (BARRI). Shapefile. 2015. https://www.sfei.org/data/baari-version-21-gis-data#sthash.7MbjqEXI.dpbs
San Francisco Estuary Institute. California Aquatic Resources Inventory (CARI). Shapefile. 2016. https://www.sfei.org/data/california-aquatic-resource-inventory-cari-version-03-gis-data#sthash.jJLS6lHY.dpbs
Bottom Left:
SCAPE / Public Sediment Team. Sea lamprey observed in headwaters of Alameda Creek. Sunol, CA, 2017.
Bottom Right:
SCAPE / Public Sediment Team. Steelhead smolt observed in headwaters of Alameda Creek. Sunol, CA, 2017.

Spread 12:
Top:
SCAPE / Public Sediment Team. Creating physical public access opportunities along Alameda Creek. Drawing. New York, NY, 2017.
California State Geoportal. California Schools. Shapefile. 2017. https://gis.data.ca.gov/datasets/f7f818b0aa7a415192eaf66f192bc9cc/explore?location=37.557775%2C-122.010398%2C12.95
California School Campus Database. California School Campus Database. Shapefile. 2017. https://www.californiaschoolcampusdatabase.org/#download
Alameda County Flood Control District. Buildings. Shapefile. 2017.
San Francisco Estuary Institute. Bay Area Aquatic Resources Inventory (BARRI). Shapefile. 2015. https://www.sfei.org/data/baari-version-21-gis-data#sthash.7MbjqEXI.dpbs
San Francisco Estuary Institute. California Aquatic Resources Inventory (CARI). Shapefile. 2016. https://www.sfei.org/data/california-aquatic-resource-inventory-cari-version-03-gis-data#sthash.jJLS6lHY.dpbs
Bottom Left:
Brett Milligan / Public Sediment Team. Creek neighborhoods. Photograph. Fremont, CA, 2017.
Bottom Right:
Public Sediment Team. Alameda Creek Crawl public event along the flood control channel. Fremont, CA, 2017.

Spread 13:
Top:
SCAPE / Public Sediment Team. Public Landscape Typologies within the Alameda Creek Flood Control Channel. Drawing. New York, NY, 2017.
Bottom:
SCAPE / Public Sediment Team. View of a Mudroom from the Terrace Trail. Drawing. New York, NY, 2017.

Spread 14:
SCAPE / Public Sediment Team. Public access to sediment landscapes within the Alameda Creek Active Channel. Drawing. New York, NY, 2017.
Alameda County Flood Control and Water Conservation District. Alameda Creek Restoration Project (Phase 1). Drawing set for Construction. Fremont, CA, December 2016.

Spread 15:
Top:
SCAPE / Public Sediment Team. Steps to Breach Alameda Creek in the Baylands. Drawing. New York, NY, 2018.
San Francisco Estuary Institute. Bay Area Aquatic Resources Inventory (BARRI). Shapefile. 2015. https://www.sfei.org/data/baari-version-21-gis-data#sthash.7MbjqEXI.dpbs
San Francisco Estuary Institute. California Aquatic Resources Inventory (CARI). Shapefile. 2016. https://www.sfei.org/data/california-aquatic-resource-inventory-cari-version-03-gis-data#sthash.jJLS6IHY.dpbs
Bottom:
SCAPE / Public Sediment Team. Bayland Bridge overlooking the Alameda Creek breach. Drawing. New York, NY, 2018.

Spread 16:
Top:
SCAPE / Public Sediment Team. Combining landmass and gravel beach cross sections to create a pebble dune. Drawing. New York, NY, 2018.
Landmass concept provided by the Alameda Flood Control District. Gravel Beach concepts developed in collaboration with Letitia Grenier at SFEI.
Bottom:
SCAPE / Public Sediment Team. Pebble dune bay edge supporting various habitat conditions. Drawing. New York, NY, 2018.

Spread 17:
Top:
Justine Holzman and Cy Keener / Public Sediment Team. Varying levels of sediment in water samples collected along Alameda Creek. Photograph. New York, NY, 2018.
Middle:
Public Sediment Team. Collecting water samples from Alameda Creek. Photograph. Fremont, CA, 2018.
Bottom:
Justine Holzman and Cy Keener / Public Sediment Team. Conceptual sensing tools to monitor sediment levels in the water. Drawing. New York, NY, 2017.

Spread 18:
SCAPE / Public Sediment Team. Sediment flowing into the baylands through a breach along Alameda Creek. Drawing. New York, NY, 2018.

Spread 19:
SCAPE / Public Sediment Team. Sediment Supply and Demand after unlocking sediment flows in Alameda Creek. Drawing. New York, NY, 2018.

Note on Methodology:
Sediment needs were derived from a spatial elevation analysis of the most recent Adapting to Rising Tides LiDAR data. Existing elevations were subtracted from target marsh plain elevations to obtain sediment needs. Those needs were multiplied by the bayland area obtained from SFEI BARRI data to produce a sediment need volume. Using bulk density conversions, sediment volumes for marsh and mudflat habitat types were converted into mass to obtain numbers that reflect a need that considers soil composition.

SCAPE / Public Sediment Team. Supplemental Sediment Supply Strategies to Sustain and Grow Baylands with Sea Level Rise. Drawing. New York, NY, 2018.

Spread 20:
SCAPE / Public Sediment Team. Experimental pilot projects to test various sediment strategies. Drawing. New York, NY, 2018.
California State Geoportal. California Schools. Shapefile. 2017. https://gis.data.ca.gov/datasets/f7f818b0aa7a415192eaf66f192bc9cc/explore?location=37.557775%2C-122.010398%2C12.95
San Francisco Estuary Institute. Bay Area Aquatic Resources Inventory (BARRI). Shapefile. 2015. https://www.sfei.org/data/baari-version-21-gis-data#sthash.7MbjqEXI.dpbs
San Francisco Estuary Institute. California Aquatic Resources Inventory (CARI). Shapefile. 2016. https://www.sfei.org/data/california-aquatic-resource-inventory-cari-version-03-gis-data#sthash.jJLS6IHY.dpbs

Spread 21:
SCAPE / Public Sediment Team. Public Sediment. Drawing. New York, NY, 2017.